Communications
in Computer and Information Science 118

Yanchun Zhang Alfredo Cuzzocrea
Jianhua Ma Kyo-il Chung Tughrul Arslan
Xiaofeng Song (Eds.)

Database Theory and Application, Bio-Science and Bio-Technology

International Conferences, DTA and BSBT 2010
Held as Part of the Future Generation
Information Technology Conference, FGIT 2010
Jeju Island, Korea, December 13-15, 2010
Proceedings

Springer

Volume Editors

Yanchun Zhang
Victoria University, Melbourne, Australia
E-mail: yanchun.zhang@vu.edu.au

Alfredo Cuzzocrea
University of Calabria, Rende, Cosenza, Italy
E-mail: cuzzocrea@si.deis.unical.it

Jianhua Ma
Hosei University, Tokyo, Japan
E-mail: jianhua@hosei.ac.jp

Kyo-il Chung
Electronics and Telecommunications Research Institute
Daejeon, Korea
E-mail: kyoil@etri.re.kr

Tughrul Arslan
Edinburgh University, Edinburgh, UK
E-mail: t.arslan@ed.ac.uk

Xiaofeng Song
Nanjing University of Aeronautics and Astronautics
Nanjing, Jiangsu, China
E-mail: xfsong@nuaa.edu.cn

Library of Congress Control Number: 2010940178

CR Subject Classification (1998): I.6, I.2, J.3, F.1, F.2, C.3

ISSN 1865-0929
ISBN-10 3-642-17621-6 Springer Berlin Heidelberg New York
ISBN-13 978-3-642-17621-0 Springer Berlin Heidelberg New York

springer.com

Printed in Germany

Typesetting: Camera-ready by author, data conversion by Scientific Publishing Services, Chennai, India
Printed on acid-free paper 06/3180

Preface

Welcome to the proceedings of the 2010 International Conferences on Database Theory and Application (DTA 2010), and Bio-Science and Bio-Technology (BSBT 2010) – two of the partnering events of the Second International Mega-Conference on Future Generation Information Technology (FGIT 2010).

DTA and BSBT bring together researchers from academia and industry as well as practitioners to share ideas, problems and solutions relating to the multifaceted aspects of databases, data mining and biomedicine, including their links to computational sciences, mathematics and information technology.

In total, 1,630 papers were submitted to FGIT 2010 from 30 countries, which includes 175 papers submitted to DTA/BSBT 2010. The submitted papers went through a rigorous reviewing process: 395 of the 1,630 papers were accepted for FGIT 2010, while 40 papers were accepted for DTA/BSBT 2010. Of the 40 papers 6 were selected for the special FGIT 2010 volume published by Springer in the LNCS series. 31 papers are published in this volume, and 3 papers were withdrawn due to technical reasons.

We would like to acknowledge the great effort of the DTA/BSBT 2010 International Advisory Boards and members of the International Program Committees, as well as all the organizations and individuals who supported the idea of publishing this volume of proceedings, including SERSC and Springer. Also, the success of these two conferences would not have been possible without the huge support from our sponsors and the work of the Chairs and Organizing Committee.

We are grateful to the following keynote speakers who kindly accepted our invitation: Hojjat Adeli (Ohio State University), Ruay-Shiung Chang (National Dong Hwa University), and Andrzej Skowron (University of Warsaw). We would also like to thank all plenary and tutorial speakers for their valuable contributions.

We would like to express our greatest gratitude to the authors and reviewers of all paper submissions, as well as to all attendees, for their input and participation.

Last but not least, we give special thanks to Rosslin John Robles and Maricel Balitanas. These graduate school students of Hannam University contributed to the editing process of this volume with great passion.

December 2010

Yanchun Zhang
Alfredo Cuzzocrea
Jianhua Ma
Kyo-il Chung
Tughrul Arslan
Xiaofeng Song

DTA 2010 Organization

Organizing Committee

Steering Co-chairs	Tai-hoon Kim (Hannam University, Korea) Wai-chi Fang (National Chiao Tung University, Taiwan)
General Co-chairs	Yanchun Zhang (Victoria University, Australia) Alfredo Cuzzocrea (University of Calabria, Italy)
Program Co-chairs	Jianhua Ma (Hosei University, Japan) Kyo-il Chung (ETRI, Korea)
Publication Co-chairs	Rosslin John Robles (Hannam University, Korea) Maricel Balitanas (Hannam University, Korea)

Program Committee

Anne James
Alfredo Cuzzocrea
Aoying Zhou
Chunsheng Yang
Damiani Ernesto
Daoqiang Zhang
David Taniar
Djamel A. Zighed
Emiran Curtmola
Feipei Lai
Fionn Murtagh
Gang Li
Guoyin Wang
Haixun Wang
Hans-Joachim Klein
Hiroshi Sakai
Hiroyuki Kawano
Hui Yang
Jason T. L. Wang
Jesse Z. Fang
Jian Lu
Jian Yin
Jixin Ma
Joel Quinqueton
Joshua Z. Huang
Jun Hong
Junbin Gao
Karen Renaud
Kay Chen Tan
Kenji Satou
Keun Ho Ryu
Krzysztof Stencel
Lachlan McKinnon
Ladjel Bellatreche
Laura Rusu
Li Ma
Longbing Cao
Lucian N. Vintan
Mark Roantree
Masayoshi Aritsugi
Miyuki Nakano
Ozgur Ulusoy
Pabitra Mitra Mitra
Pang-Ning Tan
Peter Baumann
Richi Nayak
Sanghyun Park
Sang-Wook Kim
Sanjay Jain
Shu-Ching Chen
Shyam Kumar Gupta
Stephane Bressan
Tadashi Nomoto
Takeru Yokoi
Tao Li
Tetsuya Yoshida
Theo Härder
Tomoyuki Uchida
Toshiro Minami
Vasco Amaral
Veselka Boeva
Vicenc Torra
Weining Qian
William Zhu
Xiaohua Hu
Xiao-Lin Li
Xuemin Lin
Yan Wang
Yang Yu
Yang-Sae Moon
Yiyu Yao
Young-Koo Lee
Zhuoming Xu

BSBT 2010 Organization

Organizing Committee

General Co-chairs
- Tughrul Arslan (Edinburgh University, UK)
- Wai-chi Fang (National Chiao Tung University, Taiwan)

Program Co-chairs
- Xiaofeng Song (Nanjing University of Aeronautics and Astronautics, China)
- Tai-hoon Kim (Hannam University, Korea)

International Advisory Board
- Saman Halgamuge (University of Melbourne, Australia)
- Joseph Kolibal (University of Southern Mississippi, USA)
- Philip Maini (University of Oxford, UK)
- Byoung-Tak Zhang (Seoul National University, Korea)
- Aboul Ella Hassanien (Cairo University, Egypt)

Publicity Co-chairs
- Muhammad Khurram Khan (King Saud University, Saudi Arabia)
- Aboul Ella Hassanien (Cairo University, Egypt)

Publication Co-chairs
- Rosslin John Robles (Hannam University, Korea)
- Maricel Balitanas (Hannam University, Korea)

Program Committee

A.Q.K. Rajpoot
Adrian Stoica
Ajay Kumar
Arun Ross
Asai Asaithambi
Bob McKay
Carlos Juiz
Cesare Alippi
Dana Lodrova
Davide Anguita
Dong-Yup Lee
Emilio Corchado
Farzin Deravi
Francisco Herrera
George A. Gravvanis
Hujun Yin
Janusz Kacprzyk
Jason T.L. Wang
Javier Ortega-Garcia
Jim Torresen
Jongwook Woo
José Alfredo Ferreira
José Manuel Molina
Juan Manuel Corchado
Kayvan Najarian
Kenji Mizuguchi
Kevin Daimi
Li Xiaoli
Liangjiang Wang
Lusheng Wang
Martin Drahansky
Matthias Dehmer
Meena K. Sakharkar
Michael E. Schuckers
Pong C. Yuen
Qing-Zhong Liu
R. Ponalagusamy
Rattikorn Hewett
Saman Halgamuge

Stan Z. Li
Stephen Cameron
Suash Deb
Tatsuya Akutsu
Tommaso Mazza
Tony (Xiaohua) Hu
Waleed Abdullah
Wei Zhong
Witold Pedrycz
Yaoqi Zhou
Yong Shi
Yu Zheng
Zeyar Aung
Zhenan Sun
Zizhong Chen

Table of Contents

Particle Swarm Optimization for Digital Image Watermarking

Hai Tao[1], Jasni Mohamad Zain[1], Ahmed N. Abd Alla[2], and Qin Hongwu[1]

[1] Faculty of Computer Systems and Software Eng., University Malaysia Pahang, Malaysia
[2] Faculty of Electrical and Electronic Engineering, University Malaysia Pahang, Malaysia
taotao27@gmail.com, jasni@ump.edu.my, waal85@yahoo.com, qhwump@gmail.com

Abstract. The trade-off between the imperceptibility and robustness is one of the most difficult challenges in digital watermarking system. To solve the problem, an optimal algorithm for image watermarking is proposed. The algorithm embeds the watermark by quantizing the wavelet packets coefficients of the image. In the proposed watermarking system, to protect the originality of the watermark image, a scrambled binary watermark embeds in the host image against intentional and unintentional attacks and each bit of the permuted watermark is embedded optimally into selected sub-bands of the decomposed host image by quantization of sub-bands coefficients. From experimental results, it demonstrates the robustness and the superiority of the proposed hybrid scheme.

1 Introduction

With the popularization and development of multimedia technologies and the spread of high-speed communication networks, it is proliferated that various digital multimedia products such as image, audio, video and three-dimensional model are more vulnerable to illegal possession, duplication and dissemination than analog data. Consequently, multimedia digital content owners are skeptical of putting their content on the Internet due to lack of intellectual property protection available to them. Digital watermarking is the process of embedding or hiding digital information called watermark into a multimedia product, and then the embedded data can later be extracted or detected from the watermarked product, for protecting digital content copyright and ensuring tamper-resistance, which is indiscernible and hard to remove by unauthorized persons [1].

These years, the wavelet packet transform (WPT) approach remains one of the most effective techniques that is a generalization of 2D discrete wavelet transform for image watermarking [2-5]. In [2], multiple copies of the watermark image are inserted into the host image and also the uses of different permutations of the watermark image cause its pixels to be distributed all over the host image. Lee et al [3] presents a genetic watermarking scheme based on the wavelet packet transform. Genetic algorithm is used to select an appropriate basis from permissible bases of wavelet packet transform to increase the robustness of the embedded watermarks. In [4], the original image is decomposed by discrete wavelet packet transform and the dominant wavelet coefficients are selected for watermark embedding from each sub-band

Y. Zhang et al. (Eds.): DTA/BSBT 2010, CCIS 118, pp. 1–8, 2010.

except the lowest frequency one. Then, each watermark bit is adaptively embedded with different strength into the selected wavelet packet coefficients based on the odd or even value after quantization.

In this paper, a novel digital image watermarking approach is introduced by optimal selection of wavelet packet coefficients. The original image is decomposed into m-level sub-bands by discrete wavelet packet transform and the wavelet coefficients are selected for watermark embedding from each sub-band. Then, to obtain the highest possible robustness without losing the transparency, each scrambled watermark bit is embedded optimally into the selected wavelet packet coefficients based on particle swarm optimization.

2 Preliminaries

2.1 Wavelet Packet Transform (WPT)

The wavelet packet transform (WPT) proposed by Coifman and Wickerhauser can be interpreted as an important extension of the wavelet transform where the image is passed through more filters than the DWT. In the image processing, four domains with different frequency characteristics are generated. Both discrete wavelet packets transform (DWPT) and discrete wavelet transform (DWT) have the similar structure and framework of hierarchical multi-resolution representation. The main difference in the two techniques is that, it is assumed that lower frequencies contain more important information than higher frequencies in DWT and only the outputs of the low frequency subband LL(coarse approximation) are further repeatedly applied for the lowest frequency subband among the very previous subbands at the next level, while the outputs of the higher-frequency subband remain as the final results, such as LH (horizontal details), HL (vertical details) and HH (diagonal details) which represent detailed information. However, in the wavelet packets, DWPT analyzes an image simultaneously at different resolution levels and orientations and allows further processing of not only the outputs of the low frequency subband but also those of higher-frequency subband for further wavelet decomposition at the next level, which provides the flexibility to produce arbitrary decomposition of the input image.

2D discrete wavelet packet transform can be described as follows, where an image W($N \times M$ pixels) is decomposed into one approximation and three detail images represented in following notation:

$$\begin{aligned}
W_{4k,(i,j)}^{p+1} &= \textstyle\sum_n \sum_m u(m)u(n)\, W_{k,(m+2i,n+2j)}^{p} \\
W_{4k+1,(i,j)}^{p+1} &= \textstyle\sum_n \sum_m u(m)v(n)\, W_{k,(m+2i,n+2j)}^{p} \\
W_{4k+2,(i,j)}^{p+1} &= \textstyle\sum_n \sum_m v(m)u(n)\, W_{k,(m+2i,n+2j)}^{p} \\
W_{4k+3,(i,j)}^{p+1} &= \textstyle\sum_n \sum_m v(m)v(n)\, W_{k,(m+2i,n+2j)}^{p}
\end{aligned} \quad (1)$$

Where u and v are The standard low-pass filter function and a high-pass filter function, respectively. $W_{0,(i,j)}^{0}$ is the pixel value of coordinates (i,j) of image W. At each step, the p-level subbands of coefficients W_k^p is decomposed into four quarter-size images of (p+1)-level which coefficients are W_{4k}^{p+1}, W_{4k+1}^{p+1}, W_{4k+2}^{p+1} and W_{4k+3}^{p+1}. Two-level wavelet packet decomposition is illustrated in Fig. 1.

In the process of inverse DWPT, the original image can be reconstructed from these DWPT coefficients. At different resolution levels and orientations, the coarse approximation images (low-frequency) and their detailed images (higher-frequency) can be used to redintegrate the reference image of higher resolution. The advantages of the wavelet packet framework are its universality in adapting the transform to a signal without training or assuming any statistical property of the signal, especially non-stationary signals because the same frequency bandwidths can provide good resolution regardless of high and low frequencies.

2.2 Particle Swarm Optimization (PSO)

The basic idea of the classical particle swarm optimization (PSO) algorithm is the clever exchange of information about the global and local best values mentioned above. Let us assume that the optimization goal is to maximize an objective function$f(r)$. Each particle will examine its performance through the following two views. Each potential solution is also assigned a randomized velocity, and the potential solutions, called particles, correspond to individuals. Each particle in PSO flies in the *D*-dimensional problem space with a velocity dynamically adjusted according to the flying experiences of its individuals and their colleagues. The location of the ith particle is represented as $X_i = [x_{i1}, x_{i2}, \ldots\ldots, x_{iD}]$, where $x_{id} \in [l_d, u_d]$, $d \in [1, D]$. l_d and u_d are the lower and upper bounds for the dth dimension, respectively. The best previous position (which gives the best fitness value) of the ith particle is recorded and represented as $P_i = [p_{i1}, p_{i2}, \ldots\ldots, p_{iD}]$, which is also called $pbest$. The index of the best particle among all the particles in the population is represented by the symbol. The location p_g is also denoted by $gbest$. The velocity of the i_{th} particle is represented by $V_i = [v_{i1}, v_{i2}, \ldots\ldots, v_{iD}]$ and is clamped to a maximum velocity $V_{max} = [v_{max1}, v_{max2}, \ldots\ldots, v_{maxD}]$ which is specified by the user. The particle swarm optimization concept consists of, at each time step, regulating the velocity and location of each particle toward its $pbest$ and $gbest$ locations according to (1) and (2), respectively.

$$v_{id}^{n+1} = wv_{id}^{n} + c_1 r_1^n (p_{id}^n - x_{id}^n) + c_2 r_2^n (p_{gd}^n - x_{id}^n) \quad (2)$$

$$x_{id}^{n+1} = x_{id}^{n} + v_{id}^{n+1} \quad (3)$$

Where w is the inertia weigh; c_1, c_2 are two positive constants, called cognitive and social parameter respectively; $d = 1,3, \ldots\ldots, D$; $i = 1,3, \ldots\ldots, m$ and m is the size of the swarm; r_1^n, r_2^n are two random sequences, uniformly distributed in [0,1]; and $n = 1,3, \ldots\ldots, N$ denotes the iteration number, *N* is the maximum allowable iteration number.

3 The Proposed Scheme

3.1 Watermark Embedding

Suppose that original image *I* is a gray-level image and $I_M \times I_N$ is the width by height of I, respectively. The watermark W is a binary image and $W_J \times W_K$ is the width by height of W, respectively. First, the original image is decomposed into the wavelet packet representation of m-level and obtain multi-resolution presentation (LH_m, HL_m, HH_m) and approximation(LH_3) as shown in Fig.1. $P_m^n(i,j)$ is a frequency coefficient in coordinate (i,j), where $n \in \{LL, LH, HL, HH\}$ represents the orientation and $m \in \{1,2,3\}$. In addition, in consideration of the visual quality and the robustness, the proposed algorithm that a binary watermark is embedded into the selected subbands. Fig.1 shows the selected sub-bands in our simulations.

1. In the embedding process, the original image is decomposed into m-level subbands to obtain a total of k sub-bands of wavelet coefficients $P_m^n(i,j)$. Each subband of the four selected subbands in Fig.1 are decomposed into non-overlapping blocks M_k with size 2×2, and k=1, 2,...,$W_J \times W_K$.

2. The watermark information $(W_K \times W_J)$ need be pretreated in order to eliminate the correlation of watermark image pixels and enhance system robustness and security. For the advantages of lowing computed complexity and obtaining inverse transform easily comparing with Arnold transform, the watermark image is pretreated through affine scrambling. The affine scrambling is showed as equation,

$$\cdot \begin{pmatrix} u' \\ v' \end{pmatrix} = \begin{pmatrix} a & b \\ c & d \end{pmatrix} \begin{pmatrix} u \\ v \end{pmatrix} + \begin{pmatrix} e \\ f \end{pmatrix}, \; where \left(\begin{vmatrix} a & b \\ c & d \end{vmatrix} \neq 0 \right) \tag{4}$$

For enhancing the statistical imperceptible through embedding watermark, series of $\{-1, 1\}$ values substitute for $\{0, 1\}$ which is the value of watermark image by scrambling, respectively. The new watermark is generated $(w_i' = w_i \cdot p_i)$, according to a sequence of the binary pseudo-random p_i modulating the watermark, where $p_i \in \{-1, 1\}$ and $0 \le i < W_K \times W_J$.

3. After pretreated watermark sequence is partitioned into four parts, they are embedded into the four selected frequency bands respectively. In the four selected subbands, embedding each bit watermark into each M_k is motivated by experiment. In each block, max { $P_3^n(i,j)$, $P_3^n(i+1,j)$, $P_3^n(i,j+1)$, $P_3^n(i+1,j+1)$ } and min{ $P_3^n(i,j)$, $P_3^n(i+1,j)$, $P_3^n(i,j+1)$, $P_3^n(i+1,j+1)$ } are calculated, and sub-bands coefficients are then modified according to the equation,

$$P_3'^n(i,j) = \begin{cases} max\{P_3^n(i,j), P_3^n(i+1,j), P_3^n(i,j+1), P_3^n(i+1,j+1)\} + \alpha_l \mathrm{w_k}, if \;\; \mathrm{w_k} = 1 \\ min\{P_3^n(i,j), P_3^n(i+1,j), P_3^n(i,j+1), P_3^n(i+1,j+1)\} - \alpha_l \mathrm{w_k}, if \;\; \mathrm{w_k} = 0 \end{cases} \tag{5}$$

Where α_l are the scaling factors.

4. Watermark bits are embedded into the original image and m level inverse wavelet packets transform of the sub images is performed. Then, the watermarked image can be obtained.

3.2 Watermark Extraction

The watermark extraction is the reverse procedure of the watermark embedding. It can be summarized as follows:

1. In the extracting process, the watermarked image is decomposed into 3-level using WPT to obtain a series of high-frequency and a high-energy subband.

2. The four selected subbands are decomposed into non-overlapping blocks M_k with size 2×2, In each block, $x = max\{P_3'^n(i,j), P_3'^n(i+1,j), P_3'^n(i,j+1, P_3'^n(i+1,j+1\}$ and $y = min\{P_3'^n(i,j), P_3'^n(i+1,j), P_3'^n(i,j+1), P_3'^n(i+1,j+1)\}$ are calculated. Then, $Average(i,j) = 0.5(x+y)$ is defined.

$$w_k = \begin{cases} 1, & Average(i,j) \le P_3^{\prime n}(i,j) \\ 0, & Average(i,j) > P_3^{\prime n}(i,j) \end{cases} \quad (6)$$

3. A complete watermark sequence w' is obtained and inverse affine transform perform on the sequence, then binary watermark image has been extracted.

4. After extracting the watermark, normalized correlation coefficients to quantify the correlation between the original watermark and the extracted one is used. A normalized correlation (NC) between w and w' is defined as:

$$NC = \frac{\sum_{k=1}^{W_K \times W_J} w_k w_k'}{\sqrt{\sum_{k=1}^{W_K \times W_J} {w_k}^2 \sum_{k=1}^{W_K \times W_J} {w_k'}^2}} \quad (7)$$

where w_k and w_k' denote an original watermark and extracted one, respectively.

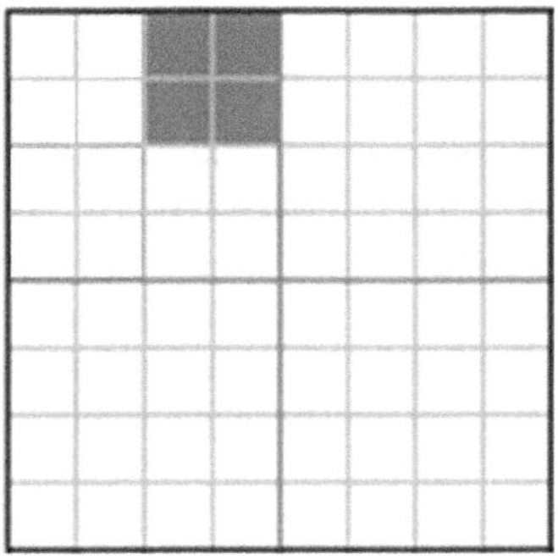

Fig. 1. Wavelet packets transform

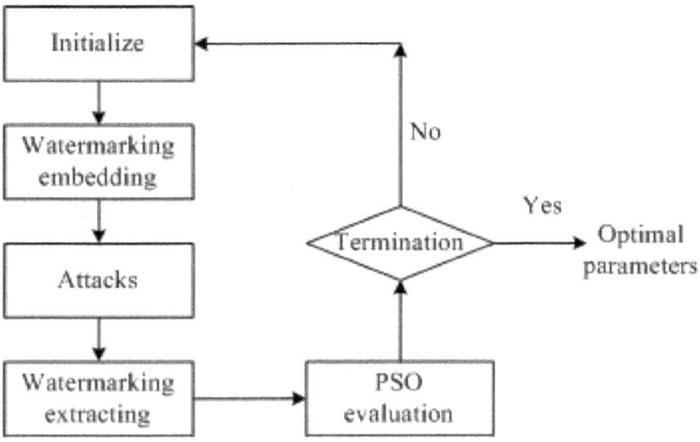

Fig. 2. Diagram for proposed scheme

3.3 Proposed Optimization Process

In order to achieve the optimal performance of a digital image watermarking algorithm, the developed technique employs PSO algorithm to search for optimal parameters. In the optimization process, the parameters are the scaling factors (α_l) which are obtained for optimal watermarking depending on both the transparency and the robustness factors. In every swarm, each member vector or particle in the particle represents a possible solution to the problem and hence is comprised of a set of scaling factors. To start the optimization, PSO use randomly produced initial solutions generated by random number generator between 0 and 1. In the proposed scheme, for solving optimization problem for multiple parameters, $\alpha_l$in Eq.(5) is a weight of each watermarked bit and embedded each modulating watermarked bit into the four selected subbands by WPT transform and therefore, all of α_l represent the multiple scaling factors. After modifying the four selected subbands coefficients of the decomposed host image by employing the scaling factors, the watermarked images of the current generation are calculated according to the watermark embedding procedure explained in Section 3.1.

In order to evaluate the feasibility of extracted watermark, both a universal quality index (UQI) [7] and NC values evaluate the objective function as performance indices. Due to UQI's role of imperceptibility measure, it is used as output image quality performance index. Similarly, NC is used as a watermark detection performance index because of its role of robustness measure. The maximum of objective value V can be calculated with

$$V = UQI \times NC \tag{7}$$

The attacks that were utilized in the process of the objective function evaluation were: median filter, Gaussian noise and rotation for obtaining the optimal scaling factors with calculating the values of UQI and NC. In simulations of PSO, a set of parameter values are identified. In the proposed scheme, the size of the initial particle for PSO is 30, c_1 =1, c_2 =1. The PSO process is repeated until the scaling factors are optimally found. Optimization diagram for digital image watermarking using PSO is shown in Fig.2.

4 Experimental Results

In this section to evaluate the performance of the proposed watermarking scheme had been tested on the grayscale 8-bit image of size 512×512 "Lena" and the 3-level wavelet packet decomposition Daubechies 9/7 filter coefficients are used. A 32×32 binary image "UMP" is used as the watermark W and in order to eliminate the correlation of watermark image pixels and enhance system robustness and security through affine scrambling. As are shown in Fig .3. (a) and (b),respectively. And watermarked image is shown in Fig.3.(c).

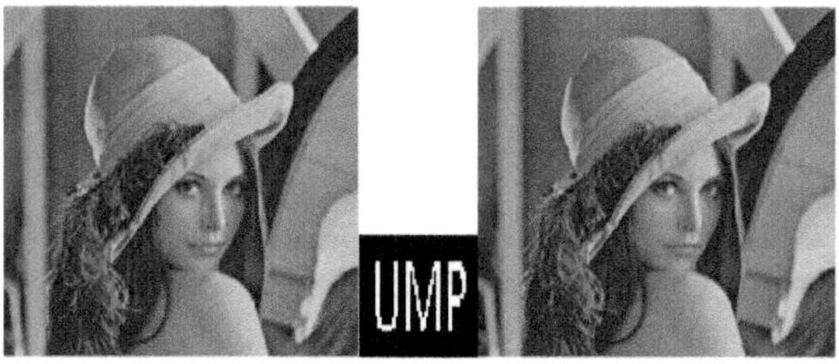

a b c

Fig. 3. (a) Original image, (b) Watermark and (c) Watermarked image

A good watermark scheme should be robust against different kind of attacks. In order to illustrate the robust nature of our watermarking scheme, robust test for various signal processing such as Gaussian filtering (0,0.003) (GF), median filtering (3×3) (MF), sharpening(SP), translating(30 pixel)(TR), rotating(30°)(RT) and cropping(30%) (CP). Table 1 presents the UQI and NC value of the detailed experiment results.

In addition, the NC value of watermark embedding at JPEG compression is evaluated over various compression factors. By comparison with several existing schemes, it is evident that the proposed scheme has the better performance than [3] and [4] as shown in Table 2.

Table 1. The experimental results under different attacks

Attacks	GF	TR	MF	RT	SP	CP
				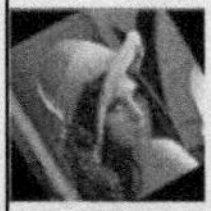		
Watermark extraction						
UQI	0.9657	0.9845	0.9261	0.9913	0.9875	0.9879
NC	0.8512	0.9214	0.7941	0.7743	0.9011	0.8791

Table 2. The comparison results with existing schemes

JPEG Quality	**[2] scheme**	**[4] scheme**	**Proposed scheme**
90%	0.9124	0.9913	1
70%	0.7743	0.8851	0.8945
60%	0. 6731	0.7643	0.8022
40%	0.5922	0.6865	0.7067

5 The References Section

Digital watermarking technique to obtain the highest possible robustness without losing the transparency is still one of the most difficult challenging issues. This paper presents an optimal robust image watermarking technique based on WPT. In this scheme, firstly, the watermark is embedded into selected sub-bands by quantization of sub-bands coefficients in wavelet packets domain, and subsequently, the scaling factors are trained by PSO which represents the intensity of embedding watermark instead of heuristics. The experimental results demonstrated that the proposed optimal watermarking scheme has strong robustness to a variety of signal processing and distortions. This simultaneously proves the more effective implementation of the novel scheme in comparison with existing schemes.

References

1. Cox, I.J., Matthew, L.M., Jeffrey, A.B., et al.: Digital Watermarking and Steganography, 2nd edn. Morgan Kaufmann Publishers, Elsevier, Burlington, MA (2007)
2. Soheili, M.R.: Redundant watermarking using wavelet packets. In: Proceedings of the 2008 IEEE/ACS International Conference on Computer Systems and Applications, pp. 591–598 (2008)

3. Chen, Y.H., Huang, H.C.: Genetic Watermarking Based on Wavelet Packet Transform. In: Proceedings of the 2009 Ninth International Conference on Hybrid Intelligent Systems, vol. 1, pp. 262–265 (2009)
4. Pun, C.M., Kong, I.K.: Adaptive Quantization of Wavelet Packet Coefficients for Image Watermarking. In: 8th WSEAS Int. Conf. on Multimedia Systems and Signal Processing, pp. 199–204 (2008)
5. Tsai, M.J., Yu, K.Y., Chen, Y.Z.: Wavelet packet and adaptive spatial transformation of watermark for digital images authentication. IEEE ICIP, 450–453 (2000)
6. Wei, Z.H., Qin, P., Fu, Y.Q.: Perceptual digital watermark of images using wavelet transform. IEEE Trans. Consumer Electron (44), 1267–1272 (1998)
7. Wang, Z., Bovik, A.C.: A Universal Image Quality Index. IEEE Signal Processing Letters 9, 81–84 (2002)

A Hardware Design for Portable Continuous Wave Diffuse Optical Tomography

Shih Kang, Shih-Yang Wu, Chih-Chung Fu, Ericson Chua, Yuan-Huang Hsu, and Wai-Chi Fang

Department of Electronics Engineering & Institute of Electronics
National Chiao Tung University
1001 Ta Hsueh Road, Hsinchu, Taiwan (R.O.C.)
wfang@mail.nctu.edu.tw

Abstract. In recent years, the rapid development of diffuse optical tomography (DOT) technology has made possible many successful applications in the field of biomedicine, such as breast cancer detection and observation of oxygenated hemoglobin distribution in the brain. In this work, we build an inexpensive and portable real-time continuous wave near-infrared (CW-NIR) DOT system hardware suitable for use in system on a chip (SOC) applications. With greatly reduced system volume, the system can pave the way for practical developments in the clinical setting. The proposed system processes digitized biomedical signals acquired from a front-end sensor circuit, and can operate in either continuous or discontinuous mode according to user settings. Finally, we demonstrate the improvements in image reconstruction associated with the two-dimensional (2D) post-processing technique employed in the proposed system.

Keywords: Diffusion Optical Tomography, CW-DOT System, Signal Processing, Portable.

1 Introduction

In recent years, DOT (Diffuse Optical Tomography) technology, due to its non-invasive and real-time capability, has been widely employed in the detection of tumors in the breast and imaging of the brain. Recent research efforts invested on DOT technology have allowed rapid progress and development and have finally paid off in recent years. DOT can be used to detect oxygenated hemoglobin (HbO) and deoxygenated hemoglobin (Hb) concentrations and volumes using bi-wavelength near-infrared. Therefore, in clinical applications, the main uses of DOT includes the monitoring of blood flow, blood volume and oxygen saturation, as well as detecting tumors within the brain and cancers of the breast [1]. By measuring different characteristics of the diffused near-infrared, DOT can be generally divided into three main categories: the Continuous Wave (CW), Frequency Domain and Time Domain. Table 1 shows the characteristics of the different DOT systems. The CW system provides advantages such as low cost, high portability, low power consumption and computation overhead, although lacking in depth information [2]. The volume of the CW-DOT system can be miniaturized which is its most attractive

Y. Zhang et al. (Eds.): DTA/BSBT 2010, CCIS 118, pp. 9–18, 2010.

advantage compared to the other algorithms. Therefore, there exists the possibility of implementing the hardware architecture for CW systems. However, few literatures have been published on the hardware architecture of CW-DOT signal processing. Most CW-DOT systems, such as [3] and [4], post-process the signal offline on a computer. Due to the bulk of such systems, the feature of portability couldn't be realized. In this paper, we propose a system on chip design for CW-DOT systems, focusing on the implementation of the signal post-processing hardware architecture. More specifically, this study demonstrates the reduction in system volume and power consumption, as well as improvements in system stability made possible by using VLSI technology.

The proposed system allows basic parameters such as detection depth, medium of reflection, scattering and absorption parameters to be configured according to user settings. Furthermore, the data acquisition scheme and delay time can be set externally. Thus, the highly configurable system will allow a more flexible application in the clinical setting. An appropriate circuit design algorithm is selected for the image reconstruction scheme, called the sub-frame image reconstruction technique [5]. This method has high performance in terms of accuracy and time consumption compared to the whole-frame technique. Therefore, the novel use of this image reconstruction algorithm to reduce the complexity of computation is a key enabling technology in the proposed CW-DOT system. However, the reduction in image reconstruction complexity causes a trade-off condition resulting in image artifacts such as discontinuities in picture edges. Consequently, the post-image processing becomes necessary before outputting the image result. The mean square error (MSE) is calculated to evaluate the accuracy between pre-processing and post-processing reconstruction images compared to the original image.

This paper is organized as follows: In section 2, a theoretical background of the CW-DOT algorithm and its image reconstruction method based on VLSI technology is given. Section 3 introduces the top-level system architecture, and illustrates the function of each module and their components. The results and comparisons are provided in Section 4 and finally a conclusion is given in Section 5.

2 Theoretical Background

DOT image reconstruction can be divided into two critical processes: determination of the forward model and inverse resolution. The forward model describes how photons are scattered and diffused in a highly scattering medium, while inverse resolution describe the optical characteristics of the medium.

2.1 Forward Model

The interaction between photons and biological matter can be classified into two types, absorption and diffusion. Typically, the effect of diffusion is larger than absorption for near-infrared in most biological media. Therefore, the transmission of photons is generally considered diffusive. Diffusion optical tomography has been developed in the past in [6], and the corresponding diffusion equation was proposed for higher scatter and lower absorption medium. For the case of CW light source, the behavioral model was represented in the literature by Eq.(1):

$$-D\nabla^2\Phi(r) + \upsilon\mu_a\Phi(r) = \upsilon S(r) \tag{1}$$

In (1) $S(r)$ is the power of optical source, $\Phi(r)$ is the photon density respect to the location, and D and μ_a are the diffusion and absorption coefficients respectively. We can understand how the photon density is distributed in biological tissue through examine the diffusion equation, so that spatial variation for diffusion and absorption can be derived. Since the absorption coefficient is more sensitive, absorption exhibits clearer variation than the diffusion coefficient. Therefore, we focus on the former, and further reconstruct its distribution in the tissue. In order to obtain the solution of diffusion equation, the Raytov approximation is used to linearize (1) [7].

To get the alterations of the absorption coefficients for more than two pairs of light source and receiver, we can express the linearized equations in matrix form as $b = Ax$, expressed in (2):

$$\begin{bmatrix} \Phi_1(r_{s1}, r_{d1}) \\ \Phi_2(r_{s1}, r_{d2}) \\ \vdots \\ \Phi_m(r_{si}, r_{dj}) \end{bmatrix} = \begin{bmatrix} a_{11} & a_{12} & \cdots & a_{1n} \\ a_{21} & a_{22} & \cdots & a_{2n} \\ \vdots & \vdots & \ddots & \vdots \\ a_{m1} & \cdots & \cdots & a_{mn} \end{bmatrix} \begin{bmatrix} \delta\mu_a(r_1) \\ \delta\mu_a(r_2) \\ \vdots \\ \delta\mu_a(r_n) \end{bmatrix} \tag{2}$$

$$\Phi_m(r_{si}, r_{dj}) = -\ln\left[\frac{\Phi(r_{si}, r_{dj})}{\Phi_{incident}(r_{si}, r_{dj})}\right] \tag{3}$$

In (2) a_{mn} represents the weighting in different locations, $\Phi_m(r_{si}, r_{dj})$ represents the light density received from different source-detector pairs, and $\delta\mu_a(r_n)$ is the change of absorption coefficient in each observed voxel.

2.2 Inverse Solution

In the inverse problem, the first step is to model the photon transmission behavior, as expressed in (2). With the $b=Ax$ matrix model, the next step is to get the variation of absorption coefficient x from known variation of light density b and weighting function matrix A, which is equivalent to solving for $x = A^{-1}b$. Some problems are introduced by the inverse process, such as the loss of light intensity information, which results in either an ill-posed problem, non-unique solution or no solution at all [8].

Therefore, matrix A generates a wide range of singular values. The Singular Value Decomposition (SVD) algorithm has proven to have very good performance in dealing with pseudo inverse, least squares fitting of data, matrix approximation and other similar ill-posed problems. The rectangular matrix $A_{m\times n}$ can be decomposed into 3 special matrices as in (4) by SVD. The columns of $U_{m\times m}$ and $V^T{}_{n\times n}$ are the eigenvectors of AA^T and A^TA . The diagonal elements of $\Sigma_{m\times n}$ are singular values of matrix A .

$$A_{m \times n} = U_{m \times m} \Sigma_{m \times n} V^T{}_{n \times n} \tag{4}$$

And Equation (5) can be written as the following form:

$$A = U \Sigma V^T \Rightarrow U^T A V = \Sigma \tag{5}$$

A practical Jacobi Singular Value Decomposition (JSVD) algorithm called the Kogbetlianz method performs two orthogonal side plane rotations to generate matrix U and V [5]. In every iteration, A_{i+1} becomes more diagonal than A_i. After n times of iteration, the matrix A becomes a diagonal matrix. The rotation equation is shown in (6). J_i^l and J_i^r are 2×2 Jacobi rotation matrices and are generated by eliminating the off-diagonal elements of A (7).

$$A_n = (J_n^l)^T (J_{n-1}^l)^T \ldots (J_0^l)^T A_0 J_0^r \ldots J_{n-1}^r J_n^r \tag{6}$$

$$J(p,q,\theta) = \begin{bmatrix} 1 & \cdots & 0 & \cdots & 0 & \cdots & 0 \\ \vdots & \ddots & \vdots & & \vdots & & \vdots \\ 0 & \cdots & \cos\theta_{pp} & \cdots & \sin\theta_{pq} & \cdots & 0 \\ \vdots & & \vdots & \ddots & \vdots & & \vdots \\ 0 & \cdots & -\sin\theta_{qp} & \cdots & \cos\theta_{qq} & \cdots & 0 \\ \vdots & & \vdots & & \vdots & & \vdots \\ 0 & \cdots & 0 & \cdots & 0 & \cdots & 1 \end{bmatrix}, \; p < q \tag{7}$$

$$\begin{bmatrix} \cos\theta & \sin\theta \\ -\sin\theta & \cos\theta \end{bmatrix}^T \begin{bmatrix} a_{pp} & a_{pq} \\ a_{qp} & a_{qq} \end{bmatrix} \begin{bmatrix} \cos\phi & \sin\phi \\ -\sin\phi & \cos\phi \end{bmatrix} = \begin{bmatrix} \sigma_1 & 0 \\ 0 & \sigma_2 \end{bmatrix} \tag{8}$$

In addition, JSVD can be performed by 2 x 2 SVD for all (p,q) pairs. The JSVD is very suitable for hardware implementation, and it can be beneficial in portable system applications.

2.3 Algorithm Technique of Image Reconstruction

Since there are six sources, twelve detectors, and ninety-six pixels for each defined image layer, a matrix size of 72 by 96 is required to completely specify the forward model. In order to avoid the computation complexity, hardware cost and the ill-posed problem of matrix inversion during calculation of absorption variation, a sub-frame image reconstruction technique is utilized. The image reconstruction problem is divided into six image blocks, and the six solutions are then recombined to form a complete reconstruction image. The size of the forward model in each sub-frame becomes 4 by 16, which means only sixteen variations of absorption coefficient have to be solved. The computation complexity is lower compared to the complete forward model, with a reasonably small loss in mean square error (MSE) accuracy [5]. Figure 1 shows the relationship between the sub-frame and whole-frame techniques.

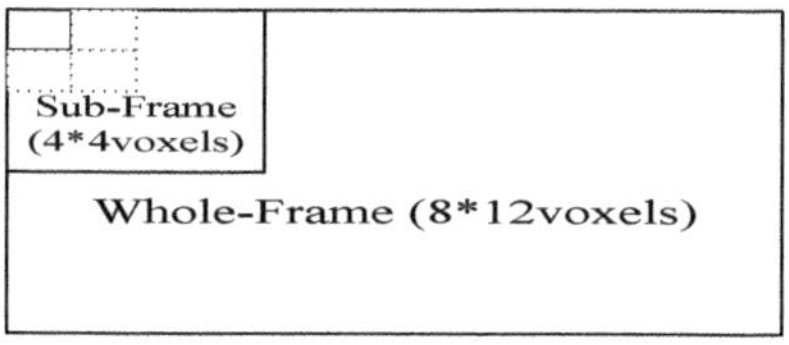

Fig. 1. The relation between sub-frame and whole-frame

3 Proposed CW-DOT System Architecture

The complete DOT system is shown in Figure 2. It consists of six processing units; a front-end interface control unit, a system control unit, a forward model processor, a Jacobi SVD engine, an image reconstructor and an image post-processor.

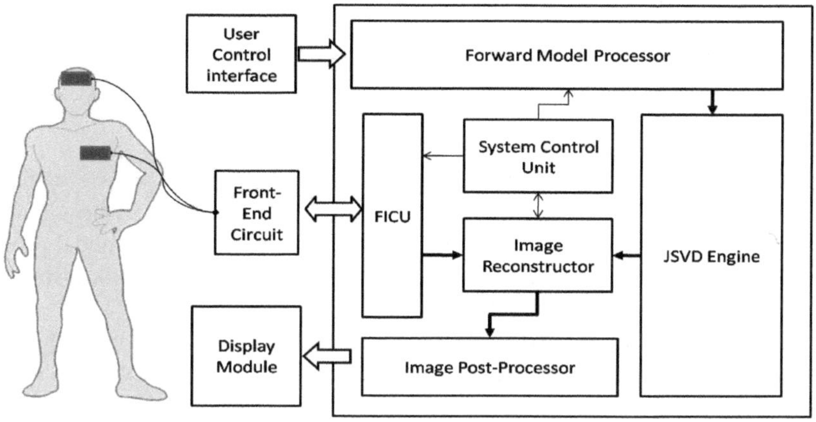

Fig. 2. Architecture of CW-DOT system

System Control Unit:

The system control unit is used to control both the off-chip sensor circuit (includes the front-end interface control unit) and the on-chip DOT system. Because the sensor circuit has six sources and twelve detectors, a ninety-six pixel image reconstruction requires a forward model with matrix size of 72 by 96. In order to reduce the computation hardware cost and avoid the ill-posed problem of matrix inversion during acquisition of absorption variation, we use a sub-frame image reconstruction technique [5]. In this system we implement two different operation modes as follows.

In mode 0, after setting the relatively parameter, the system starts to emit near infrared wave length for continuous reconstruction. This mode is used to obtain continuous images, and is a more convenient method for practical observation in clinical applications.

On the other hand, mode 1 is the preferred operating mode when the parameters are unknown such as depth of target or background coefficient of medium. The method of detection used in mode 1 is discontinuous, and uses only one image as reference after detection. To overcome this problem, the user can adjust and set parameters using the result of these images through software at a later time. In addition, using the software to change the parameters of the system can be done very quickly, and thus eventually allows the immediate application of continuous mode (mode 0) detection for clinical applications. The flowchart of the system control unit is shown in Figure 3.

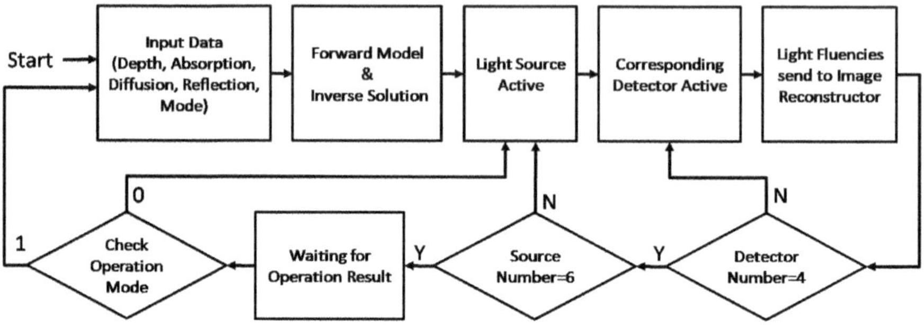

Fig. 3. Flowchart of System Control Unit

Forward Model Processor:

The forward model is used to build a theoretical model of photon transfer behavior in highly scattered tissue. The parameters of the forward model will be changed by different target or observation depth. The architecture of the forward model processor, shown in Figure 6(a), consists of four parts; a control unit, a pre-processing unit, a look-up table (storage) and a calculation module.

The control unit is a finite state machine, and controls the data flow in the forward model processor. The pre-processing module is used to modify the calculation parameter for different detective condition, while the look-up table is used to store constant physical parameters such as speed of light or distance between the light source and detector, etc.

The calculation module mainly performs the final calculation for the forward model coefficiencts. It consists of eleven operator modules, including two floating point adders, two floating point subtractors, two floating point multipliers, two floating point squaring units, one floating point divider, one floating point radical unit and one floating point exponential unit. Each coffecient of the forward model can be calculated within twenty three cycles.

The forward model processor uses a number representation based on the IEEE-754 format, and is composed of one sign bit, 8 exponent bits and 6 significant bits. Additional complexity results from the required number transformation modules between the forward model processor and Jacobi SVD engine, since other modules uses fixed-point calculations.

Jacobi SVD Engine:

The proposed JSVD processor is targeted mainly for biomedical signal processing applications, particularly portable instruments. Therefore, design requirements such as high precision, low area and low power consumption must be satisfied. The architecture of the JSVD follows from these requirements, and the JSVD is used to solve the inverse problem in the DOT system. The Architecture is shown in figure 6(b): It contains four main parts: CORDIC engine, CORDIC control unit, memory control and dual port memories. The CORDIC Engine is implemented using the basic structure in [9]. Before computation of the data, the product of $\cos\theta_r, \sin\theta_r, \cos\theta_l, \sin\theta_l$ is determined first, and the results of the products are stored as ***x, y, z,*** and ***s***. The schedule of the data flow to renew A_i is presented in Figure 4 The renewal process of U_i and V_i employ the same methodology and shares four multipliers. With this method, renewal of the two matrix elements can proceed in parallel, and memory access can be performed efficiently.

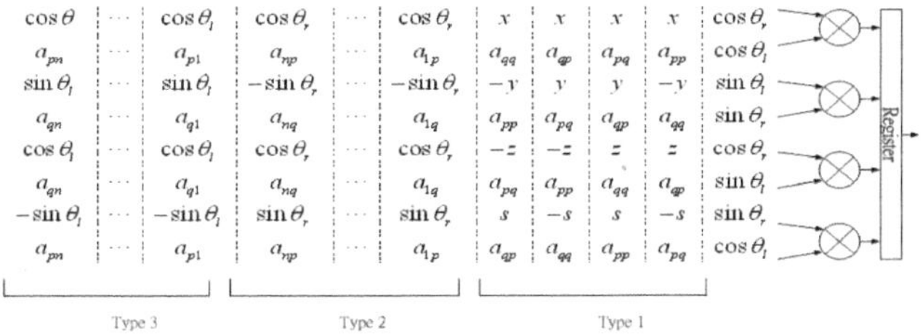

Fig. 4. First renewal flow of input data

Image Reconstructor:

Figure 6(c) shows the architecture of the image Reconstructor. The purpose of this module is to perform the inner product between the solved inverse solution and collected data to form a reconstructed image. The behavior of the module is controlled by setting the operation mode. For example, the inverse solution component is cleared each time after a complete operation when working in the first mode. The image reconstructor, based on the sub-frame algorithm, however, even though the sub-frame algorithm is effective in reducing the processing time [5], its result may be less accurate than that of the whole-frame algorithm. Therefore, the trade-off between computation effort and accuracy is an important consideration.

Image Post-Processor:

Since the system is based on the sub-frame algorithm, the resulting image has discontinuities and is of low resolution. Therefore, the image post-processor is employed to expand and smooth the reconstructed image so that the result will have better observability. Figure 6(d) shows the architecture of the image post-processor, which includes an input buffer, a control unit, a weighting array, and a central operations unit. The processor operates as follows. First, the original image is stored into the input buffer array. Next, the process control initiates and stores the weighting variables into the weighting array. Finally, the image processor reads in the input image and weight data and produces the continuous improved image output. VLSI DSP reduction techniques, such as effective reuse of components, were employed to achieve a low area complexity.

Figure 5 compares the image quality before and after image post-processing. The original image, shown in Figure 5(a), has a lot of edge discontinuities exhibited by low resolution image blocking effect. As a result, it is difficult to observe in a clinical setting. Figure 5(b) shows the result after post-processing. As we can see, the object target is better-defined with smoother edges and definite volume. The processed result has improved image resolution and will be more useful for clinical observation.

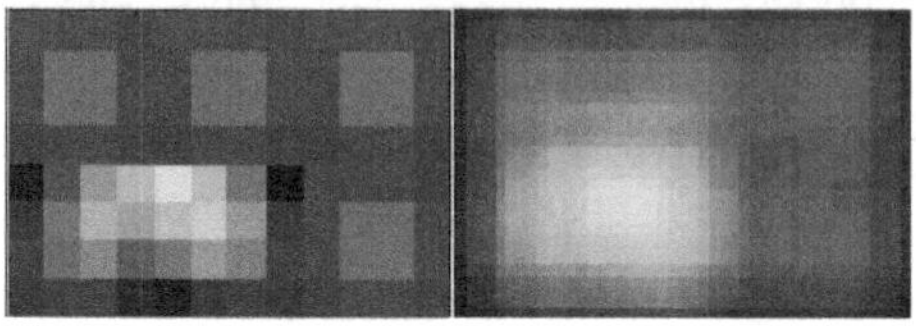

(a) Pre-Processing (b) Post- Processing

Fig. 5. Reconstructed images after pre-processing and post-processing

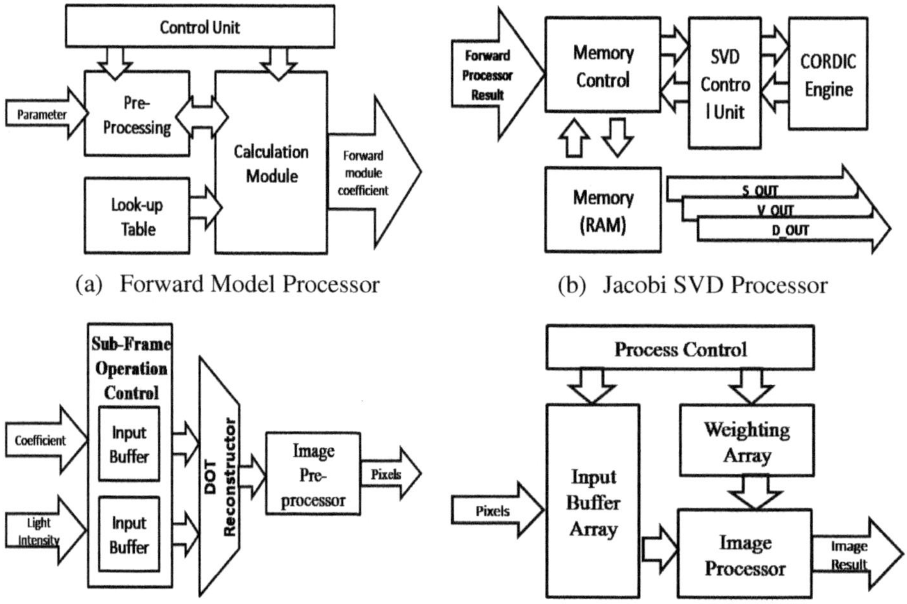

(a) Forward Model Processor (b) Jacobi SVD Processor

(c) Image Reconstructor (d) Image Post-Processor

Fig. 6. Architecture of each component

4 Results and Discussion

To test the proposed DOT system, the area of each frame is $4\times 6cm^2$ and the voxel size is $(0.25cm)^3$. The background medium is homogenous with $\mu_a^{bg} = 0.05cm^{-1}$ and the reduced scattering coefficient is $\mu_s^{'bg} = 10cm^{-1}$. Two types of inhomogeneous media were embedded at a depth of 0.5 cm below the surface. The right-side inhomogeneous medium is $\mu_a^{Anomaly1} = 0.21cm^{-1}$ and left-side medium is $\mu_a^{Anomaly2} = 0.5cm^{-1}$. The experiment result is shown in Figure 7(b) and Figure 7(c), and the original result by MATLAB simulation is shown in Figure 7(a). Table 1 shows the reconstruction image accuracy of pre-processing and post-processing compared to original image. Clearly, the original image and reconstructed image after post-processing (system result) is seen to be positively correlated with one another. These findings indicate that post-processing achieves superior reconstruction performance compared to pre-processing alone.

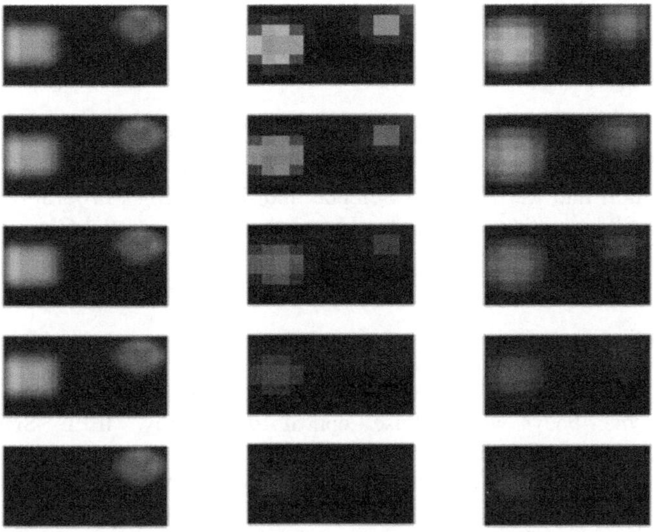

(a) Original (b)Pre-Processing (c)Post-Processing

Fig. 7. Image reconstruction results for object depths of 0.2cm, 0.4cm, 0.6cm, 0.8cm and 1cm (top to bottom)

Table 1. Image reconstruction performance of pre-processing and post-processing compared to original image using mean square error (MSE)

Depth	Preprocessing	Post-processing
(cm)	MSE	MSE
-0.2	3.191	2.599
-0.4	2.502	2.099
-0.6	3.354	2.830
-0.8	4.673	4.042
-1.0	3.412	3.252

5 Conclusion

In this paper, we confirm that the proposed CW-DOT system can be used to detect foreign target matter in a uniform medium. We highlight the novel implementation of the system architecture in VLSI technology, noting that currently there is still no such development feasible for use in portable CW-DOT systems. We propose the DOT system based on the sub-frame algorithm to be combined with other biomedical signal processors and integrated into a single SOC. With such breakthrough, the SOC will allow development of next generation portable and wireless biomedical devices, benefitting more doctors, patients, and researchers alike.

Finally, the chip area and working frequency of the proposed CW-DOT system architecture implemented in UMC 90nm technology are 900um x 900um and 650KHZ, respectively, with an estimated power consumption of 1.13mW using Synopsys Prime Power.

Acknowledgement

This work was performed at NCTU System-on-Chip research center under the grants NSC99-2220-E-009-028 and NSC99-2220-E-009-030 sponsored by the National Science Council and the National Science and Technology Program for System-on-Chip, Taiwan. The authors would also like to thank the National Chip Implementation Center for CAD tool support.

References

1. Boas, D.A., Brooks, D.H., Miller, E.L., DiMarzio, C.A., Kilmer, M., Gaudette, R.J.: Imaging the body with diffuse optical tomography. IEEE Signal Processing Magazine 18(6), 57–75 (2001)
2. Strangman, G., Boas, D.A., Sutton, J.P.: Non-invasive neuroimaging using near-infrared light. Biological Psychiatry 52, 679–693 (2002)
3. Siegel, A., Marota, J.J., Boas, D.: Design and evaluation of a continuous-wave diffuse optical tomography system. Opt. Express 4, 287–298 (1999)
4. Lin, Y., Lech, G., Nioka, S., Intes, X., Chance, B.: Noninvasive, low-noise, fast imaging of blood volume and deoxygenation changes in muscles using light-emitting diode continuous-wave imager. Review of Scientific Instruments 73(8), 3065–3074 (2002)
5. Hsu, Y.-H., Fu, C.-C., Fang, W.-C., Sang, T.-H.: A VLSI-inspired image reconstruction algorithm for continuous-wave diffuse optical tomography systems. In: IEEE/NIH Life Science Systems and Applications Workshop, LiSSA 2009, April 9-10, pp. 88–91 (2009)
6. Cheng, X., Boas, D.A.: Diffuse optical reflection tomography with continuous-wave illumination. Med. Biol. 42, 841–854 (1997)
7. Farrell, T.J., Patterson, M.S., Wilson, B.: A diffusion theory model of spatially resolved, steady-state diffuse reflectance for the noninvasive determination of tissue optical properties in vivo. Medical physics 19, 879 (1992)
8. Arridge, S.: Optical tomography in medical imaging. Inverse problems 15, 41–41 (1999)
9. Cavallaro, J.R., Luk, F.T.: CORDIC arithmetic for an SVD processor. Journal of parallel and distributed computing 5, 271–290 (1988)

An XML-Based Data Model for Moving Object Database

Xiajun Jiang, Yuping Zhang, and Dechang Pi

Nanjing University of Aeronautics and Astronautics, College of Information Science & Technology, 210016, Nanjing
{xiajunja,ypzhang,dc.pi}@nuaa.edu.cn

Abstract. An XML-based data model to represent moving points and moving regions is provided in this paper. The algorithms of two important types of queries are designed for the model. They are snapshot query and trajectory query. The moving object dataset stored in our XML-based model is generated, and experiments on the storage and query methods of spatio-temporal attributes are done on the basis of the dataset. The generation algorithm of moving regions is researched in details in this paper, and the shape change of regions is considered in the algorithm. Results show that using XML technology to store and query spatio-temporal data is especially efficient and flexible.

Keywords: Moving object database, XML, Data model, Query processing.

1 Introduction

Moving objects are entities with spatio-temporal information, such as persons, vehicles, ships and aircrafts in the battlefield. Moving object database (MOD) is especially useful for recording the location and location changes of objects, and it also provides functions related to storage, query and update of spatio-temporal information. MOD systems often have too many entities, and every entity contains its whole historical informations and current information, so the data volume is very large. The motivation of MOD applications is to know the movement condition of entities well, to optimize the path of entities and to solve problems encountered while moving, etc.

Spatial databases and temporal databases are the basis of MOD. But previous works (such as [1,2,3]) show that simply extending a spatial data model to include temporal data, or vice versa, will result in inefficient representations for spatio-temporal data. The positions of moving objects change frequently. The values of other attributes such as entity identifiers or names are constant or change rarely. Representing all the information in relational databases will result in denormalization, or will result in too many joining operations among tables while querying. Reference [4] considers that XML will be the most useful and important language for semi-structured data in the future. So XML is a better choice for modeling MOD. Furthermore, technologies such as DOM or XPath will help us to implement the operations of database systems.

2 Related Work

The spatial databases and temporal databases have been both researched extensively before the MOD being brought forward. Reference [5] lists a comprehensive

Y. Zhang et al. (Eds.): DTA/BSBT 2010, CCIS 118, pp. 19–28, 2010.

bibliography on spatio-temporal databases. But up to now, models designed for spatio-temporal databases are still in their infancy[6]. ROSE algebra[7, 8] designs the data structures, algorithms and implementations for spatial data types. Based on them, reference [9] completes the systematic design and formal definition of data types and operations at the abstract level for moving objects. We represent the moving objects in our XML-based model according to the algebra.

Trajectory query is very important to many applications of MOD[10], such as battlefield simulations[11] and transportation systems[12]. Reference [13,14,15] study the storage and query methods of trajectory data, reference [16] gives some kinds of spatio-temporal pattern queries based on the trajectory, and reference [17] discusses the data security for trajectory query. On the basis of XML model, we will focus our work on the algorithm of trajectory query.

Reference [18] considers that XML database has two types: Native XML database and XML-Enabled database. The former designs special logic models and storage methods. It's unnecessary to design algorithms to transform data format. The examples of native XML database systems include Tamino[19], Timber[20], Lore[21], Natix[22], etc. The latter is based on the traditional relational model or object-oriented model. Data in XML database are translated firstly according to the format of bottom database, and then stored to files of tables. For the complication of spatial and temporal data, the native XML database is more appropriate to MOD. In fact, native XML database has already been used in some GIS applications.

3 XML-Based Data Model for Moving Objects

In our XML model, we only take into account the space and time information of moving objects, and neglect the rest attributes. The time information is treated as attribute of an element in XML document, and the space information is element.

3.1 The Representation of Moving Points

In a 2-D space, a point is represented as (X, Y). The location of a moving point is function of time. In our model, line segments are used as the basic structure for a moving point's trajectory.

Definition 1. The following XML segment shows XML-based representation of moving points.

```
<MPoint ID="Entity-Identifier" s="Point-StartTime " e="Point-EndTime">
 <MemberuPoints>
  <MemberuPoint>
   <uPoint s="UnitPoint-StartTime" e="UnitPoint-EndTime">
    <Point0>
     <x>X</x>
     <y>Y</y>
    </Point0>
    <Point1>
     <x>X</x>
     <y>Y</y>
    </Point1>
   </uPoint>
  </MemberuPoint>
  ......
 </MemberuPoints>
</MPoint>
```

The elements and attributes must satisfy the following conditions:

(1) $$[\text{Point-StartTime}, \text{Point-EndTime}) \supseteq \bigcup_{\forall \text{MemberuPoint} \in \text{child(MemberuPoints)}} [\text{UnitPoint-StartTime}, \text{UnitPoint-EndTime})$$

The [UnitPoint-StartTime, UnitPoint-EndTime) can't intersect each other. The function child(MemberuPoints) gets the child element set of <MemberuPoints>.

(2) As child elements of a <uPoint>, <Point0> represents the location of the moving point at UnitPoint-StartTime, and <Point1> represents the location of the moving point at UnitPoint-EndTime. The trajectory of the point between UnitPoint-StartTime and UnitPoint-EndTime is a line determined by <Point0> and <Point1>.

(3) As child elements of a <MemberuPoints>, two neighbor <MemberuPoint> have spatial relation. The descendant element <Point1> of a <MemberuPoint> meets with the descendant element <Point0> of the succedent <MemberuPoint>.

3.2 The Representation of Moving Regions

A regular region can contain holes. In this paper, we assume that a set of line segments form the boundary of a region, and a region has no holes. A region can change its shape when moving. For the reason of simplicity, we assume that the shape does not change in this section.

Definition 2. The following XML segment shows XML-based representation of moving regions.

```
<MRegion ID="Entity-Identifier" s="Region-StartTime" e="Region-EndTime">
 <OuterMemberCycle>
  <MemberuPoints s="UnitPoints-StartTime" e="UnitPoints-EndTime" >
   <MemberuPoint>
    <Point1>
     <x>X</x>
     <y>Y</y>
    </Point1>
    <Point2>
     <x>X</x>
     <y>Y</y>
    </Point2>
   </MemberuPoint>
   ......
  </MemberuPoints>
  ......
 </OuterMemberCycle>
</MRegion>
```

The elements and attributes must satisfy the following conditions:

(1) $$[\text{Region-StartTime}, \text{Region-EndTime}) \supseteq \bigcup_{\forall \text{MemberuPoints} \in \text{child(OuterMemberCycle)}} [\text{UnitPoints-StartTime}, \text{UnitPoints-EndTime})$$

The [UnitPoints-StartTime, UnitPoints-EndTime) can't intersect each other. The function child(OuterMemberCycle) gets the child element set of <OuterMemberCycle>.

(2) As descendant elements of a <MemberuPoints>, all <Point1> constitute the vertex set of a region at UnitPoints-StartTime. Lines connecting <Point1> and < Point2> in the same <MemberuPoint> of the <MemberuPoints> constitute the outer cycle of the region at UnitPoints-StartTime.

(3) As child elements of a <OuterMemberCycle>, two neighbor <MemberuPoints> have spatial relation. <MemberuPoints> represents the boundary of a region at UnitPoints-StartTime, and the boundary at UnitPoints-EndTime is determined by the next <MemberuPoints>. The region's positions and shapes between UnitPoints-StartTime and UnitPoints-EndTime are calculated by interpolation functions.

Figure 1 shows an example of a moving region. The trajectory in the figure is actually the center's trajectory of the region. Each key frame of the region is mapped to <MemberuPoints>.

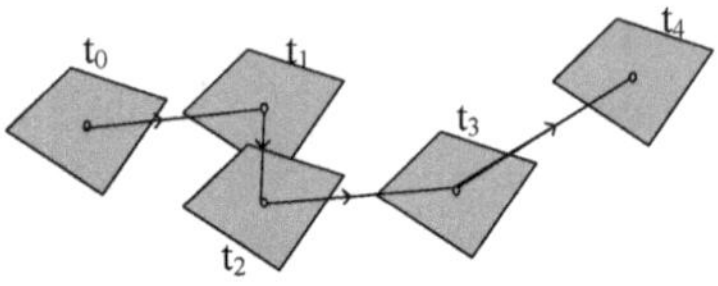

Fig. 1. The trajectory of a moving region

4 Algorithms for Spatio-temporal Queries in XML MOD

We focus our attention on operations related to space and time information. We only discuss two typical types of spatio-temporal queries in this paper. Following describes the meaning of the queries.

Q1: Querying the locations of all moving objects at time *t* in a system. We also call it snapshot query.

Q2: Querying the trajectory of a moving object at a time interval. We also call it trajectory query.

4.1 Snapshot Query

Q1 takes only one argument, namely the time *t*, and outputs two object sets containing location and shape information (only for moving regions) of all live moving objects at time *t*.

Procedure: **SnapshotQuery(*t*)**

1. Check all child elements of <MODEntities>. If $t \in$[*-StartTime,*-EndTime), go step 2 (* stands for Point or Region according to the object type in XML document).

2. Check the type of moving objects. If the element name is <MPoint>, go step 3. If it is <MRegion>, go step 4.

3. For points, check all <uPoint>. If $t \in$[UnitPoint-StartTime,UnitPoint-EndTime), get the value of <x> and <y>, and give the value to X_{Point0}, X_{Point1}, Y_{Point0} and Y_{Point1} respectively. Calculate the location of the moving point at time *t* as follows.

$$X_t = X_{\text{Point0}} + \frac{X_{\text{Point1}} - X_{\text{Point0}}}{\text{UnitPoint-EndTime} - \text{UnitPoint-StartTime}} \times (t - \text{UnitPoint-StartTime})$$

$$Y_t = Y_{\text{Point0}} + \frac{Y_{\text{Point1}} - Y_{\text{Point0}}}{\text{UnitPoint-EndTime} - \text{UnitPoint-StartTime}} \times (t - \text{UnitPoint-StartTime})$$

(X_t, Y_t) is the location of the moving point at time *t*. Put (X_t, Y_t) into the result set (SnapshotPointSet).

4. For regions, check all <MemberuPoints>. If $t \in$ [UnitPoints-StartTime, UnitPoints-EndTime), count the <MemberuPoint> to n. n is the vertex number of the region at time t. The location of i'th vertex is calculated as follows.

(1) Get the value of <x> and <y> of < Point1>, which father element is the i'th child element of <MemberuPoints>. Give the value to *PreX* and *PreY* respectively.
(2) For the next <MemberuPoints>, get the value of <x> and <y> of <Point1>, which father element is the i'th child element of <MemberuPoints>. Give the value to *X* and *Y* respectively.
(3) Calculate the position of the i'th vertex of the region at time *t* as follows.

$$X_{t,i} = \mathrm{Pr}eX + \frac{X - \mathrm{Pr}eX}{\text{UnitPoints - EndTime} - \text{UnitPoints - StartTime}} \times (t - \text{UnitPoints - StartTime})$$

$$Y_{t,i} = \mathrm{Pr}eY + \frac{Y - \mathrm{Pr}eY}{\text{UnitPoints - EndTime} - \text{UnitPoints - StartTime}} \times (t - \text{UnitPoints - StartTime})$$

The vertex set $\{(X_{t,i}, Y_{t,i})\}$ constitutes the targeted region at time t. Put the vertex set $\{(X_{t,i}, Y_{t,i})\}$ into the result set(SnapshotRegionSet).

5. Return the SnapshotPointSet and SnapshotRegionSet.

4.2 Trajectory Query

Q2 takes three arguments, namely *EntityID*, *StartTime* and *EndTime*. Q2 outputs an object set, which contains location and shape information (only for moving regions) of *EntityID*'s trajectory at [StartTime, EndTime).

Procedure: **TrajectoryQuery(*EntityID, StartTime, EndTime*)**

1. Check all child elements of <MODEntities>. If the name is MPoint and the ID equals to EntityID, go step 2. If the name is MRegion and the ID equals to EntityID, go step 3.

2. The following process queries the moving point's trajectory at [StartTime, EndTime).

(1) Check all <uPoint>, which are descendants of the <MPoint>, and find the <uPoint> on condition that *StartTime*$\in$[UnitPoint-StartTime,UnitPoint-EndTime). Get the value of <x> and <y>, and give the value to X_{Point0}, X_{Point1}, Y_{Point0} and Y_{Point1} respectively. Calculate the start position (X_{t0}, Y_{t0}) of the moving point's trajectory as follows.

$$X_{t0} = X_{\text{Point0}} + \frac{X_{\text{Point1}} - X_{\text{Point0}}}{\text{UnitPoint - EndTime} - \text{UnitPoint - StartTime}} \times (StartTime - \text{UnitPoint - StartTime})$$

$$Y_{t0} = Y_{\text{Point0}} + \frac{Y_{\text{Point1}} - Y_{\text{Point0}}}{\text{UnitPoint - EndTime} - \text{UnitPoint - StartTime}} \times (StartTime - \text{UnitPoint - StartTime})$$

Put (X_{t0}, Y_{t0}) into moving point's result set (TrajectoryPointSet).

(2) Start from the <uPoint> located in step (1), get the child element <Point0>'s coordinate (X_{ti}, Y_{ti}) of <uPoint> in turn, and put them into result set (TrajectoryPointSet), until find the <uPoint> on condition that *EndTime*$\in$[UnitPoint-StartTime, UnitPoint-EndTime). Calculate the end position (X_{tn}, Y_{tn}) of the trajectory at *EndTime* as step (1). Put (X_{tn}, Y_{tn}) into moving point's result set (TrajectoryPointSet).

3. The following process queries the moving region's trajectory at [StartTime, EndTime).

(1) Check all <MemberuPoints>, which are descendants of <MRegion>, and find the <MemberuPoints> on condition that $StartTime \in$ [UnitPoints-StartTime,UnitPoints-EndTime). Calculate the i'th vertex of the start region as follows.

Find <Point1> of the i'th child element of <MemberuPoints>, get the value of <x> and <y> of < Point1>, and give the value to $PreX_i$ and $PreY_i$ respectively.

Find <Point1> of the i'th child element of the next <MemberuPoints>, get the value of <x> and <y> of < Point1>, and give the value to $X_{i,\ Point1}$ and $Y_{i,\ Point1}$ respectively.

The (X_i,Y_i) of i'th vertex of the start region is

$$X_i = \mathrm{Pr}eX_i + \frac{X_{i,\mathrm{Point1}} - \mathrm{Pr}eX_i}{\text{UnitPoints-EndTime} - \text{UnitPoints-StartTime}} \times (\text{StartTime} - \text{UnitPoints-StartTime})$$

$$Y_i = \mathrm{Pr}\,eY_i + \frac{Y_{i,\mathrm{Point1}} - \mathrm{Pr}\,eY_i}{\text{UnitPoints - EndTime} - \text{UnitPoints - StartTime}} \times (\text{StartTime} - \text{UnitPoints - StartTime})$$

The vertex set $\{(X_i,Y_i)\}$ constitute Region_{t0}. Put Region_{t0} into result set (Trajectory-RegionSet).

(2) Start from the <MemberuPoints> located in step (1), get the descendant element < Point1>'s coordinate (X_i, Y_i) of <MemberuPoints> in turn. These (X_i, Y_i) constitute a vertex set $\{(X_i, Y_i)\}$ of Region_{tj}, which is the j'th boundary of the trajectory. Put the vertex set $\{(X_i, Y_i)\}$ into result set (TrajectoryRegionSet). Until find the <MemberuPoints> on condition that $EndTime \in$ [UnitPoints-StartTime,UnitPoints-EndTime), calculate the last boundary of the moving region's trajectory as step (1). Put Region_{tn} into result set (TrajectoryRegionSet).

4. Return the result set of the moving point's or moving region's trajectory.

5 Dataset Generation of Moving Objects

The datasets are generated by simulation. We assume that the simulation starts at time 0, and all moving objects are alive in the simulation without any object adding. The shapes of moving regions are constant. The velocity and orientation of moving objects and the adjustment algorithm refer to [23].

If all vertexes are generated, the shape of a region is determined. Figure 2 shows the principle of vertex generation. If a region contains n vertexes, we give every vertex a number from 1 to n. First, we produce reference lines for every vertex. The center of a region is the origin of the auxiliary coordinate system, and the reference lines are real lines starting from the origin. The angle of reference line i to x axis is determined by the following formula: $(i\text{-}1)*360/n$. Second, we determine the angles between lines from origin to every vertex and the x axis as follows: generate a value of *Angle* at interval (-DeltaAngle, DeltaAngle) satisfying uniform distribution, so the angle of i'th vertex can be calculate by $(i\text{-}1)*360/n+Angle$. Last, the distance from origin to every vertex distributes at a certain interval.

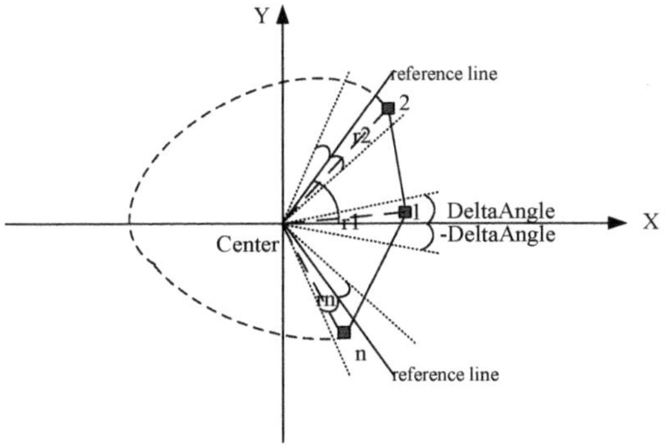

Fig. 2. Vertex generation of a moving region

The regions can rotate when moving, and sometimes they even can change their shapes. We neglect the rotation of regions in this paper, and in our experiment we will consider the simplest shape change.

6 Experiment

The shape of a moving region is determined by its vertexes. The number of <MemberuPoint> contained in a <MemberuPoints> corresponds to the vertex number of a region. Actually, all <Point1> or <Point2> contained in a <MemberuPoints> are vertexes for the region at time determined by the attribute "s" of the same <Memberu Points>. If shapes of the region change, the vertex number of its neighbor <Memberu Points> is different. We add attribute "add" or "delete" of the <MemberuPoints> to determine the changes of region shape. In our model, if no vertex changes at time determined by the value of attribute "s", the value of attribute "add" or "delete" equals -1. The value of the attribute "add" or "delete" other than -1 also determines the position of a vertex to be added or deleted. We assume that every time only one vertex will be added or deleted, and the probability of addition or deletion is little (about 10 percent of all <MemberuPoints> in the experiment).

As shows in figure 3, a vertex is added to a region which ID is 7 at time 65, and the position number of the vertex added is 0 (add="0"). At time 87, a vertex which position number is 1 (delete="1") in the former <MemberuPoints> is deleted. The position of the vertex to be added or deleted is generated stochastically. In this example, the region at interval (57, 65) or (83, 87) will be calculated by the interpolation algorithm. The number of vertexes of a region is the same as the child element number of <MemberuPoints> that has more vertexes.

Figure 4 shows some results in our experiment. At time 65, a new vertex which position is determined by its neighbor vertexes is added. At time 87, a vertex is deleted. The dashed rectangles in figure are MBRs of the regions.

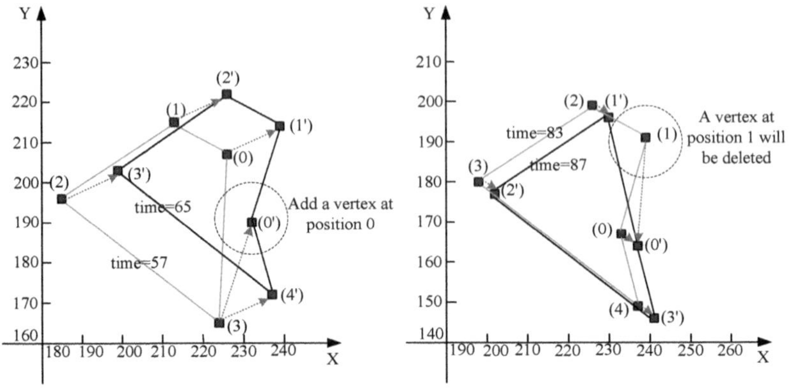

Fig. 3. The determination of vertex between two neighbor <MemberuPoints> considering adding or deleting

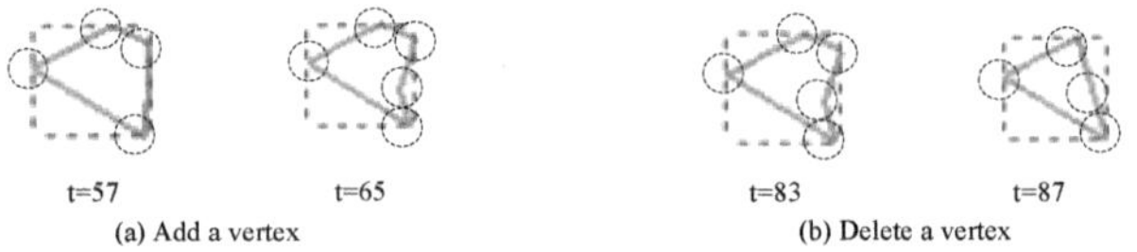

Fig. 4. The results of the experiment after vertex adding or deleting

Figure 5 shows the trajectory query results in the experiment. The dashed circles contain a new added vertex or a vertex to be deleted. The dashed lines are trajectories of the centers of the regions' MBR.

Fig. 5. The trajectory of region 7 at some intervals

7 Conclusion and Future Work

The XML-based representation of moving objects is presented in this paper. We only consider two types of moving objects: points and regions. The algorithms of snapshot query and trajectory query are designed. Compared with traditional databases, XML-based MOD are very flexible, especially for the representation of space and time information. The type of coordinates can change arbitrarily in a document, such as geocentric coordinate system, longitude and latitude system, etc. Information of a moving object cluster together. It's very efficient for querying objects' all life information. Like other database systems, we can also design XML indices to improve query efficiency.

Other than the methods designed in this paper, the motion equation can be recorded in XML documents, altogether with the trend estimation equation and its adjustment algorithms. We will further study the MOD dataset generation algorithms especially for moving regions and their shape changes when moving. We will also design the query optimization methods and spatio-temporal data mining algorithms for our XML MOD in the future. XML index is one of the important aspects among them.

Acknowledgments. This research was supported by NUAA Research Funding, NO.NS2010099.

References

1. Peuquet, D.J.: Making Space for Time: Issues in Space-time Data Representation. GeoInformatica 5(1), 11–32 (2001)
2. Peuquet, D.J.: It's about Time: A Conceptual Framework for the Representation of Temporal Dynamics in Geographic Information Systems. Annals of the Association of American Geographers 84, 441–461 (1994)
3. Sellis, T.K.: Research Issues in Spatio-temporal Database Systems. In: Güting, R.H., Papadias, D., Lochovsky, F.H. (eds.) SSD 1999. LNCS, vol. 1651, p. 5. Springer, Heidelberg (1999)
4. Quanzhong, L., BongKi, M.: Indexing and Querying XML Data for Regular Path Expressions. In: Proceedings of the 27th Intl. Conf. on VLDB, Roma, Italy, pp. 361–370 (2001)
5. Al-Taha, K., Snodgrass, R.T., Soo, M.D.: Bibliography on Spatio-Temporal Databases. International Journal of Geographical Information Science 8(1), 95–103 (1994)
6. Erwig, M., Gueting, R.H., Schneider, M., Vazirgiannis, M.: Spatio-temporal Data Types: An Approach to Modeling and Querying Moving Objects in Databases. GeoInformatica 3(3), 269–296 (1999)
7. GuÈting, R.H., Schneider, M.: Realm-Based Spatial Data Types: The ROSE Algebra. VLDB Journal 4, 100–143 (1995)
8. GuÈting, R.H., de Ridder, T., Schneider, M.: Implementation of the ROSE Algebra: Efficient Algorithms for Realm-Based Spatial Data Types. In: Proc. of the 4th Intl. Symposium on Large Spatial Databases, Portland, pp. 216–239 (1995)
9. GuÈting, R.H., BoÈhlen, M.H., Erwig, M., Jensen, C.S., Lorentzos, N., Schneider, M., Vazirgiannis, M.: A Foundation for Representing and Querying Moving Objects. ACM Transactions on Database Systems 25(1), 1–42 (2000)
10. Wolfson, O., Xu, B., Jiang, L., Chamberlain, S.: Moving Objects Databases: Issues and Solutions. In: 10th International Conference on Scientific and Statistical Database Management, Capri, Italy, pp. 111–123 (1998)
11. Chamberlain, S.: Automated Information Distribution in Bandwidth-constrained Environments. In: Military Communications Conference, Fort Monmouth, NJ, USA, pp. 537–541 (1994)
12. OmniTRACS. Communicating Without Limits, `http://www.qualcomm.com/ProdTech/Omni/prodtech/omnisys.html`
13. Pfoser, D., Jensen, C.J., Theodoridis, Y.: Novel Approaches in Query Processing for Moving Objects Trajectories. In: Proceedings of the 26th Intl. Conf. on VLDB, San Francisco, CA, USA, pp. 395–406 (2000)

14. Prasad Chakka, V., Adam Everspaugh, A., Patel, J.M.: Indexing Large Trajectory Data Sets With SETI. In: 1st Biennial Conf. on Innovative Data Systems Research, Asilomar, CA, USA (2003)
15. Tao, Y., Papadias, D.: MV3R-Tree: A Spatio-Temporal Access Method for Timestamp and Interval Queries. In: Proceedings of the 27th Intl. Conf. on VLDB, Rome, Italy, pp. 431–440 (2001)
16. Zhang, W., Li, J., Zhang, W.: Spatio-temporal Pattern Query Processing based on Effective Trajectory Splitting Models in Moving Object Database. In: Proceedings of the First International Multi-Symposiums on Computer and Computational Sciences, Washington, DC, USA, pp. 540–547 (2006)
17. GkoulalasDivanis, A., Verykios, V.S.: A Privacy-aware Trajectory Tracking Query Engine. ACM SIGKDD Explorations Newsletter archive 10(1), 40–49 (2008)
18. Meng, X.F., Luo, D.F., Lee, M.L.: OrientStore: A Schema Based Native XML Storage System. In: Proc. of the 29th Intl. Conf. on VLDB, Berlin, German, pp. 1057–1060 (2003)
19. Tamino, S.H.: A DBMS Designed for XML. In: Proc. 17th Int. Conf. on Data Engineering, Heidelberg, Germany, pp. 149–154 (2001)
20. Jagadish, H.V., Khalifa, S.A.: TIMBER: A Native XML Database. The VLDB Journal 11(4), 274–291 (2002)
21. McHugh, J., Abiteboul, S., Goldman, R.: Lore: A Database Management System for Semistructured Data. SIGMOD Record 26(3), 54–66 (1997)
22. Fiebig, T., Helmer, S., Kanne, C.C.: Anatomy of a Native XML Base Management System. The VLDB Journal 11(4), 292–314 (2002)
23. Theodoridis, Y., Silva, J.R.O., Nascimento, M.A.: On the Generation of Spatiotemporal Datasets. In: Güting, R.H., Papadias, D., Lochovsky, F.H. (eds.) SSD 1999. LNCS, vol. 1651, pp. 147–164. Springer, Heidelberg (1999)

Improving Response Time, Availability and Reliability through Asynchronous Replication Technique in Cluster Architecture of Web Server Cluster

Aznida Hayati Zakaria[1], Wan Suryani Wan Awang[1], Zarina Mohamad[1], Ahmad Nazari Mohd Rose[1], and Mustafa Mat Deris[2]

[1] Faculty of Informatics, Universiti Sultan Zainal Abidin, Gong Badak Campus, Gong Badak, 21300, Kuala Terengganu, Terengganu, Malaysia
{aznida,suryani,zarina,anm}@udm.edu.my
[2] Fakulti Teknologi Maklumat & Multimedia, Universiti Tun Hussein Onn Malaysia, Beg Berkunci 101 Parit Raja, Batu Pahat, 86400 Johor Malaysia
mmustafa@uthm.edu.my

Abstract. In Web Server Cluster (WSC), replication techniques are necessary when accessing data from multiple locations whether it is from a local area network environment or geographically distributed worldwide. In WSC, for the sake of availability and reliability when sharing or backing-up data, the best alternative would be to replicate a copy of the data in the server across network. Thus providing reliability and efficient services are our primary goals in designing a cluster architecture data replication scheme. However, the primary concern for the technique is not only to provide high availability and reliability but also to boost up the performance of data replication: to minimize the response times and to maximize the throughput. This paper proposes an algorithm of data replication scheme based on asynchronous replication in order to improve web server cluster system reliability. The proposed technique also manages to reduce to a very minimal response time while providing maximum throughput.

Keywords: web server cluster, data replication, reliability, response time, throughput.

1 Introduction

Applications in finance and telecommunications are areas that generate an ever increasing interest in distributed and replicated data management [1]. The boost in this popularity has resulted in large bandwidth demands and notoriously high latencies experienced by end users. To combat these problems the benefits of data replication were early recognized by the research community especially data replication in cluster systems. Beside data replication, one of the main operational aspects of any distributed system is the demand for a reliable web server. The need for reliability in web server cluster also likely to increase due to the ever increasing applications in the world wide web (WWW) [2]. When components fail, a reliable cluster system should still be able to continue executing user requests [3]. As such, in order to provide reliable services, WSC needs to maintain the availability of some

Y. Zhang et al. (Eds.): DTA/BSBT 2010, CCIS 118, pp. 29–36, 2010.

data replicas while preserving one-copy consistency among all replicas [3]. Furthermore, data replication is also concerned with its performance. For example, an application might access a local database at local server rather than at remote server in order to achieve minimum response time and maximum throughput. Improving the performance of the replication technique is therefore a major task in order for WSC systems to become more efficient.

Replication is more than copying. It refers to the act or result of reproducing. As such, any type of data processing object can be implemented. The definition describes replication as the act of reproducing. Therefore replication is much more than simply the copying of any object; it must also address the management of the complete copying a process [4]. Furthermore replication should encompass the administration and monitoring of a service that guarantees data consistency across multiple sites in a distributed environment. Reliability refers to the probability that the system under consideration does not experience any failure in a given time interval. Thus, a reliable WSC is one that can continue to process user's requests even when the underlying system is unreliable [5]. When components fail, it should still be able to continue executing the requests without violating the database consistency.

The most common approaches to replication are the synchronous and asynchronous replications. The synchronous replication is the replication of data from each transaction at the same time updating all copies elsewhere. This technique ensures that a remote copy of data, which is identical to primary copy, is created at the time the primary copy is updated. This means that the latency between data consistency is zero. Data in all nodes/replicas is always the same, no matter from which replica the updated originated. However, synchronous replication has drawbacks in practice [6]: the response time to execute an operation is high. While, the asynchronous replication works by taking a copy of the data and replicate it to a remote site, on a continuous basis. A remote copy of the data is created at another time after the primary copy is updated. Thus, the response time is lower than that of the synchronous replication technique.

This paper proposes an algorithm of data replication scheme based on asynchronous replication in order to improve web server cluster system reliability. The proposed technique also manages to reduce to a very minimal response time while providing maximum throughput.

The rest of the paper is organized as follows. Section 2 reviews asynchronous data replication technique. Section 3 describes asynchronous data replication algorithm. In Section 4 we present an analytical model of the asynchronous replication technique and performance analysis. Section 5 summarizes our findings and concludes the paper.

2 Asynchronous Data Replication Technique

In the design of the WSC, a client on the Internet will notice only one IP address coming from the cluster, not those individual servers in the cluster. The cluster (with only one IP address visible to the public) is composed of a node called Request Distributor Agent (RDA) and a group of servers. The servers are logically connected to each other in the form of a grid structure, each of which is connected to RDA. RDA is designed to perform two jobs. Firstly, RDA will communicate with client whereby in this case the RDA will forward legitimate Internet requests to the appropriate servers in the cluster. It returns any replies from the servers back to the clients. Secondly, the RDA act as a manager to control the six servers in the cluster.

As a manager, the RDA will communicate with one of the primary server depending on the request needs. For example if a client need to access a file x in server-2, the RDA will forward request to server-2. Since file x is also available in other servers, in case of failure in server-2, file x is still accessible. In the implementation proposed the RDA through rsync will monitor the file changes in each server and needs to know the source and destination for the copy operation in this cluster server environment.

One advantage of a server cluster over a monolithic server is its high security. If a monolithic server is used, it is reachable from the Internet and therefore vulnerable for vicious [5]. Only the RDA has the IP address visible from the Internet, and all other stations of the cluster bear only private IP address. Therefore, all cluster-server stations are not reachable directly from the outside. A firewall system may be installed on the RDA to protect the whole cluster. To attack one of the cluster server stations, one has to first land on RDA and launches an attack from there. Network Address Translation is used on RDA to translate the destination IP address of incoming packets to an internal IP address, and that of the outgoing packets to the IP address on Internet where the requests.

3 An Analytical Model and Algorithm of the Asynchronous Data Replication Technique

This section presents the mathematical implementation of closed queuing network model of distributed database based on asynchronous data replication technique. The basic entities in queuing network models are service centers, which database nodes, and customers, which represent users or transactions. Table 1 shows the parameters and example of parameter values for three (3) transaction types.

Table 1. Model Parameters

Parameter	Meaning
n	Number of sites
τ	Number of transaction types
a_i	Percentage of transactions of type i
q_i	Function to distinguish between queries and updates
λ_i	Transaction arrival rate
t_i	Mean service time for a transaction of type i
k	Coherency index
Bps	Network bandwidth (bps)
ℓ_i	Probability of local transaction execution
$t_c^{send\_i}$	Mean time to send a transaction of type
$t_c^{return\_i}$	Mean time to return query results

3.1 Transaction Processing and Arrival Rates

The arrival of the update and query transactions from every node is assumed to be a Poisson process, and then their sum is also Poisson. Update transactions are assumed to be propagated asynchronously to the secondary copies. Furthermore, transactions are also assumed to be executed at a single site, either the local or a remote site.

The performance of replicated databases can be improved if the requirement of mutual consistency among the replicas of a logical data item is relaxed. Various concepts of relaxed coherency can be denoted by coherency conditions which allow to calculate a coherency index $k \in [0;1]$ as a measure of the degree of allowed divergence. Small values of k express high relaxation, $k = 0$ models suspended update propagation, and for $k = 1$ updates are propagated immediately.

Taking locality, update propagation, and relaxed coherency into account, the total arrival rate of transactions of type i $(1 \le i \le \tau)$, at a single site amounts to

$$\lambda_i^{total} = \ell_i \lambda_i + (n-1) \cdot (1-\ell_i) \cdot \lambda \cdot \frac{1}{(n-1)}$$

The first term describe a share of ℓ_i of the incoming λ_i transactions can be executed locally, and whereas the remaining $(1-\ell_i) \cdot \lambda_i$ transactions are forwarded to nodes where appropriate data is available. The other n-1 nodes also forward $(1-\ell_i)$ of their λ_i transactions, which are received by each of the remaining databases with equal probability 1/(n-1). The above formula simplifies to

$$\lambda_i^{total} = \lambda_i \quad \text{where}$$

$$\lambda^{total} = \sum_{i=1}^{\tau} \lambda_i^{total} = \sum_{i=1}^{\tau} \lambda_i$$

3.2 Performance Criteria

This paper considers the average response times and the transaction throughput as performance criteria. Similar to the calculation of $\overline{W}_c$, the mean waiting time $\overline{W}$ at a local database is found to be:

$$\overline{W} = \frac{\sum_{i=1}^{\tau} \lambda_i^{total} \cdot t_i^2}{1 - \sum_{i=1}^{\tau} \lambda_i^{total} \cdot t_i}$$

The mean waiting time at local database site is the time that user or transaction spends in a queue waiting to be serviced. Meanwhile, the response time is the total time that a job spends in the queuing system [7]. In other word, the response time is equal to the summation of the waiting time and the service time in the queuing system. On average, a transaction needs to wait for $\overline{W}$ seconds at a database node to receive a service of t_i seconds. Additionally, with probability (1- ℓ_i) a transaction needs to be forwarded to a remote node that takes $\overline{W}_c$ seconds to wait for plus the time to be sent and returned. Thus, the response time is given by

$$\overline{R}_i = \overline{W} + t_i + (1 - \ell_i) \cdot (\overline{W}_c + t_c^{send\_i} + t_c^{return\_i})$$

and the average response time over all transactions types results in

$$\overline{R} = \sum_{i=1}^{\tau} a_i \cdot \overline{R}_i$$

3.3 Asynchronous Data Replication Algorithm

The Data Replication Scheme is a scheme identifying neighborhoods of the cluster. The main function of maintaining the original data of any neighborhood community will be undertaken by the primary server. If for some certain reasons, the primary server is unavailable, then the secondary server will be given the main role of the neighborhood.

The *Rsync* utility is used in the implementation pertaining to data replica algorithm. *Rsync* is an open source utility that provide incremental file transfer capabilities. It is a utility for Unix Systems and provides a fast mechanism for synchronizing remote file systems. It is not a real time mirroring software utility, but it must be called by the user or a program either periodically (via cron job) or manually, after updates have been done. Generally *Rsync* is good for replicating data in WSC environment. Latency cost is

```
Procedure Initialization
Get n  #n is a number of row
Row = 0
Cow = 0
while Not EOF(NRDT.conf)
   row +=1
   metric = row.cow
   copy IP# to ARRAY(1).IP#
   copy ADD# to ARRAY(2).ADD#
   assign 2 to ARRAY(3).Status
   PUSH all data into @NRDT(row,colum)
   If (row =n)
      Row = 0
      Cow +=1
end-while
End Procedure
Procedure communication-module
Life = 1
  Pinghost = 'ping -c 1 IP'
  If (pinghost =~/100\% packet loss/)
       Life = 0
  Return life
End procedure
```

```
Procedure Neighbor-module
 Foreach  array (@NRDT)
    row = substri[1,1](metric)
    cow = substri[2,1](metric)
    Get IP
    count = 0
    prow = row + 1
    mrow = row -1
    pcow = cow + 1
    mcow = cow -1
    Foreach  array (@NRDT)
      nrow = substri[1,1](metric)
      ncow = substri[2,1](metric)
     if ( (((mrow==nrow) || (prow == nrow))
    &&    (cow == ncow)) || ( (row == nrow)
  && ((mcow ==ncow) || (pcow ==  ncow))))
           get IP_nei
           neighbor{IP}{count} = Ip_nei;
           count+=1
     end foreach
   end foreach
 write  array to neighbor.txt file
End procedure
```

```
Main
  Call Initialization-module
  Read NRDT.conf
  While(1)
    Call communication-module
    If life = 0
      Call neighbor-module until all
neighbors
      If life = 1
        Call recovering-module
End Main

Procedure recovering-module
    Read DNS text fail
    Flag = 0
    While(1)
          If line ==data (host and ip)
            If (line_host == host) & (line_ip
== ip)
                Line_ip = ip_nei
              Flag = 1
    End While
  If flag = 1
    Re-run DNS software
End Procedure

Main
  Call Initialization-module
  While(1)
    Call Check-history-status/*log file
    If life=1 && history-status = 1
      Call Copy1 module
      If life =1 &&history-status=1
  Call Update module
      If life=0 &&(history-status =1
       || history- status=0)
      While life-neighbor=1
        Call Neighbor-module
       Call Copy2 module
      End While
   End If
  End While
  End Main

Procedure Primary to Neighbor
 For each neighbor to the primary
   Rsync primary -> neighbor
 End for
End procedure

Procedure Neighbor to Primary
  For each neighbor to the primary
    Call check-server
   Rsync neighbor->primary
 End for
End procedure

Procedure Neighbor to Neighbor
For each neighbor to the primary
   Call check-server
   Rsync neighbor -> neighbor
End for
End procedure
```

minimized as *Rsync* uses pipelining files transfer. It can also use any transparent remote shell such as remote shell (rsh) and secure shell (ssh). Furthermore *Rsync* does not require any root privileges, it supports anonymous and authenticated *Rsync* servers. At the same time it can also copy selected file recursively into any subdirectories.

We have designed a special algorithm for asynchronous data replication technique. The algorithm functioned independently of each other but they complement each other in achieving high availability in WSC. Six modules are used in this schemes : the communication module, initialization module, neighboring module, primary to neighbor module, neighbor to primary module and neighbor to neighbor module.

4 Results and Performance Analysis

The results are based on the discussed model. The example in Table 2 summarizes the parameters of the analytical model including the base settings used to obtain the following graph. The example considers 3,5 and 10 transaction types, which are short queries, updates, and long queries. The values of the parameter are estimated using measurements in a real system and considering assumptions of the other models.

Based on the parameter values given in Table 2, the average response time for 3 transaction types is 0.9431. Table 3 and Figure 1 show the detail results of the average response time for 3, 5, and 10 transaction types.

Table 2. Example of parameter values for 3 transaction types

Parameter	Base setting
n	6
λ	3
τ	3
q_i	q_1=1, q_2=0, q_3=1
a_i	a_1=0.85, a_2=0.1, a_3=0.05
loc_i	1 ($loc_i \in [0;1]$)
f_plcmt_i	0.55
sel_i	1
r_2	4
K	0.01639
t_i	t_1=0.06, t_2=0.125, t_3=0.5
Bps	64000
$size_c^{send\_i}$	400
$size_c^{return\_1}$	2000
$size_c^{return\_3}$	3500

Table 3. Detail results for 3, 5, and 10 transactions types

Number of Transaction Types	Average Response Time
3	0.9431
5	3.2878
10	10.8660

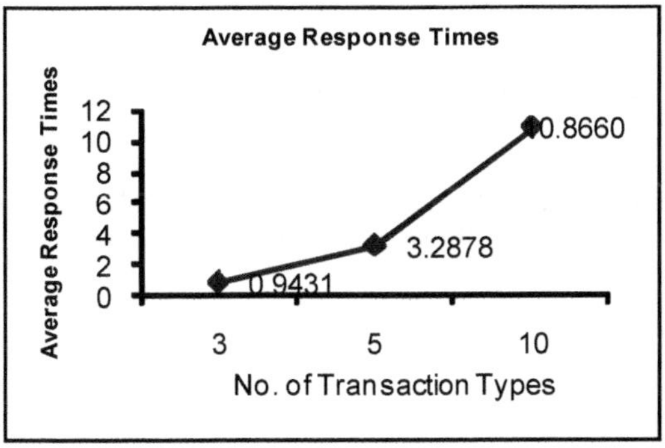

Fig. 1. Average Response Times

Figure 1 shows the average response time versus the number of transaction types. The curve shows a positive relationship where an increasing in the average response time is associated with an increasing in the number of transaction types. In this situation, the average response time can be reduced if the number of transaction types is decreased. The increasing numbers of transaction types caused the services take much longer time to response to all requests. This reflects directly with the definition of the response time, which refer to the elapsed time from the initiation to the completion of the transaction, and with the inclusion of the waiting time and service time taken.

5 Conclusion

In distributed database systems such as web server cluster, data replication is useful and very important technique. Through this technique, data object will be accessed from multiple locations or geographically distributed. This paper proposed an algorithm and analytical model based on asynchronous approach to improve the response time, reliability and availability in WSC. The provision of high reliability in this model is by imposing a neighbor logical structure on data copies. Data from one server will be replicated to its neighboring server and vice versa in the face of failures. The algorithm of data replication scheme was presented. The performance of the asynchronous data replication model has been analyzed and successfully proven mathematically.

References

1. Gallersdorfer, R., Nicola, M.: Improving Performance in Replicated Databases through Relaxed Coherency. In: Proceedings of the 21th International Conference on Very Large Databases, pp. 445–456 (1995)
2. Zhou, W., Goscinski, A.: Managing Replicated Remote Procedure Call Transactions. The Computer Journal 42(7), 592–608 (1999)
3. Mat Deris, M., Mamat, A., Seng, P.C., Ibrahim, H.: Three Dimensional Grid Structure for Efficient Access of Replication Data. Int'l Journal of Interconnection Network 2(3), 317–329 (2001)
4. Buretta, M.: Data replication: Tools and Techniques for Managing Distributed Information. John Wiley, New York (1997)
5. Liu, J., Xu, L., Gu, B., Zhang, J.: Scalable, High Performance Internet Cluster Server. In: IEEE 4th Int'l Conf. HPC-ASIA, Beijing (2000)
6. Moissis, A.: Sybase Replication Server: A practical Architecture for Distributing and Sharing Information, Technical Document, Sybase Inc., (1996)
7. Barbara, D., Garcia-Molina, H.: How Expensive is Data Replication? An Example. In: 2nd International Conference on Distributed Computing Systems, pp. 263–268 (1982)

Change Detection in Ontology Versioning: A Bottom-Up Approach by Incorporating Ontology Metadata Vocabulary

Heru-Agus Santoso[1], Su-Cheng Haw[2], and Chien-Sing Lee[2]

[1] Faculty of Computer Science, Dian Nuswantoro University, 50131 Semarang - Indonesia
[2] Faculty of Information Technology, Multimedia University, 63100 Cyberjaya - Malaysia
{heru.agus.santoso08,schaw,cslee}@mmu.edu.my

Abstract. In ontology versioning, change detection should be related to dependent database or metadata, and vice-versa. Therefore, there is the need to detect change based on a bottom-up approach. This paper discusses bottom-up change detection by incorporating Ontology Metadata Vocabulary (OMV) as the interface to detect change from a database or metadata perspective. With respect to ontology versioning tasks, the approach emphasizes the use of change detection as an effort to preserve the compatibility serialization between ontology and knowledge-based systems (KBS). Our proposed framework and algorithm are presented.

Keywords: ontology change detection, ontology metadata vocabulary, ontology versioning, ontology change management.

1 Introduction

In most Semantic Web applications, ontology is at the heart of the application [1] because of its role in representing knowledge and in addressing heterogeneity in collaborative environments. This kind of system usually encourages the community to contribute new information to the existing knowledge stored in the database. On the other hand, another user may integrate their application's ontology with specific domain ontology, which is available on the Web into the knowledge-based system (KBS). As such, the support for ontology change management becomes extremely important in Semantic Web applications.

As one of important phases in ontology change management, change detection should provide information about the differences among the ontology versions, as well as information about dependent metadata and resources affected by these changes. Ontology change management is also an initial step to keep the knowledge up-to-date so that it meets the user requirement. Besides, it is also crucial to provide information about compatibility serialization between ontology and dependent resources (e.g. database and ontology metadata).

In general, ontology is dynamic and subject to continuous change affected by the changes of its environment. For instance, an implementation of new regulation or law, discovering new scientific facts, and so on may produce new concept entities. This

Y. Zhang et al. (Eds.): DTA/BSBT 2010, CCIS 118, pp. 37–46, 2010.

type of change needs further analysis regarding the conceptualization of the domain. Once new concepts are added to the ontology, it should be followed by the processes of finding and characterizing the effect of changes for further action. Generally, finding and characterizing changes in ontology has to provide the following functions [2]: (1) reading changes, (2) identification of change consistently and unambiguously, (3) the effects of change analysis, and (4) change documentation.

In ontology change management, it is necessary to understand how ontology can be changed and how to detect these changes. Ontology has two types of changes, [3] i.e. bottom-up change and top-down change. Top-down change is an explicit change made by engineer, which needs to be reflected into the related resources used in the system, such as metadata or database. On the other hand, bottom-up ontology change happens when change in the related database or metadata needs to be reflected in the ontology model. This paper focuses on bottom-up change detection and consequently ontology versioning in the context of change management. Ontology versioning is a process to manage changes of different versions of ontology, how to minimize the side effect of the changes to related ontology, application or other elements [4]. Our approach based on analysis of the legacy database and ontology metadata is an effort to preserve compatibility serialization between ontology and the legacy database.

This paper consists of six sections. Section 2 presents related works about change detection. Section 3 discuses the case study of ontology change in Amino Acid Ontology (AAO). Section 4 discusses the proposed framework. Section 5 discusses the bottom-up change detection algorithm. The conclusion is presented in Section 6.

2 Related Work

Eder and Wiggisser [5] proposed change detection of ontology based on Directed Acyclic Graph (DAG) whereby vertices represent classes and edges represent properties of an ontology. The approach compares versions based on the graph structure which represents an ontological structure. In change detection, concept name is employed as a key using a heuristic approach.

Plessers and Troyer [6] proposed change detection of ontology based on version log. Since instances, properties and classes generate related information or instance in the version log, it provides lifetime information of the ontology. The value comparison of the version log is used to detect the change in ontology.

Maynard et al. [7] proposed a bottom-up approach based on the change ontology metadata. The metadata provides information about the number of classes, properties and instances. By using metadata comparison, the approach detects change at the concept, property and instance level, captures the effect of the change and gives information what action need to be done once the change has been detected.

The algorithm in Promptdiff [8] consists of two main parts, the heuristic matcher which exploits the structural property of ontology, and the fixed-point algorithm which uses the matcher to generate structural differences in an ontology. It identifies complex changes, such as moving class by combining hierarchical information with structural information of the ontology.

Tury and Bielikova [9] proposed a change detection approach to provide information about the currentness of an ontology. The approach uses a mapping approach by comparing the changes at the structural level and content level. At the structural level, the approach identified the differences of concept name and subclass equivalences, whereas at the content level, it identified the changes of slot value and individual name.

Table 1 shows the summary on the features supported by the ontology change detection approaches discussed earlier.

Table 1. The comparison of the current ontology change detection approaches

	Eder, J. & Wiggisser K.[5]	Plessers & De Troyers [6]	Maynard et al. [7]	Noy, N. F., et al. [8]	Tury & Bielikova [9]	Proposed approach
Approach	Bottom-up	Top-down	Bottom-up	Top-down	Top-down	Bottom-up
Change management	-	Ontology evolution	Metadata evolution	Ontology evolution	Ontology evolution	Ontology Versioning
Ontology Language	OWL	OWL	OWL	OKBC/RDFS /OWL	OWL	OWL
Technique	Tree comparison algorithm using RDAG	Heuristic approach (Version log comparison)	Mapping (metadata comparison)	Heuristic-algorithm based on structural diff	Mapping based on structural and instance diff	Metadata comparison & RDB-OWL mapping
Key Input	Concept name	Version log	Metadata of ontology	Structural properties of ontology	Structural properties & slot values	RDB components & OMV

We propose change detection based on the analysis of legacy database and ontology metadata. Since the ontology is built using a learning approach on top of the relational database (RDB), the changes of legacy database components such as instances, column metadata, and constraints can serve as the trigger for ontology change detection in the versioning process. Analyzing the changes, which occur on the source of the database and on the metadata, to the underlying ontology, can be reflected as change detection of concepts, object properties, datatype properties, restrictions and instances of the ontology.

3 Case Study: Ontology Changes in Amino Acid Ontology (AAO)

In ontology versioning approach, ontology should be related to which dependent data source or metadata to be adapted [10]. On the other hand, change management should enable changes of the dependent data or metadata to be reflected to the underlying ontology [7]. The changes of metadata can be captured due to change occurred at corresponding conceptual structure and resources mapping process [7]. In that sense, changes of domain, e.g. new discovery in scientific field need to be reflected to ontology metadata (i.e. OMV) and ontology.

Change detection is a process to find out whether several changes have occurred, and what kind of changes could be identified for further analysis. For case study of ontology versioning, we use Amino Acid Ontology[1] (AAO). AAO is ontology about Amino Acids and their molecules characteristics such as size and polarity. There are four latest versions of AAO available on the Web which are: Version 1.0 dated 10/11/2005, Version 1.1 dated 05/15/2006, Version 1.2 dated 05/18/2006 and Version 1.3 dated 16/02/2009. Fig. 1. shows the two versions of AAO. There is a change from Version 1.1 to be Version 1.2 as highlighted in Fig. 1. As can be observed from the figure, in case of concept addition, once the RefiningFeature concept is added, the change does not only mean a new class addition, but it also change in the hierarchical structure of the ontology. Thus, further analysis is required to determine the effect of the change.

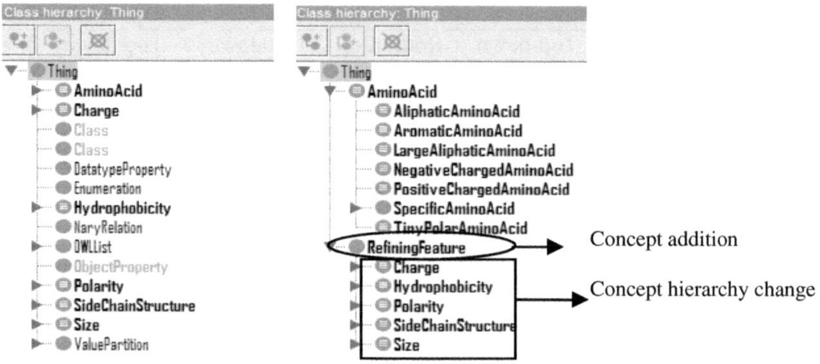

AAO Version 1.1 AAO Version 1.2

Fig. 1. The changes in the version 1.1 and 1.2 of AAO

Table 2 presents several versions of AAO. From Table 2, there are several changes occurred from Version 1.1 to be Version 1.2. On the other hand, there is no particular change detected on Version 1.2 to be Version 1.3. Nevertheless, several annotations are added in Version 1.2. to be Version 1.3., although, there is no change in the number of concept, object properties, subclasses axiom, equivalent classes axiom and disjoint classes axiom.

Table 2. The metadata of four versions of Amino Acid Ontology

Version	∑ concepts	∑ object properties	∑ subclasses axiom	∑ equivalent classes axiom	∑ disjoint classes axiom
1.0	55	14	245	15	199
1.1	55	14	245	14	199
1.2	46	5	238	12	199
1.3	46	5	238	12	199

Given the two versions of ontology, V_1 and V_2, the differences between V_1 and V_2 can be identified as addition and subtraction ontology components [11], [12], denoted

[1] http://www.co-ode.org/ontologies/amino-acid/2009/02/16/

by Δ_1 and Δ_2 respectively. Addition ontology component (Δ_1) is a set of required added components which should be added in V_1 to be V_2. Subtraction ontology component (Δ_2) is a set of required deleted component which should be deleted in V_1 to be V_2. Given two set of ontology components, OC_1 be a set of V_1 components and OC_2 be a set of V_2 components, the deltas can be computed as $\Delta_1 = OC_2 \setminus (OC_2 \cap OC_1)$ and $\Delta_2 = OC_1 \setminus (OC_1 \cap OC_2)$. For instance, the properties subtraction of Version 1.1 mean to be Version 1.2, Version 1.1 requires nine deleted properties.

In bottom-up approach, the changes regarding new information in specific domain need to be reflected as the change of ontology metadata and ontology [3][7], for instance, how to reflect new information in Amino Acid Quantitative Structure Property Relationship Database (AA-QSPR Db) [13]. AA-QSPR Db is a web-based platform of Amino Acid database which encourages communities to contribute information about new molecules and new information about existing molecules. In choosing database strategy, AA-QSPR Db uses relational database and XML database [13]. They provide simple web form which community can submit new information and store them to the database. Since instances and database schema can be associated and mapped to particular ontology components, and vice versa [14], the changes of relational database instances and schema have possibility to be reflected as the change of OMV and ontology as well. For example, an addition of new molecules should be reflected as an addition of numberOfIndividuals attribute of the OMV ontology entity, and OWL individual as well. For further details on the mapping strategy from relational database to OWL ontology can be found in [15], [16].

4 Our Proposed Framework

Our approach emphasizes the use of change detection in the context of ontology versioning in the effort to preserve compatibility serialization between ontology and KBS. Considering ontology versioning should be related to the dependent data source or metadata [10], Ontology Metadata Vocabulary (OMV) can be used as metadata interface to provide compatibility information between ontology and the legacy database. Once application ontology is built from the schema and instance mapping of the legacy database, OMV may also contain particular information about the legacy database. OMV attributes such as the number of classes, properties and individuals can be used to provide compatibility information of ontology and KBS. For this reason, there is the need to derive change detection from a bottom-up approach. Bottom-up change detection can use the differences between OMV and metadata of the legacy database as a trigger for change detection of the ontology, although further analysis is required to possibly reflect the change in the ontology.

Fig. 2. illustrates the proposed framework for the change detection approach. In the framework, version identification is important since ontology versioning can only be applied to different versions of the same ontology. OMV components such as URI, version and resourceLocator can be combined with the version information stored in the Web Ontology language (OWL) to perform version identification.

The change detection can be broken down into several steps such as hierarchical change detection, concept-driven change detection, and instance driven change detection. Hierarchical change detection detects the change based on the hierarchical structure of concepts. On the other hand, concept and instance-driven change detection analyze based on the change of concept and instance, e.g. instance addition or subtraction. The algorithm in Section 4 discusses the bottom-up approach to derive change detection caused by addition or subtraction components.

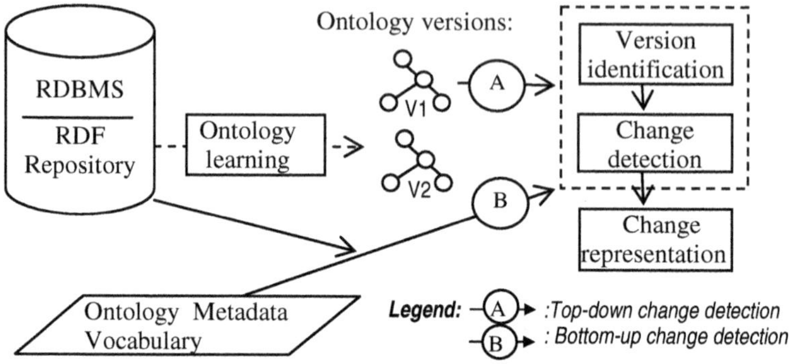

Fig. 2. The change detection framework

OWL provides tags to store version information at the <owl:Ontology rdf:about=""/> element. This tag contains annotations about the version of ontology. The <owl:priorVersion> is used to indicate the earlier version of the ontology. <owl:backwardCompatibleWith> is related to version compatibility as it identifies specific prior ontology and is backward compatible with the current version. The <owl:incompatibleWith> is used to define version incompatibility among ontology versions. When the granularity of ontology version on the ontology level is not sufficient, user may keep the version information at the level of class and property without creating a new version as a different ontology file. In order to apply versioning on the class or property level, version annotation can be placed after rdf:Id and rdf:Resource attribute in the <owl:class> or <owl:DatatypeProperty> tag of the new version.

OMV [17] is a vocabulary standard, which captures information about ontology. OMV also provides version information of ontology, which is stored in the Ontology component. Fig. 3. illustrates the general overview of OMV. Metadata means machine-processable information on the Web[2]. Hence, ontology metadata is machine-processable information about ontology. The main purpose of ontology metadata is to provide additional information about ontology in order to improve accessibility, sharing and reuse of ontology [17]. As ontology versioning should provide transparent access [10] and should include metadata [18], change detection in the context of ontology versioning must take into account ontology metadata. This approach will be focused on employing the *ontology* entity of OMV. The ontology attributes, which are useful are *URI, Version, creationDate, modificationDate, numberOfClasses, numberOfProperties, numberOfIndividuals and numberOfAxioms.*

[2] http:/www.w3.org/metadata

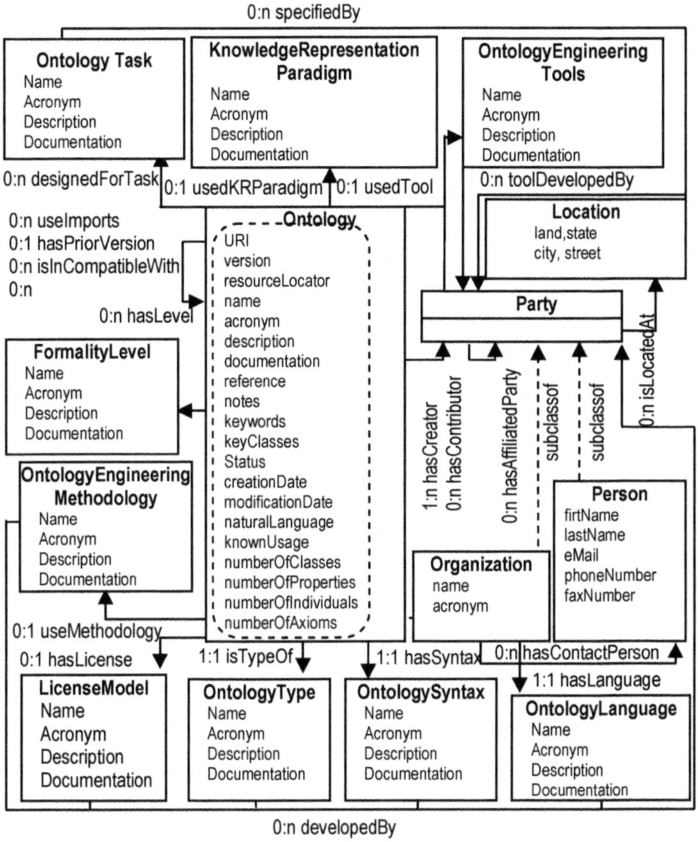

Fig. 3. OMV core entities

5 Change Detection Algorithm

Implementing change detection in RDB (which can be reflected as change detection of OMV and ontology) needs further study on the relation between RDB and the ontology component. Based on [14], [15] and [16], RDB can be mapped with the OWL ontology. RDB components which are tables, attributes, rows and restrictions, denoted by T, A, R and RES respectively, can be mapped to concept, datatype property, instances and OWL restriction. Let V_1 be the ontology which is built on top of RDB. RDBC is the particular RDB components used as the source to build V_1, so that $RDBC \subseteq RDB$. According to OMV core entities, the Ontology's Attributes (denoted by OA for brevity), which can be associated with RDBC are *numberOfClass*, *numberOfProperties*, *numberOf Individuals and numberOfAxiom.*

We use the term *initial difference* in which the changes have not been detected. The initial difference, denoted by δ_0, is the difference between the number of RDBC metadata and the values of OA. Identified as the state before the changes occurred, the number of RDBC metadata and the values of OA may have either the same or different value. Let δ_0 be the initial difference between the numbers of RDBC metadata and OA values, δ_1 be the difference after components addition, δ_2 be the difference after components subtraction. $RDBC_0$ is RDBC in which the changes have not been occurred, $\delta RDBC_1$ is addition component in RDBC, and $\delta RDBC_2$ is subtraction component in RDBC.

Given C as a set of changes occurring in RDBC then $RDBC_1 = RDBC_0 + C$ and $RDBC_2 = RDBC_0 - C$. The difference between the number of RDBC metadata and OA values after components addition can be computed as $\delta RDBC_1 = RDBC_2 \setminus (RDBC_2 \cap RDBC_1)$. The difference between the number of RDBC metadata and OA values after components subtraction can be computed as $\delta RDBC_2 = RDBC_1 \setminus (RDBC_1 \cap RDBC_2)$.

Fig. 4 depicts the algorithm of change detection. The counting of RDBC metadata is based on the tables associated with the concepts in the ontology.

```
1: Preparation:
   -If exist (δ0) then load δ0
   -Load numberOfClass, numberOfProperties,
    numberOfIndividuals, numberOfAxiom from OMV
2: For all tables in RDBC
           Count RDBC metadata
   End For;
3: Compute: δRDBC1, δRDBC2 , δ1 and δ2;
4: Change detection:
   If δ1∪δ2≠0 then ∃ change in RDBC such that δ1>0 or
      δ2>0
   If δ1>0 then change_type:= addition;
   If δ2>0 then change_type:= subtraction;
5: End
```

Fig. 4. The algorithm for change detection

The algorithm can be explained as follows: In the preparation step, δ_0 (which is created after the creation of the previous version of ontology) is loaded, and the values of the *ontology* attributes are loaded as well. Since RDBC is a subset of RDB, then RDBC metadata can be computed based on the tables, which are associated as the mapping sources of the ontology concepts. Column metadata (data type, primary key, mandatory, etc), constraints (unique, check), and rows, which are present in each table are calculated as well.

δ_1 is the difference between the numbers of RDBC metadata and the values of OA caused by the addition of T (tables), A (attributes), R (rows) and RES (constraints or restrictions) in RDBC, hence $\delta_1 = \delta_0 + \sum(\delta RDBC_1)$. δ_2 is the difference between the numbers of RDBC metadata and the values of OA caused by the subtraction of T, A, R and RES in RDBC, hence $\delta_2 = \delta_0 - \sum(\delta RDBC_2)$.

The changes in RDB which should be reflected as the changes of ontology can be detected if $\delta_1 \cup \delta_1 \neq 0$. In other words, the change occurring in RDB can be detected as additional components of ontology if $\delta_1 > 0$, whereas the change occurring in RDB can be detected as subtraction components of ontology if $\delta_2 > 0$. For instance, the addition of table t, where $t \in$ RDBC, since t can be mapped to concept c which is needed in V_2 regarding the change of legacy database, then we can say that concept c is required added concept which should be added in V_1 to be V_2. On the other hand, the subtraction of table t, where $t \in$ RDBC, and t is associated with concept c in V_1, then we can say that concept c is required deleted concept in V_1 to be V_2.

6 Conclusion and Future Work

Ontology tends to change over time. In order to keep it up-to-date to meet user needs, change management is indeed essential. Change management approach in ontology versioning concerns how to manage the changes in the versioning process of ontology. Thus, finding and characterizing the changes are crucial for further action in ontology management.

This paper emphasizes on detecting changes in the context of ontology management due to the need to reflect changes, which occurs to the related database or metadata to the underlying ontology. This approach incorporates OMV, which is focused on the utilization of the Ontology entity of OMV. OMV originally created for providing ontology metadata to improve accessibility and reusability of ontology, but it also contains particular properties, which can be exploited for bottom-up change detection approach. We have presented our framework and algorithm for change detection. Our future work will involve simulation of this algorithm on real data.

References

1. Davies, J., Studer, R., Warren, P.: Semantic Web Technologies: Trends and Research in Ontology-based Systems. John Wiley and Son Ltd., West Sussex (2006)
2. Klein, M., Kiryakov, A., Ognyanov, D., Fensel, D.: Finding and Characterizing Changes in Ontology. LNCS, pp. 79–89. Springer, Heidelberg (2002)
3. Stojanovic, L., Meadche, A., Motik, B., Stojanovic, N.: User-driven Ontology Evolution Management. In: 13th International Conference on Knowledge Engineering and Knowledge Management (EKAW 2002), Siguenza, Spain, pp. 285–300 (2002)
4. Flouris, G., Manakanatas, D., Kondylakis, H., Plexousakis, D., Antoniou, G.: Ontology Change: Classification and Survei. The Knowledge Engineering Review 00(01), 1–19, 1–29 (2007)
5. Eder, J., Wiggisser, K.: Change Detection in Ontologies Using DAG Comparison. In: Meersman, R., Tari, Z., Herrero, P. (eds.) OTM 2006 Workshops. LNCS, vol. 4277, pp. 42–43. Springer, Heidelberg (2006)
6. Plessers, P., De Troyer, O.: Ontology Change Detection using a Version Log. LNCS, pp. 578–592. Springer, Heidelberg (2005)
7. Maynard, D., Peters, W., d'Aquin, M., Sabon, M.: Change Management for Metadata Evolution. In: Proceeding International Workshop on Ontology Dynamics (IWOD 2007)-4th European Semantic Web Conference (ESWC 2007), Innsbruck, Austria (2007)
8. Noy, N.F., Kunnatur, S., Klein, M., Musen, M.A.: Tracking Change During Ontology Evolution. In: McIlraith, S.A., Plexousakis, D., van Harmelen, F. (eds.) ISWC 2004. LNCS, vol. 3298, pp. 259–273. Springer, Heidelberg (2004)

9. Tury, M., Bielikova, M.: An Approach to Detection Ontology Changes. In: ICWE 2006, ACM, New York (2006)
10. Klein, M., Fensel, D.: Ontology Versioning on the Semantic Web. In: Proceedings of the International Semantic Web Working Symposium (SWWS), Stanford, California, USA, pp. 75–91 (2001)
11. Papavassiliou, V., Flouris, G., Fundulaki, I., Kotzinos, D., Christophides, V.: On Detecting High-Level Changes in RDF/S KBs. In: Bernstein, A., Karger, D.R., Heath, T., Feigenbaum, L., Maynard, D., Motta, E., Thirunarayan, K. (eds.) ISWC 2009. LNCS, vol. 5823, pp. 473–488. Springer, Heidelberg (2009)
12. Volkel, M., Groza, T.: Semversion: An RDF-based Ontology Versioning System. In: IADIS International Conference WWW/Internet 2006, Murcia, Spain, vol. VI (2), pp. 195–202 (2006)
13. Lu, Y., Bulka, B., DesJardins, M., Freeland, S.J.: Amino Acid Quantitative Structure Property Relationship Database: a Web-based Platform for Quantitative Investigation of Amino Acid. In: Protein Engineering Design & Selection, pp. 1–5. Oxford University Press, Oxford (2007)
14. Perez de Laborda, C., Conrad, S.: Relational.OWL - A Data and Schema Representation Format Based on OWL. The Second Asia-Pasific Conference on Conceptual Modelling (APCCM). Newcastle, Australia (2005)
15. Myroshnichenko, I., Murphy, M.C.: Mapping ER Schemas to OWL Ontology. In: IEEE International Conference on Semantic Computing, pp. 324–329. Berkeley, CA (2009)
16. Albarrak, M.K., Sibley, E.H.: Translating Relational and Object-Relational Database Models into OWL model. In: IEEE IRI 2009, Las Vegas, Nevada (2009)
17. Hartmann, J., Palma, R., Sure, Y., Suarez-Figueroa, M., Haase, P., Gomez-perez, A., Studer, R.: Ontology Metadata Vocabulary and Application. In: Meersman, R., Tari, Z., Herrero, P. (eds.) OTM-WS 2005. LNCS, vol. 3762, pp. 906–915. Springer, Heidelberg (2005)
18. Klein, M., Fensel, D., Kiryakov, A., Ognyanov, D.: Ontology Versioning and Change Detection on the Web. In: Proceedings of the 13th International Conference on Knowledge Engineering and Knowledge Management (EKAW 2002), Siguenza, Spain (2002)

Controlling Multi Algorithms Using Round Robin for University Course Timetabling Problem

Khalid Shaker and Salwani Abdullah

Data Mining and Optimisation Research Group (DMO),
Center for Artificial Intelligence Technology,
Universiti Kebangsaan Malaysia, 43600 Bangi, Selangor, Malaysia
{khalid,salwani}@ftsm.ukm.my

Abstract. The university course timetabling problem (CTTP) involves assigning a given number of events into a limited number of timeslots and rooms under a given set of constraint. The objective is to satisfy the hard constraints (essential requirements) and minimise the violation of soft constraints (desirable requirements). In this study, we apply three algorithms to the CTTP problem: Great Deluge, Simulated Annealing and Hill Climbing. We use a Round Robin Scheduling Algorithm (RR) as a strategy to control the application of these three algorithms. The performance of our approach is tested over eleven benchmark datasets: one large, five medium and five small problems. Competitive results have been obtained when compared with other state-of-the-art techniques.

Keywords: Timetabling, Simulated Annealing, Great Deluge, Hill Climbing, Round Robin.

1 Introduction and Problem Description

In university course timetabling problem (CTTP), a set of courses are scheduled into a given number of rooms and timeslots across a period of time. This usually takes place within a week and the resultant timetable is replicated for as many weeks as the courses run. Also, students and teachers are assigned to courses so that the teaching delivery activities can take place. The course timetabling problem is subject to a variety of hard and soft constraints. Hard constraints need to be satisfied in order to produce a *feasible* solution. The hard and soft constraints introduced by Socha *et al.* [15] are as follows:

The hard constraints:

- *No student can be assigned to more than one course at the same time.*
- *The room should satisfy the features required by the course.*
- *The number of students attending the course should be less than or equal to the capacity of the room.*
- *Not more than one course is allowed to be assigned to a timeslot in each room.*

Y. Zhang et al. (Eds.): DTA/BSBT 2010, CCIS 118, pp. 47–55, 2010.

The soft constraints:

- *A student has a course scheduled in the last timeslot of the day.*
- *A student has more than 2 consecutive courses.*
- *A student has a single course on a day.*

The problem has

- A set of *N* courses, $e = \{e_1,\ldots,e_N\}$.
- 45 timeslots.
- A set of *R* rooms.
- A set of *F* room features.
- A set of *M* students.

In this paper we used the problem description as in Socha *et al.* [15] to test the efficiency of our proposed approach.

Existing meta-heuristic optimisation algorithms designed to solve the CTTP can be categorised as single-solution and population-based approaches. Some examples of single-solution methods are dual simulated annealing [4]; variable neighbourhood search [2]; graph-based hyper heuristic [8]; and non-linear great deluge [9]; extended great deluge [11]. Population-based methods applied to CTTP include genetic algorithm [18]; ant algorithm [15]; artificial immune system [19] and harmony search [6]. Interested readers are referred to Lewis [13] for a comprehensive survey of the university timetabling approaches in recent years, and Ross and Corne [17] for a comparison between genetic algorithm, simulated annealing and stochastic hill climbing. In addition, other related papers on enrolment-based course time tabling problems can be found in McCollum [7], Lu and Hao [10], Qu et al. [12], and Schaerf [14].

2 The Proposed Algorithm

The search algorithm consists of two processes. The first process is concerned with producing an initial solution where a complete feasible solution is found using a Least Saturation Degree Heuristic. The second process is to optimise the soft constraint violations using three heuristic algorithms: Great Deluge, Simulated Annealing and Hill Climbing.

2.1 Constructive Heuristic

This approach starts with generating an initial solution that satisfies all the hard constraints. Least saturation degree is used to generate initial solutions that start with an empty timetable [11], which involves two phases. In phase I, those events with less rooms available and more difficult to be scheduled will be attempted first without taking into consideration the violation of any soft constraints, until the hard constraints are met. If a feasible solution is found, the algorithm stops. Otherwise, Phase II is executed with an attempt to move from an infeasible to feasible solution. In this phase, a neighbourhood moves (coded as N_1 and N_2) are applied. N_1 is applied for a certain number of iterations. If a feasible solution is met, then the algorithm stops. Otherwise the algorithm continues by applying a N_2 neighbourhood structure for a certain number of iterations. The details of neighbourhood structure are discussed in subsection 3.3. Across all instances tested, solutions are made feasible before the improvement algorithm is applied.

2.2 Improvement Algorithm

During the optimisation process, the neighbourhood structures outlined in subsection 3.3 are applied in all three algorithms: Hill Climbing, Great Deluge and Simulated Annealing (see subsections 3.4, 3.5 and 3.6). The hard constraints are never violated during the timetabling process. The general pseudo code of the improvement algorithm is given in Fig.1.

```
Set the initial solution Sol by employing a constructive heuristic;
Calculate initial cost function f(Sol);
Set best solution Solbest ← Sol;
Set quantum time, q_time;
Set initial value to counter_qtime;

do while (not termination criteria)
   Set a sequence algorithms in a queue which is ordered as ALG_i where i ∈ {1,...,K} and
   K = 3;
   do while (q_time not met )
      Select an algorithm ALG_i in the queue where i ∈ {1,...,K} where K = 3;
  A: Apply ALG_i  on current solution, Sol to generate new solution, Sol*;
     if Sol* < Solbest
        update Solbest, Sol;
        repeat label A
     else
        reset ALG_i parameters;
        insert ALG_i into the queue;
       counter_qtime = q_time;
   end do
end do
```

Fig. 1. The pseudo code for the improvement algorithm

The Round Robin Scheduling (RR) algorithm is employed to control the three algorithms, which are ordered in sequence. In this work, the algorithms are ordered as ALG_1 (Hill Climbing), ALG_2 (Great Deluge) and ALG_3 (Simulated Annealing). A time quantum is assigned for each algorithm in equal portions and in a circular order. The algorithm is dispatched in a FIFO manner at a given quantum denoted as q_time which is set to 15 minutes. We did experiment with several time quantum, namely, 5, 10, 15, 20, 30 minutes. Finally we chose a fixed value (q_time =15 minutes) used for medium and large data sets and 10 seconds for small datasets (the details of the datasets are shown in Table 1). Note that in this paper, all parameters used are based on a number of preliminary experiments.

After the completion of the time quantum of a current algorithm, the preemption is given to the next algorithm waiting in a queue which will start with the best solution

(*Solbest*). The pre-empted algorithm is then placed at the back of the queue and reset as parameters. Note that the parameters involved in the Great Deluge algorithm are the estimated quality (coded as *Optimalrate*) and time to spend (coded as *NumOfIteGD*) as in Fig. 2. When the algorithm ALG_i is unable to generate a better solution during the given quantum time, the algorithm will be added into the queue. In the next iteration (in Fig. 1), the first algorithm in the queue will be used to generate a new solution.

2.3 Neighbourhood Structures

Two neighbourhood structures employed in this approach are as follows:

N_1: Choose a single course at random and move to a feasible timeslot that can generate the lowest penalty cost.

N_2: Select two courses at random from the same room (the room is randomly selected) and swap timeslots.

2.4 Hill Climbing Algorithm

Hill Climbing algorithm is used in order to find the local optimum. Initially, the generated initial solution *Sol* is assigned as current solution (denoted Sol_{Hill}) and best solution (denoted $Sol_{bestHill}$). A change in the assignment of the current solution Sol_{Hill} in each iteration is proposed by applying two neighborhood structures on Sol_{Hill} to generate new two solutions and select the best among them, called $TempSol_{Hill}$. The $f(TempSol_{Hill})$ is compared to the $f(Sol_{bestHil})$. The generated solution moves on $TempSol_{Hill}$ will be accepted when it does not worsen the overall solution value (the weighted sum of violated soft constraints). The current and best solutions within the Hill Climbing algorithm operations are then updated ($Sol_{bestHill} \leftarrow TempSol_{Hill}$, $Sol_{Hill} \leftarrow TempSol_{Hill}$). The process is repeated until the termination criterion is met, and then returns $Sol_{bestHill}$.

2.5 Great Deluge Algorithm

Great Deluge algorithm [11] uses a bound *level* that is imposed on the overall value of the current solution that the algorithm is working with. The generated solution is only accepted when the value of the solution after applying the neighbourhood does not exceed the *level*.

Fig. 2 shows the pseudo code for Great Deluge algorithm. Here *SolGD and SolbestGD* set to be *Sol*. The *level* starts at a value (*level* = *f*(*SolGD*)), where *SolGD* is the current solution of Great Deluge algorithm where *SolbestGD* is the overall value of the best solution so far.

The *level* is decreased after each iteration (*NumOfIteGD*) by ΔB, which is calculated using ΔB = ((*f*(*SolGD*)–*Optimalrate*)/(*NumOfIteGD*), where *Optimalrate* is the estimated quality of the final solution that a user requires. Two of neighbourhoods are applied to *SolGD* to obtain $TempSolGD_i$. The best solution among $TempSolGD_i$ is identified, called, *SolGD**. The *f*(*SolGD**) is compared to the *f*(*SolbestGD*). If it is better, then the current and best solutions are updated. Otherwise *f*(*SolGD**) will be compared against the level. If the quality of *SolGD** is less than the level, the current solution, *SolGD* will be updated as *SolGD**. Otherwise, the *level* will be increased with a certain number (which is set in between 1 and 3 in this experiment) in order to allow some flexibility in accepting worse solutions. The process is repeated until the termination criterion is met.

```
Initialization

SolGD ← Sol;
SolbestGD ← Sol;
f(SolGD) ← f(Sol);
f(SolbestGD)← f(Sol)
Set optimal rate of final solution, Optimalrate;
Set number of iterations, NumOfIteGD;
Set initial level: level ← f(SolGD);
Set decreasing rate ΔB = ((f(SolGD)–Optimalrate)/(NumOfIteGD);
Set iteration ← 0;
Set not_improving_counter ← 0, not_improving_ length_GDA;

Improvement

do while (not termination criteria)
   Apply neighbourhood structure Ni where i ∈ {1,2} on  SolGD,TempSolGDi;
   Calculate cost function f(TempSolGDi);
   Find the best solution among TempSolGDi where i ∈ {1,2} call new solution
SolGD*;
   if (f(SolGD*) < f(SolbestGD))
      SolGD ← SolGD*;
      SolbestGD ← SolGD*;
      not_improving_counter ← 0;
      level = level - ΔB;
   else
      if (f(SolGD*)≤ level)
         SolGD ← SolGD*;
         not_improving_counter ← 0;
      else
        not_improving_counter++;
        if (not_improving_counter == not_improving_length_GDA)
            level= level + random(0,3);
   Increase iteration by 1;
end do;
return SolbestGD;
```

Fig. 2. The pseudo code for the Great Deluge

2.6 Simulated Annealing Algorithm

The Simulated Annealing algorithm uses a temperature *temp*. Here, current solution (*SolSA*) and the best solution (*SolbestSA*) are set to be *Sol*. The same parameters as those employed in Abdullah et al. [4] are used where the initial temperature T_0 is equal to 1000; the final temperature T_f is equal to 0.5. Two of neighbourhoods outlined in subsection 3.3 are applied to *SolSA* to obtain $TempSolSA_i$ and the best among $TempSolSA_i$ called, *SolSA** is chosen. A generated solution is accepted when it does not worsen the overall value of the current solution or with the following probability [4] otherwise.

$$P_{accept} = exp\ (-\frac{f(SolSA*) - f(SolSA)}{f(SolSA*)temp})$$

f(SolSA) is the value of the current solution and *f(SolSA*)* is the value of the new solution after a number of non-improvements (worse solution). At every iteration, *temp* is decreased by α, defined as $\alpha = (\log(T_0) - \log(T_f)/Iter\_max)$. The process is repeated until the termination criterion is met. The pseudo code for the Simulated Annealing algorithm is shown in Fig. 3.

```
SolSA ← Sol;
SolbestSA ← Sol
f(SolSA) ← f(Sol);
f(SolbestSA)← f(Sol)
Set initial temperature T0
Set final temperature Tf;
Set decreasing rate α = (log (T0) - log (Tf)/Iter_max);
Set not_improving_counter ← 0, not_improving_ length_SA;
   do while (not termination criteria)
  Define a neighbourhood N_i where i ∈ {1, 2} on SolSA to generate TempSolSA_i;
   Calculate cost function f(TempSolSA_i);
   Find the best solution among TempSolSA_i where i ∈ {1, 2} call new solution SolSA*;
      if (f(SolSA*) < f(SolbestSA))
        SolSA ← SolSA*;
        SolbestSA ← SolSA*;
     else
     not_improving_counter++;
     if (not_improving_counter == not_improving_length_SA)
          Generate a random number, RandNum in [0, 1];
          Calculate the acceptance propability of SolSA*, Paccept(SolSA*)
           if (RandNum < Paccept(SolSA*)) // Paccept(SolSA*) is a function to calculate
           the acceptance // probability of SolSA*
             SolSA ← SolSA*;
       temp ← temp/(1+ α);
  end do;
  return SolbestSA;
```

Fig. 3. The pseudo code for the Simulated Annealing

3 Experimental Results

3.1 Problem Instances and Experimental Protocol

The algorithm was implemented on a Pentium 4 Intel 2.33 GHz PC Machine using Matlab on a Windows XP Operating System. For each benchmark data set, the algorithm was run for 200,000 evaluations with 10 test-runs to obtain an average value.

We have evaluated our results on the instances taken from Socha et al. [15] which are available at http://iridia.ulb.ac.be/~msampels/tt.data/. They are divided into three categories: small, medium and large. We deal with 11 instances: 5 small, 5 medium and 1 large.

Table 1. The parameter values for the course timetabling problem categories

Category	*Small*	*Medium*	*Large*
Number of courses	100	400	400
Number of rooms	5	10	10
Number of features	5	5	10
Number of students	80	200	400
Maximum courses per student	20	20	20
Maximum student per courses	20	50	100
Approximation features per room	3	3	5
Percentage feature use	70	80	90

3.2 Simulation Results

The best results out of 10 runs obtained are presented. Table 2 shows the comparison of the approach in this paper with other available approaches in the literature. These include the genetic algorithm and local search by M1-Abdullah and Turabieh [5], the randomised iterative improvement algorithm by M2-Abdullah et al. [3], the graph hyper heuristic by M3-Burke et al. [8], the variable neighbourhood search with tabu by M4-Abdullah et al. [2], the hybrid evolutionary approach by M5-Abdullah et al. [1], the extended great deluge by M6-McMullan[11], the non linear great deluge by M7-Landa-Silva and Obit [9], the electromagnetism-like mechanism approach by M8-Turabieh et al. [16], the Dual simulated annealing by M9-Abdullah et al. [4], and M10- the Harmony search by Al-Betar et al. [6]. Note that the best results are presented in bold. It can be seen that our approach is able to produce competitive results.

From Table 1 it can be seen our approach has competitive results and better results on *medium3* and *medium4*.

Table 2. Results comparison

Data Set	Our method Best	M1	M2	M3	M4	M5	M6	M7	M8	M9	M10
small1	**0**	2	**0**	6	**0**	**0**	**0**	3	**0**	**0**	**0**
small2	**0**	4	**0**	7	**0**	**0**	**0**	4	**0**	**0**	**0**
small3	**0**	2	**0**	3	**0**	**0**	**0**	6	**0**	**0**	**0**
small4	**0**	**0**	**0**	3	**0**	**0**	**0**	6	**0**	**0**	**0**
small5	**0**	4	**0**	4	**0**	**0**	**0**	**0**	**0**	**0**	**0**
medium1	117	226	242	372	317	221	**80**	140	96	93	124
medium2	108	215	161	419	313	147	105	130	96	**98**	117
medium3	**135**	231	265	359	357	246	139	189	**135**	149	190
medium4	**75**	200	181	348	247	165	88	112	79	103	132
medium5	160	195	151	171	292	130	88	141	87	98	**73**
large	589	1012	-	1068	-	529	730	876	683	680	**424**

Fig. 4 shows the box plots of the penalty cost when solving small, medium and large instances. The results for the large dataset are less dispersed compared to medium and small (where small instance shows a worse dispersed case in these experiments). We can see that the median is closer to the best in small and medium datasets; however the worst is closer to the median in large datasets. From these results, we believe that the size of the search space may not be dependent on the problem size due to the fact that the dispersion of solution points are significantly different from one to another, even though the problems are from the same group of datasets with the same parameter values. This shows that the algorithm behaves differently on various datasets, possibly due to the different complexities of the datasets and the nature of the solution space. We believe that the algorithm might be able to obtain better results on all the datasets by introducing a mechanism that can adaptively and intelligently employ different neighbourhood structures for different situations, based on the quality of the solution in hand.

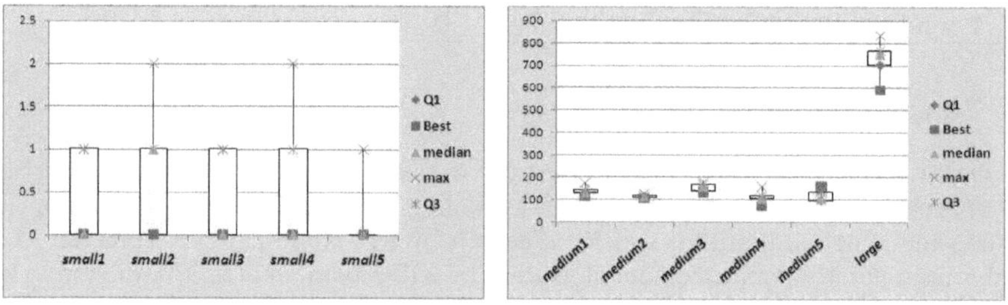

Fig. 4. Box plots of the penalty costs for *small*, *medium* and *large* datasets

4 Conclusion and Future Work

This paper presents Hill Climbing, Great Deluge and Simulated Annealing algorithms applied to the course timetabling problem. The Round-Robin algorithm is employed on these algorithms to control the selection of the algorithms, given a slice time or quantum. In order to test the performance of our approach, experiments are carried out based on course timetabling problems and compared with state-of-the-art methods from the literature. Preliminary comparisons indicate that our approach is competitive with other approaches in the literature and able to produce two best known solutions on medium3 and medium4 dataset. In future work, efforts will be made to establish and compare the approach in relation to previously reported literature. We believe that the proposed approach can be adapted to new problems, thus the ITC2007 datasets will be the subject of future work.

References

1. Abdullah, S., Burke, E.K., McCollum, B.: A hybrid evolutionary approach to the university course timetabling problem. IEEE Congres on Evolutionary Computation, 1764–1768 (2007) ISBN: 1-4244-1340-0
2. Abdullah, S., Burke, E.K., McCollum, B.: An investigation of variable neighbourhood search for university course timetabling. In: The 2nd Multidisciplinary International Conference on Scheduling: Theory and Applications (MISTA), pp. 413–427 (2005)

3. Abdullah, S., Burke, E.K., McCollum, B.: Using a randomised iterative improvement algorithm with composite neighbourhood structures for university course timetabling. In: Metaheuristics: Progress in complex systems optimisation. Operations Research/Computer Science Interfaces Series, ch. 8, Springer, Heidelberg (2007) ISBN:978-0-387-71919-1
4. Abdullah, S., Shaker, K., McCollum, B., McMullan, P.: Dual Sequence Simulated Annealing with Round-Robin Approach for University Course Timetabling. In: Cowling, P., Merz, P. (eds.) EVOCOP 2010. LNCS, vol. 6022, pp. 1–10. Springer, Heidelberg (2010)
5. Abdullah, S., Turabieh, H.: Generating university course timetable using genetic algorithms and local search. In: The Third 2008 International Conference on Convergence and Hybrid Information Technology, ICCIT, vol. I, pp. 254–260 (2008)
6. Al-Betar, M., Khader, A., Yi Liao, I.: A Harmony Search with Multi-pitch Adjusting Rate for the University Course Timetabling. In: Geem, Z.W. (ed.) Recent Advances in Harmony Search Algorithm, SCI 270, pp. 147–161. Springer, Heidelberg (2010)
7. McCollum, B., Burke, E.K., McMullan, P.: A review and description of datasets, formulations and solutions to the University Course Timetabling Problem. To be submitted to the Journal of Scheduling (April 2009)
8. Burke, E.K., Meisels, A., Petrovic, S., Qu, R.: A graph-based hyper-heuristic for timetabling problems. European Journal of Operational Research 176, 177–192 (2007)
9. Landa-Silva, D., Obit, J.H.: Great deluge with non-linear decay rate for solving course timetabling problem. In: The Fourth International IEEE conference on Intelligent Systems, Varna, Bulgaria (2008)
10. Lu, Z., Hao, J.: Solving the Course Timetabling Problem with a Hybrid Heuristic Algorithm. In: Dochev, D., Pistore, M., Traverso, P. (eds.) AIMSA 2008. LNCS (LNAI), vol. 5253, pp. 262–273. Springer, Heidelberg (2008)
11. McMullan, P.: An extended implementation of the great deluge algorithm for course timetabling. In: Shi, Y., van Albada, G.D., Dongarra, J., Sloot, P.M.A. (eds.) ICCS 2007. LNCS, vol. 4487, pp. 538–545. Springer, Heidelberg (2007)
12. Qu, R., Burke, E.K., McCollum, B., Merlot, L.T.G., Lee, S.Y.: A Survey of Search Methodologies and Automated System Development for Examination Timetabling. Journal of Scheduling 12(1), 55–89 (2009)
13. Lewis, R.: A survey of metaheuristic-based techniques for university timetabling problems. OR Spectrum 30(1), 167–190 (2008)
14. Schaerf, A.: A Survey of Automated Timetabling. Artif. Intelli. Rev. 13, 87–127 (1999)
15. Socha, K., Knowles, J., Samples, M.: A max-min ant system for the university course timetabling problem. In: Dorigo, M., Di Caro, G.A., Sampels, M. (eds.) ANTS 2002. LNCS, vol. 2463, pp. 1–13. Springer, Heidelberg (2002)
16. Turabieh, H., Abdullah, S., McCollum, B.: Electromagnetism-like Mechanism with Force Decay Rate Great Deluge for the Course Timetabling Problem. In: Wen, P., Li, Y., Polkowski, L., Yao, Y., Tsumoto, S., Wang, G. (eds.) RSKT 2009. LNCS, vol. 5589, pp. 497–504. Springer, Heidelberg (2009)
17. Ross, P., Corne, D.: Comparing genetic algorithm simulated annealing and stochastic hill climbing of timetabling problems. In: Fogarty, T.C. (ed.) AISB-WS 1995. LNCS, vol. 993, pp. 94–102. Springer, Heidelberg (1995)
18. Lewis, R., Paechter, B.: New crossover operators for timetabling with evolutionary algorithms. In: Lotfi, A. (ed.) Proceedings of the 5th International Conference on Recent Advances in Soft Computing, UK, December 16-18, pp. 189–194 (2004)
19. Malim, M.R., Khader, A.T., Mustafa, A.: Artificial Immune Algorithms for University Timetabling. In: Burke, E.K., Rudova, H. (eds.) The 6th International Conference on Practice and Theory of Automated Timetabling, Brno, Czech Republic, pp. 234–245 (2006)

Strategy Modeling and Classifier Training for Share Trading

Yain-Whar Si, Weng-Lon Lei, and Chi-Chong Chiu

Faculty of Science and Technology, University of Macau
fstasp@umac.mo, kevin.wl.lei@gmail.com,
patrickchiu517@gmail.com

Abstract. In technical analysis, a trading strategy can trigger either buy or sell signals whenever the specified conditions are satisfied. When a set of strategies are applied to a particular stock, a trader often receives conflicting recommendations from each strategy. In this paper, we propose a unified data mining approach in which the outcomes from all strategies are taken into consideration for decision making. First, we develop a framework for composing complex trading strategies. Next, we show how to perform simulation analysis on constructed strategies using extracted historical prices. The result of the simulation analysis is then used for training classifiers which can be used for recommending stock trading actions. Experiments conducted with the price data from Hong Kong Stock Market show promising results.

Keywords: Technical indicators, trading signals, trading strategies.

1 Introduction

Trading securities based on technical analysis [3][4] is considered as one of the most important techniques in today stock markets. However, the vast majority of online trading systems only provide functions for configuring common technical indicators, generating predefined sets of trading signals, and triggering buy/sell actions based on simple if-then rules. Experienced traders often find these functions too rigid since they cannot be tailored for the individual needs. In this paper, we describe a decision support system for trading securities based on use-defined strategies. Through the developed system, traders can create complex trading strategies by combing trading signals incrementally.

Optimization of technical indicators [18] and mining of trading strategies [19] based on evolutionary algorithms have been extensively studied in recent years. In these approaches, to achieve optimum results, selection of parameters for the technical indicators and composition of trading strategies are done automatically based on genetic operators through evolution process. Although automatic composition approaches are proven to be effective, manual composition and testing of complex trading strategies is equally vital for certain circumstances. For instance, a trader may have designed a set of strategies for a particular stock. However, these individual strategies can generate conflicting results (e.g. buy/sell suggestions) when they are applied to a particular stock simultaneously. In this case, the trader will need to decide which strategy to use for

Y. Zhang et al. (Eds.): DTA/BSBT 2010, CCIS 118, pp. 56–67, 2010.

actual trading. However, choosing the right strategy can be extremely complex and any wrong decision can cause devastating financial loss. To alleviate this problem, a unified approach is needed to incorporate individual strategy when making trading decisions. In this paper, we propose a novel approach for training classifiers based on the outcomes of user-defined strategies.

First, we perform a simulation analysis in which each constructed trading strategy from the trader is applied to the extracted historical prices of a selected stock. Next, different classifiers are trained with the results from the simulation analysis. Once the traning is completed, traders can use these classifiers for making buy and sell decisions during trading sessions. In this approach, any potential conflicts which can be caused by these strategies are implicitly eliminated since classifiers are trained with the outcomes (signals) from all these strategies. Experiments conducted with some of the stocks from Hong Kong Stock Market shows promising results.

The rest of the paper is organized into 7 sections. A brief introduction to technical indicators, trading signals, and user-defined strategies is given in Section 2. The proposed simulation method is discussed in Section 3. In section 4, we describe the training process for clissifiers. In section 5, we detail the experimental results obtained from testing with price data from Hong Kong Stock Market. In Section 6, we briefly review recent work before summarizing our ideas and future work in Section 7.

2 Technical Indicators, Trading Signals, and User-Defined Strategies

Trading strategies are often used to determine what kind of actions to be taken at specific point in time. In this paper, we propose an incremental approach for composing strategies from trading signals and technical indicators. The steps for composing trading strategies are described in Figure 1.

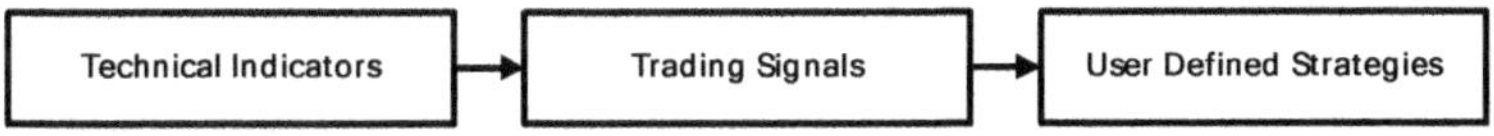

Fig. 1. Steps for composing trading strategies

2.1 Technical Indicators

Technical indicators can be derived from the price data of a security. In the following sections we briefly describe three common technical indicators which are used in this research.

Simple Moving Average (SMA): Moving average is one of the most widely used technical indicators. A simple moving average (SMA) can be constructed by adding a set of price data and then dividing by the number of observations in the period examined [3]. For example, an N-day simple moving average of closing price can be calculated as:

$$SMA = \frac{P_1 + P_2 + ... + P_N}{N} \quad (1)$$

Where P_i is the closing price of ith day and N is the number of days in the moving average (selected by the trader). In technical analysis, values for N are usually chosen as 200, 60, 50, 30, 20, and 10. The choice of N depends on the kind of movement the trader would like to analyze, such as short, intermediate, or long term. A rising SMA indicates an upward trend whereas a falling SMA indicates a downward trend.

Exponential Moving Average (EMA): In calculating SMA, older price data outside of the specified days is not considered. Such omitting can be a problem when there is a large change in the neglected part of the data. EMA assigns larger weight to the latest price data. EMA also responds to changes faster than SMA. EMA can be calculated as follows [3]:

$$EMA_{Today} = P_{Today} \times k + EMA_{Yesterday} \times (100 - k) \tag{2}$$

Where $k = \frac{2}{N+1}$, N is the number of days in the EMA (selected by the trader), P_{Today} is today's price, and $EMA_{Yeserday}$ is the EMA of yesterday.

Moving Average Convergence/Divergence (MACD): The Moving Average Convergence/Divergence indicator was developed by Gerald Appel. The MACD indicator consists of two lines: a fast line and a slow line. The fast line (also called the MACD line) is usually the difference between 12-day EMA and 26-day EMA [4]. The slow line (also called the signal line) is usually a 9-day EMA of the fast line [4]. Based on the two lines, we can plot a MACD-histogram based on the following formula.

$$MACD\text{-}histogram = Fast\ line - Slow\ line \tag{3}$$

MACD-histogram can be plotted as a series of vertical bars and often used to identify whether current trend is losing momentum or not. For instance, when the histogram is positive but starts to fall toward the zero line, it indicates that the uptrend is weakening [4]. When the histogram is negative and begins to move toward the zero line, it indicates that the downtrend is losing momentum [4].

2.2 Trading Signals

In technical analysis, trading signals are used to trigger a buy/sell action. Trading signals can be calculated from technical indicators. In the following sections, we briefly describe three common trading signals which can be constructed from SMA, EMA, and MACD.

Basic trading signals based on SMA: "Golden Cross" [5] signal appears when SMA(10) crosses up through SMA(20) from the bottom to the top. It indicates a "Buy" signal. "Dead Cross" signal appears when SMA(10) crosses down through the line SMA(20) from the top to the bottom. It indicates a "Sell" signal. Example signals based on SMA(10) and SMA(20) are depicted in Figure 2. At point 1 and 3, SMA(10) line crosses up through SMA(20) line indicating "Buy" signals. We can notice that the price of the stock rises after the signal appeared. At point 2 and 4, SMA(10) line crosses down through SMA(20) line indicating "Sell" signals. After the "Sell" signals appeared, the price of the stock drops.

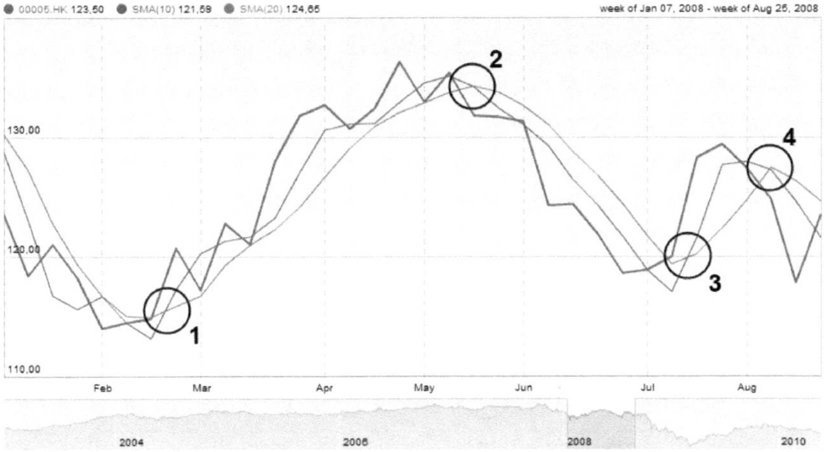

Fig. 2. SMA(10) and SMA(20) of stock 00005.HK

Basic trading signals based on EMA: When EMA(10) crosses up through EMA(20) from the bottom to the top, it indicates a "Buy" signal. When EMA(10) crosses down through EMA(20) from the top to the bottom, it indicates a "Sell" signal. Example trading signals generated from EMA(10) and EMA(20) are shown in Figure 3.

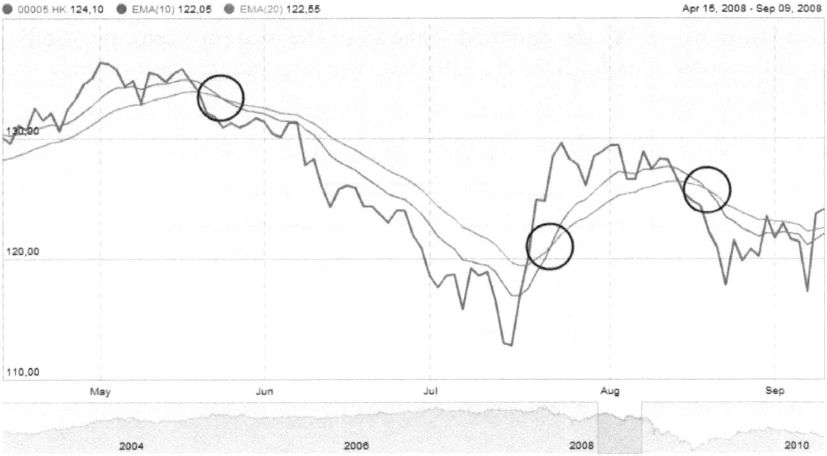

Fig. 3. EMA(10) and EMA(20) of stock 00005.HK

Basic trading signals based on MACD: In this research, we use a standard indicator setting (i.e. 12, 26, 9) for MACD. When the MACD-histogram (blue line) crosses up through the zero line, it indicates a "Buy" signal. When the MACD-histogram crosses down through the zero line, it indicates a "Sell" signal. Example signals identified from MACD(12, 26, 9) are depicted in Figure 4.

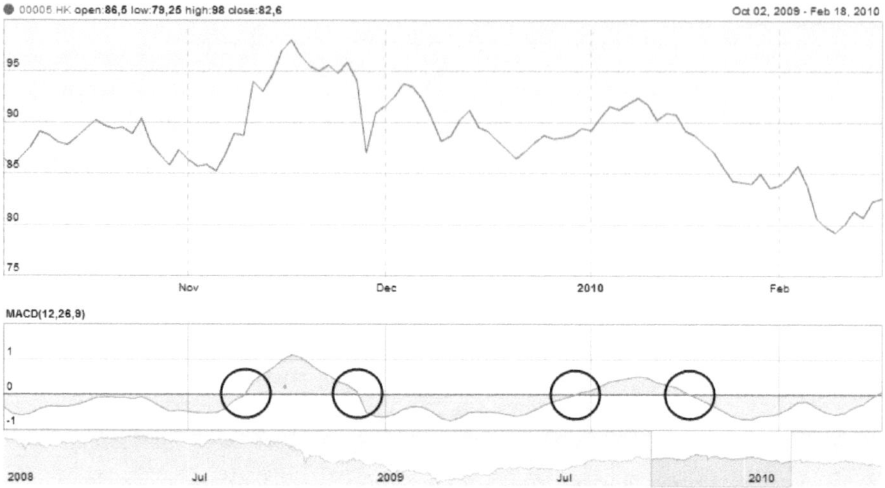

Fig. 4. MACD(12, 26, 9) of stock 00005.HK

2.3 User Defined Strategies

A user defined strategy is a predefined set of conditions for making trading decisions. A strategy contains two groups of conditions: Signal Appear Conditions (SAC) and Signal Termination Conditions (STC). The system performs "Buy" action when all the conditions for SAC are satisfied. Likewise, the system performs "Sell" action when all the conditions for STC are satisfied. The structure for user defined strategies is depicted in Figure 5.

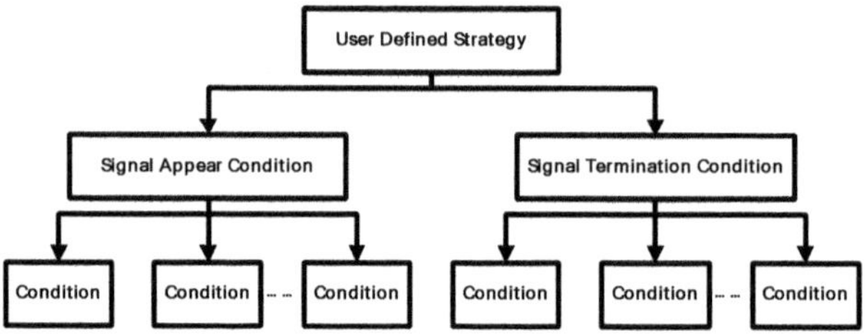

Fig. 5. The structure for user defined strategies

Operators which can be used in constructing conditions for the strategies are listed in Table 1. These operators can be categorized into either "Binary" or "Unary" depending on the indicators used in the strategies. Binary operators can be used for SMA and EMA indicators whereas only unary operators can be used for MACD indicators.

A sample strategy based on EMA(10) and EMA(20) is depicted in Table 2. In some situations, the price of the stock can be directly used in composing strategies. For instance, SAC from Table 2 can be revised as "Price Crosses up through EMA(20)".

Table 1. Operators for indicator

Operator	Type	Description
Cross up	Binary	Indicator A crosses up through Indicator B from bottom to top
Cross Down	Binary	Indicator A crosses down through Indicator B from bottom to top
Above	Binary	The value of indicator A is greater than that of indicator B
Below	Binary	The value of indicator A is less than that of indicator B
Appear Buy	Unary	When the indicator matches its buy signal condition
Appear Sell	Unary	When the indicator matches its sell signal condition
First Appear Buy	Unary	When the indicator matches its buy signal condition at the first day
First Appear Sell	Unary	When the indicator matches its sell signal condition at the first day

Table 2. A sample strategy based on EMA(10) and EMA(20)

Signal Appear Condition (SAC)	EMA(10) Crosses up through EMA(20)
Signal Termination Condition (STC)	EMA(10) Crosses down through EMA(20)

3 Simulation with Strategies

Strategies defined by the users can be tested against the historical price of the stock. To test the performance, the system automatically downloads the historical data (daily closing price) from Yahoo beginning at a specified date. The system then simulates buy and sell actions whenever the specified conditions in the strategy are satisfied. The example actions performed during a simulation are shown in green (buy) and red (sell) circles in Figure 6.

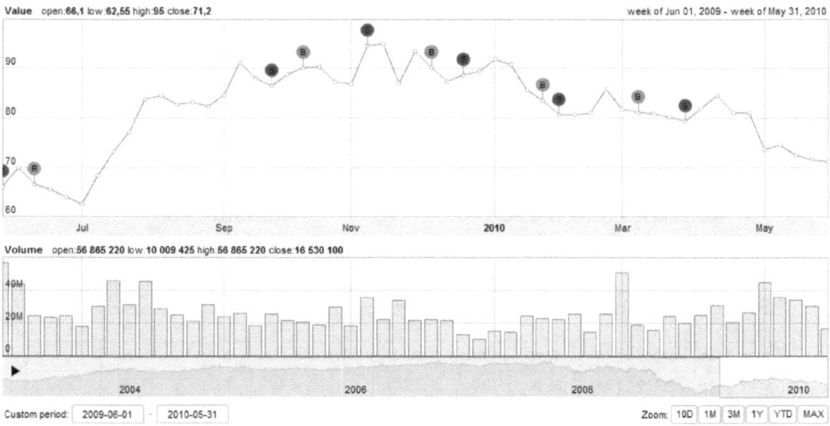

Fig. 6. The buy and sell actions triggered by a strategy

The sample simulation result of HSBC stock (00005.HK) is depicted in Figure 7. According to the simulation, the "Buy" and "Sell" signal pairs appeared 38 times during the simulation period and the signal has exceeded a predefined threshold (percentage of increase in the price) 11 times. The simulation uses the initial capital $10000 to buy and sell the stocks according to a user defined strategy. The total return is then calculated based on the initial capital.

Strategy Performance for 00005.HK				
Title	Data (Start from 2003-01-01)			
Strategy 2	Signal Happened:	38 times	Ideal Signal:	11 times
	Average Changing:	6.72%	Deviation Changing:	17.78%
	Average Duration:	26.29 days		
	Total Return:	715.92%		More

Fig. 7. The sample performance report for 00005.HK (starting form 2003-01-01)

4 Classifiers Training

In this section we describe how the results of the simulation can be used to train classification models. Recall that in Section 2, a trader can compose more than one strategy for a particular stock. However, each strategy can trigger a "Buy" or a "Sell" action whenever the SAC and STC are satisfied. If all strategies are used simultaneously for decision making, a trader may receive multiple or even conflicting buy/sell signals at the same time. Therefore, in order to resolve these conflicts, a unified approach is required to take into account the outcomes (states) of each strategy.

Based on this underpinning, we developed a data mining approach to train classification models for stock recommendation. In this approach, recommendations from each strategy and the price changes in the stocks are used for training. Our prototype system stores the training data in Attribute-Relation File Format (ARFF) from WEKA [2]. The training data contains two sets of columns: Change in Stock Price (CSP) and Outcomes from Applying Strategies (OAS). The output column is the recommended action which is either "Buy", "Sell", or "Nil". A sample training data is given in Table 3.

Table 3. A sample training data after applying BAT and SAT

Reference information				Training data						
				CSP			OAS			Class
Start Date	End Date	$P_{StartDate}$	$P_{EndDate}$	CSP1	CSP5	CSP10	SA	SB	SC	Action
2009-04-30	2009-06-11	54.90	67.15	-2.10%	-1.12%	-3.12%	A	N	N	Buy
2009-06-11	2009-07-16	67.95	67.1	12.12%	3.12%	1.10%	O	A	N	Nil
2009-07-16	2009-09-08	67.60	84.45	3.12%	4.12%	-2.10%	T	O	N	Buy
2009-09-08	2009-10-09	84.90	85.9	-2.10%	3.10%	3.10%	N	O	A	Nil
2009-10-09	2009-10-15	89.00	88.8	3.30%	12.10%	-2.10%	A	T	O	Nil
2009-10-15	2009-11-10	90.50	87.2	-2.10%	7.10%	5.12%	O	A	O	Sell
2009-11-10	2010-01-04	90.05	89.7	8.10%	6.10%	-2.10%	O	O	T	Nil
2010-01-04	2010-02-23	89.90	89.3	3.10%	12.10%	6.10%	O	O	A	Nil
2010-02-23	2010-03-09	85.60	81.75	-2.70%	5.10%	4.10%	T	O	O	Sell

CSP(i): Change in stock price for i days, **SA**: Strategy A, **SB**: Strategy B, **SC**: Strategy C
A: Appear, **O**: Ongoing, **T**: Terminated, **N**: Nil.

First, the trader composes strategies for a targeted stock and set the start date for simulation. Next, the system automatically downloads the historical prices and performs simulation based on these strategies. During the simulation, the system automatically extracts the states of each strategy and the corresponding time period (denoted as start and end date in Table 3). During that period, the states of these strategies remain invariant. Whenever one of the strategies changes its state, the system automatically records the time and begins a new period. There are four possible states for a strategy: appear (A), ongoing (O), terminated (T), and nil (N). The outcomes of applying strategies (OAS) are then indexed by the dates as shown in Table 3.

Next, the system calculates the Change in Stock Price (CSP) for 1, 5, and 10 days based on equation 4.

$$CSP(k) = \frac{P_i - P_{i-k}}{P_{i-k}} \times 100\% \tag{4}$$

Where P_i is the price of day i, and $CSP(k)$ is the change in stock price of last k days. The system then calculates the Return (R) based on the price difference between the start and the end date as follows.

$$R = \frac{(P_{EndDate} - P_{StartDate})}{P_{StartDate}} \times 100\% \tag{5}$$

Finally, the action attribute from Table 3 can be obtained by comparing the return to the Buy Action Threshold (BAT) And Sell Action Threshold (SAT). These two thresholds (see Figure 8) are set by the trader before the training begins. The recommended action is "Buy" if the return is greater than BAT, the action is "Sell" if the return is less than SAT, otherwise the action is Nil.

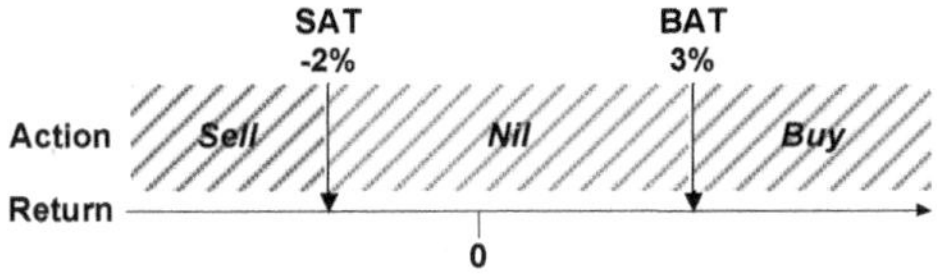

Fig. 8. Deciding action based on BAT and SAT

By varying the value of BAT and SAT, the trader can fine-tune the frequency and profit consciousness of the trading. For instance, high BAT value will produce less "Buy" actions during the generation of training data. The recorded data is then used by the prototype system for training classifiers based on J48, Multi-layer Perceptron, and Naïve Bayes algorithms.

5 Experimental Result

The trained classifiers have been tested against stocks from Hong Kong Stock Market. Due to the limited space, we select three stocks (00001.HK: CHEUNG KONG, 00005.HK: HSBC Holdings, 00941.HK: China Mobile) for discussion. For these stocks, historical price data (closing price) from 1st Jan 2005 to 28th Feb 2009 is used for training and the price data from 1st Mar 2009 to 31st May 2010 is used for testing the classifiers. Both training and testing is performed daily. Therefore, the period considered for training increases as the time progresses. In our experiment, we consider following four strategies (see Table 4).

Table 4. Four strategies used in the experiment

Strategy 1: MACD(12,26,9)	SAC	MACD(12,26,9) Signal Appear
	STC	MACD(12,26,9) Signal Terminated
Strategy 2: SMA(10,20)	SAC	SMA(10) Crosses up through SMA(20)
	STC	SMA(10) Crosses down through SMA(20)
Strategy 3: EMA(2,19)	SAC	EMA(2) Crosses up through EMA(19)
	STC	EMA(2) Crosses down through EMA(19)
Strategy 4: EMA(price, 7)	SAC	PRICE Crosses up through EMA(7)
	STC	PRICE Crosses down through EMA(7)

In these experiments, we consider three combinations of BAT and SAT values: (BAT 2%, SAT -2%), (BAT 3%, SAT -2%), and (BAT 2%, SAT -3%). J48, Multi-layer Perceptron, and Naïve Bayes classification models are used for training. The results of the testing are ranked by the total return as shown in Figure 9, 10, and 11.

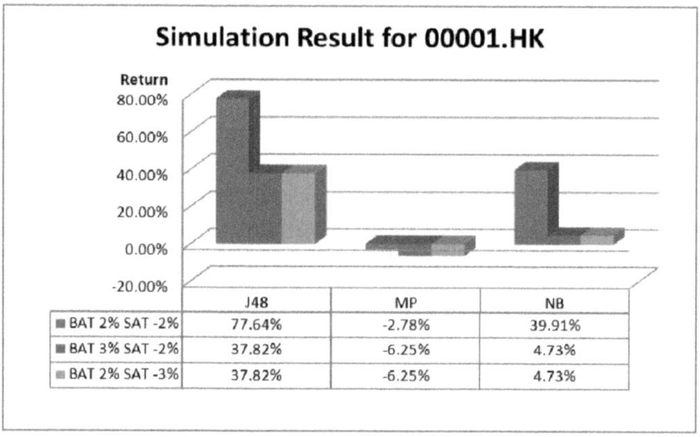

Fig. 9. Simulation result for 00001.HK

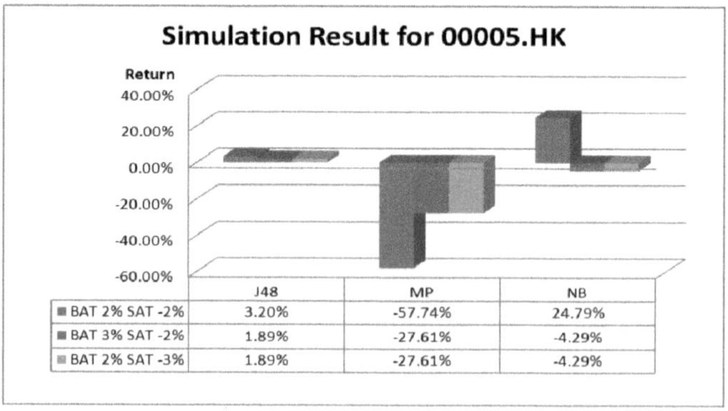

Fig. 10. Simulation result for 00005.HK

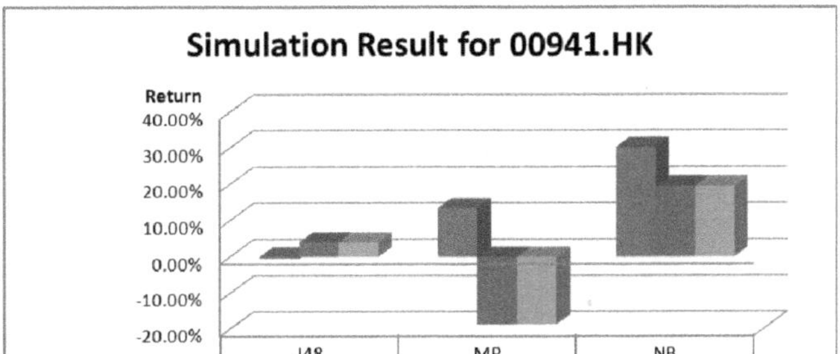

	J48	MP	NB
BAT 2% SAT -2%	-0.49%	13.34%	30.15%
BAT 3% SAT -2%	4.00%	-18.75%	19.51%
BAT 2% SAT -3%	4.00%	-18.75%	19.51%

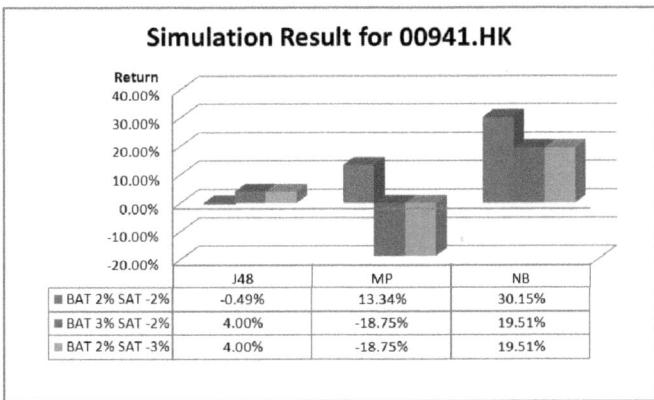

Fig. 11. Simulation result for 00941.HK

The results from the experiment show that the performance of the algorithms varies from one stock to another. For instance, J48 yields the best result for the stock 00001.HK whereas Naïve Bayes classification yields the best result for the remaining two stocks. In addition, the result also shows that the performance of the recommendation also depends on the BAT and SAT values. Larger BAT and SAT can result reduced trading frequencies whereas smaller BAT and SAT values can lead to risk-taking behavior in trading.

6 Related Work

Soft computing techniques have been widely used to forecast the stock returns. The survey by Atsalakis et al. [6] identified various studies on forecasting for more than 20 stock markets. Although our approach has been demonstrated through some of the selected stocks from Hong Kong stock market, it can be applied to any other markets.

Various input parameters have also been considered by researchers to forecast stock returns, such as stock prices [8], trading volume, dividend, yield, indexes [7], technical indicators [9], and technical analysis variables [10]. In our approach, we use the history of the stock price and the signals generated by the user-defined strategies. One of the important characteristics of our approach is that, in training classification models, our model allows a variable number of input parameters. Specifically, the number of attributes (columns) for OAS in Table 3 can increase or decrease depending on the total number of strategies defined by the trader.

Numerous modeling techniques have also been proposed for forecasting stock markets. Similar to our approach, multilayer perceptron (MLP) model was used to forecast the return on the Spanish bIbex-35Q stock index [11] and TEPIX index from Tehran's Stock Exchange [12]. In [13], C4.5 was used in predicting the direction of Taiwan Stock Index returns whereas in [14] C4.5 classifiers are used to forecast the short-term market reaction to news. Naïve Bayes classification was also used to forecasting movement direction of Hang Seng Index based on its opening price, high price, low price, S&P 500 index, and currency exchange rate between HK dollar and

US dollar [16]. In [17], Naïve Bayes classification was used to predict sentiment about stock using financial message boards. Sentiments expressed by individual authors are extracted from financial message boards to learn the correlation between the sentiments and the stock values. The learned model is then be used to make future predictions about stock values.

7 Conclusion

In this paper, we describe a framework for modeling strategies based on technical indicators and trading signals. We show how classifiers can be trained based on the outcomes of the strategies. The trained classifiers are then used to assist traders in deciding when to long or short a particular stock. A prototype system including strategy modeling framework and classifier training/testing was developed in JAVA, WEKA [2], and amCharts [1]. Experiments conducted with the price data from Hong Kong Stock Market show promising results. The main contribution of our approach is twofold. First, our approach allows incremental composition of complex trading strategies. Such capability is crucial for advanced traders who need to tailor their trading strategies. Second, based on the proposed approach, classifiers can be trained and tested against historical price data regardless of the complexity of these strategies. Instead of relying on a predefined set of technical indicators, our approach allows incorporation of all signals which can be generated from the user-defined strategies. Specifically, the data mining approach presented in this paper implictly eliminates any potential conflicting recommendations from the strategies. As for the future work, we are currently testing other data mining techniques for recommending stock trading actions. To improve the performance, we are also planning to include additional market related attributes for training classifiers.

Acknowledgments. This research is funded by the Research Committee, University of Macau.

References

1. AmCharts, `http://www.amcharts.com/` (last accessed April 27, 2010)
2. Weka, `http://www.weka.org/` (last accessed April 27, 2010)
3. Kirkpatrick, C.D.: Technical Analysis, The Complete Resource for Financial Market Technicians. FT Press (2007)
4. Murphy, J.J.: Technical Analysis of the Financial Markets, A Comprehensive Guide to Trading Methods and Applications. Institute of Finance Press, New York (1999)
5. Fyfe, C., Marney, J., Tarbert, H.: Technical Analysis Versus Market Efficiency - a Genetic Programming Approach. Applied Financial Economics 9, 183–191 (1999)
6. Atsalakis, G.S., Valavanis, K.P.: Surveying Stock Market Forecasting Techniques - Part II: Soft Computing Methods. Expert Systems with Applications 36(3), Part 2, 5932–5941 (2009)
7. Abraham, A., Philip, N.S., Saratchandran, P.: Modelling Chaotic Behaviour of Stock Indices Using Intelligent Paradigms. Neural, Parallel & Scientific Computations Archive 11, 143–160 (2003)

8. Doeksen, B., Abraham, A., Thomas, J., Paprzycki, M.: Real Stock Trading Using Soft Computing Models. In: Proceedings of the International Conference on Information Technology: Coding and Computing (ITCC 2005), vol. II, pp. 162–167 (2005)
9. Dourra, H., Siy, P.: Investment Using Technical Analysis and Fuzzy Logic. Fuzzy Sets and Systems 127, 221–240 (2002)
10. Motiwalla, L., Wahab, M.: Predictable Variation and Profitable Rrading of US equities: a Trading Simulation Using Neural Networks. Computers and Operations Research 27(11-12), 1111–1129 (2000)
11. Pérez-Rodríguez, J.V., Torra, S., Andrada-Félix, J.: STAR and ANN models: Forecasting Performance on the Spanish "Ibex-35" Stock Index. Journal of Empirical Finance 12(3), 490–509 (2005)
12. Tabrizi, H.A., Panahian, H.: Stock Price Prediction by Artificial Neural Networks: A Study of Tehran's Stock Exchange (T.S.E), `http://www.abdoh.net/dbase/upload/TSE-PP.pdf` (last accessed April 27, 2010)
13. Cheng, J.H., Chen, H.P., Lin, Y.M.: A Hybrid Forecast Marketing Timing Model Based on Probabilistic Neural Network. Rough Set and C4.5. Expert Systems with Applications 37(3), 1814–1820 (2010)
14. Robertson, C., Geva, S., Wolff, R.: Predicting the Short-Term Market Reaction to Asset Specific News: Is Time Against Us? In: Washio, T., Zhou, Z.-H., Huang, J.Z., Hu, X., Li, J., Xie, C., He, J., Zou, D., Li, K.-C., Freire, M.M. (eds.) PAKDD 2007. LNCS (LNAI), vol. 4819, pp. 1–13. Springer, Heidelberg (2007)
15. Huang, T.T., Chang, C.H.: Intelligent Stock Selecting via Bayesian Naive Classifiers on the Hybrid Use of Scientific and Humane Attributes. In: Proceedings of 8th International Conference on Intelligent Systems Design and Applications (ISDA 2008), pp. 617–621 (2008)
16. Ou, P., Wang, H.: Prediction of Stock Market Index Movement by Ten Data Mining Techniques. Modern Applied Science 3(12) (2009)
17. Sehgal, V., Song, C.: SOPS: Stock Prediction Using Web Sentiment. In: Proceedings of the 7th IEEE International Conference on Data Mining Workshops (ICDMW 2007), pp. 21–26 (2007)
18. Bodas-Sagi, D.J., Fernández, P., Hidalgo, J.I., Soltero, F.J., Risco-Martín, J.L.: Multiobjective Optimization of Technical Market Indicators. In: Proceedings of the 11th Annual Conference Companion on Genetic and Evolutionary Computation Conference: Late Breaking Papers, pp. 1999–2004 (2009)
19. Ni, J., Zhang, C.: Mining Better Technical Trading Strategies with Genetic Algorithms. In: Proceedings of the International Workshop on Integrating AI and Data Mining (AIDM 2006), pp. 26–33 (2006)

BT+-tree: A New Index for Temporal Information in Web Pages

Hong Chen, Qiang Li, and Peiquan Jin

School of Computer Science and Technology,
University of Science and Technology of China, 230027, Hefei, China
jpq@ustc.edu.cn

Abstract. With the growth of Web information, traditional search engines, which are built on the text-based search technology, are unable to meet users' demands on Web search. As many queries are time-related, and most Web pages contain time information, it has been an important issue to develop time-aware Web search engines. Based on this view, in this paper we study the indexing mechanism of the temporal information in Web pages. Our work is based on the assumption that each Web page only has one primary time, which will be utilized in time-based Web search. We present a new index structure called BT+-tree which is based on the MAP21-tree. However, unlike MAP21-tree's double-tree structure, BT+-tree only uses one tree structure. Furthermore, duplicated keys can be effectively treated in BT+-tree, while the MAP21-tree has little consideration on duplicated keys. After discussing the index structure as well as manipulation algorithms of BT+-tree, we design a testing program to measure the performance of BT+-tree. The experimental results show that BT+-tree is effective for indexing temporal information in Web pages.

1 Introduction

Web search engines such as Google and Bing have been an important part in people's life. Most people rely on Google to find useful information. The major goal of search engine is to deliver right information to right users quickly, which is generally implemented by a query processing system. In order to achieve this goal, search engines provide many effective ways for users to express their queries precisely, and also develop some efficient algorithms in ranking and indexing. However, previous research on Web search does not pay enough attention to the temporal information in Web pages. For example, it is difficult to express queries like "to find the discount information about Nike in the next week" in Google. On the other side, time is one of essential characteristics of information [1], and most Web pages are related with temporal information, e.g., business news, discount information and so on. Recently, time has been a focus in the area of Web information extraction [2]. Therefore, it is useful and meaningful to utilize temporal information in Web search to enhance traditional search engines, that is, to develop a temporal-textual Web search engine.

In this paper, we focus on the index structures for temporal-textual Web search. Our basic idea is to develop an efficient index structure to cope with temporal queries.

Y. Zhang et al. (Eds.): DTA/BSBT 2010, CCIS 118, pp. 68–78, 2010.

As MAP21-tree [3] is an efficient temporal index structure in temporal database area, we aims at developing an improved index structure based on MAP21-tree. When considering the time in Web pages, we find that the MAP21-tree has two problems. First, it uses two tree structures to index time, where one for historical time with determined start and end time instant, and another for NOW-related time. Second, it has little consideration on duplicated time. In order to improve the MAP21-tree, we propose the BT+-tree in this paper, which has the following specific features:

(1) The BT+-tree uses only one tree structure to represent both historical and NOW-related time.
(2) The BT+-tree is able to efficiently deal with duplicated time.

The remainder of this paper is organized as follows. Section 2 describes related work. Section 3 introduces the framework of our temporal-textual Web search engine and its main components. Section 4 discusses the five hybrid index structures. Section 5 provides the experimental results. Finally, we conclude the paper and discuss our future work in Section 6.

2 A Framework of Time-Related Web Search Engine

Index is the core of search engine. Indexing algorithm has a great influence on the performance of search engines. The validity of a search engine largely depends on the quality of the index.

The index position in a time-related Web search engine is shown in Fig 1.

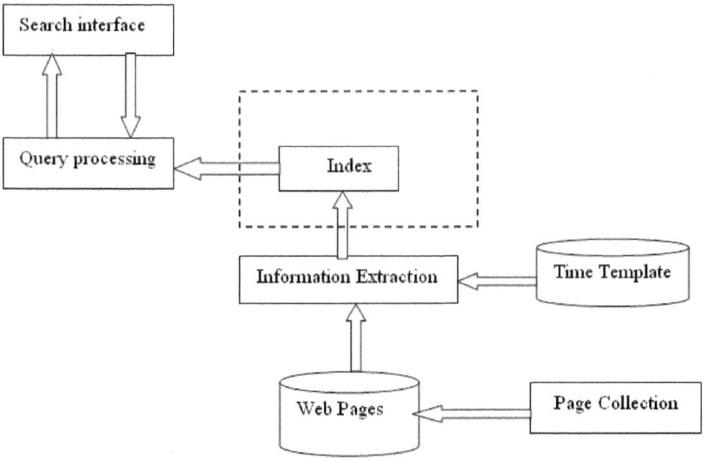

Fig. 1. The system architecture of the search engine

As shown in Fig 1, the establishment of index is based on the extracted time information from Web pages. At the same time, index provides external interface to query processing module. It mainly provides query, insertion, and deletion operations for upper modules to maintain data structures.

3 BT+-tree

3.1 Basic Idea

We suppose that each Web page only corresponds to one time interval. The time interval is the most appropriate time that describes the events of the Web page from the set of content times, and we can regard the time interval as valid time in the database. The time interval does not involve the transaction time, so the indexing structure is not suitable to use R-tree which is for Multi-dimensional indexing method. Time interval is two dimensions, which has the starting point and the end point, so we can not simple use B+-tree. In our paper, we consider mapping the two end points of a valid time range into a single value and using this one as an indexing value for the range, the idea comes from MAP21 algorithm. MAP21 algorithm is to resolve the indexing problem in temporal databases, but it has some defects to deal with the index which is about Web temporal information. In the paper, we improve the MAP21 algorithm in the application of Web temporal information indexing and propose BT+-tree to achieve the Web temporal information indexing. There are two key defects in MAP21 algorithm:

(1) If the Web temporal information contains temporal variable NOW which is the end point of a valid time range, another B+-tree will be built. The B+-tree is called OET-tree, Fig.2 shows an example. This tree only regards left end point as the key value, and the right end point is open, so we just query the range according to the left end point. The establishment of two B+-trees not only affects query performance, but also has a overhead of index size.

Data Set

ID	Dept	Vs	Ve
E11	A	0	3
E12	B	4	now
E21	B	0	5
E31	C	0	7
E32	A	8	9
E41	C	2	3
E42	A	8	now

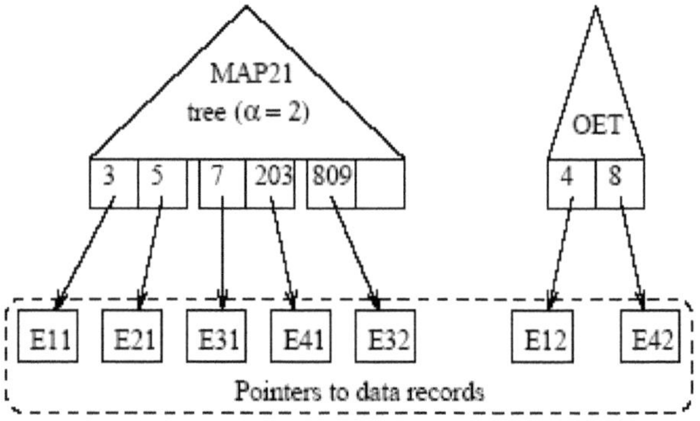

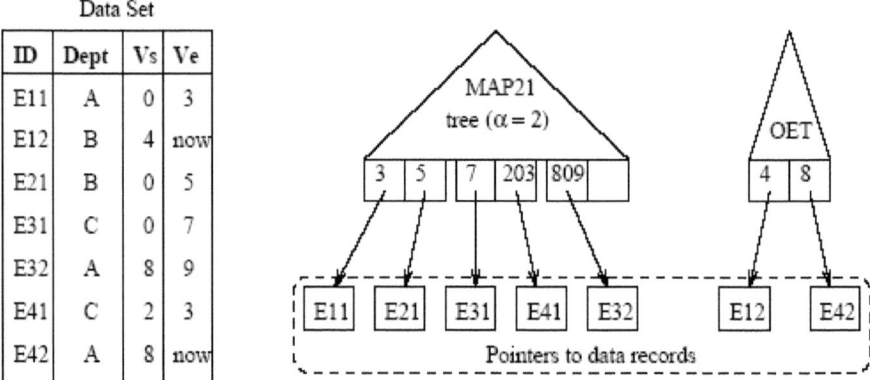

Fig. 2. An example of B+-tree in temporal databases

(2) Mapping the two end points of a valid time range into a single value and using this single value as an indexing value for the range, however, when two time intervals generate the same key value, there is only one value is inserted into one B+-tree. MAP21 algorithm does not support critical value duplicated. Because there is not the same tuple in the temporal database, MAP21 does not consider the processing

mechanism of duplicate key values. In the paper, we improve the MAP21 algorithm and give an index structure called BT+-tree which is suitable for Web temporal information. Compared with MAP21-tree, BT+-tree has two advantages:

(1) We use the system time to replace the NOW variable in time-based Web search. As a consequence, the two index trees in MAP21-tree are simplified to one tree.

(2) We use intermediate bucket to support duplicate key values, which improves the query performance on duplicated keys.

3.2 Insertion

Index insertion algorithm and the traditional B + tree insertion algorithm are similar. The algorithm is shown in Fig 3.

```
Algortihm Insert(nodepointer, start, end, newchildentry)
   // nodepointer is the insert node pointer
   //start and end are the insert interval
   // newchildentry is the pointer to the inserted node
key = start×10^α + end;
if nodepointer is a non-leaf node
     find i such that K_i ≤ key < K_{i+1};
     Insert(P_i, start, end, newchildentry);
if newchildentry is null
     return;
else if nodepointer has free space
       put newchildentry in it and set newchildentry to null;
       return;
else split nodepointer;
if nodepointer is the root
       create a new node with left child nodepointer and right child newchildentry;
        make root point to this new node;
        return;
if nodepointer is a leaf node
    if nodepointer has free space
       put newchildentry in entry and set newchildentry to NULL;
       return;
    else split nodepointer;
End Insert
```

Fig. 3. The insertion algorithm

For example, provided that the order of the tree is 3, the insertion sequence of time period is (0, 3), (0, 5), (0, 7), (8, 9), (2, 3), (5, 8), (1, 8) and (6, 7), α is set as 2, then we get the keyword sequence as 3, 5, 7, 809, 203, 508, 108, 607, and generate the tree structure as shown in Fig 4. If we continue to insert time period (1, 9) into the second leaf node, the node should be split, then the interval [2, 3] moves up to the parent node.

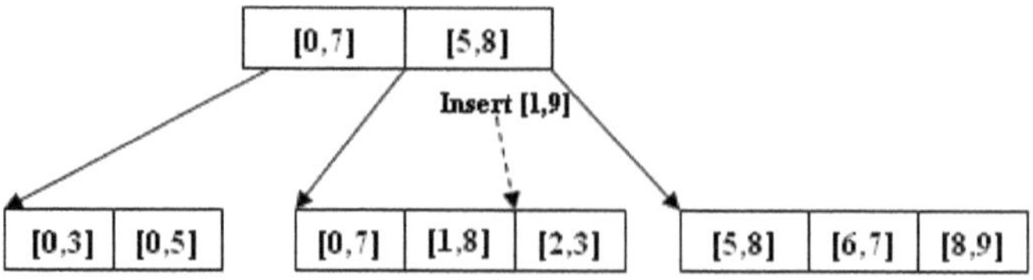

Fig. 4. Before insertion

The tree after insertion is shown in Fig 5.

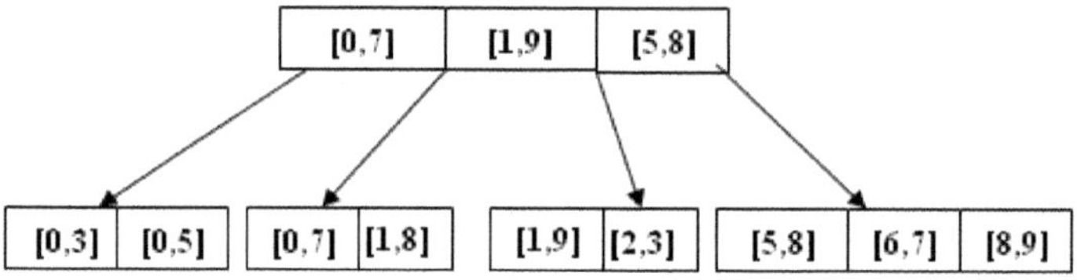

Fig. 5. After insertion

3.3 Search

The search algorithm of BT+-tree is shown in Fig 6.

Algorithm Search(*M, max, start, end*) //*max* is the maximum intervals of all time periods
if end = 'NOW'
  finish=get systemtime();
else finish=end;
$P = \text{NULL}$;
Traverse *M* down to the leaf entry with value $\lceil (start - \max)\times 10^{\alpha} + start \rceil$;
While $key \le finish \times 10^{\alpha} + finish + max$;
  Traverse (in ascending order) each leaf entry of *M*
  if $key \% 10^{\alpha} \ge start$
    $P = P \cup$ {pointers associated to this leaf entry};
return *P*;

Fig. 6. Search algorithm

For instance, when the input time is 1(0, 3), 2(0, 5), 3(0, 7), 4(8, 9), 5(2, 3), since $key = start \times 10^{\alpha} + end$ (α is 2 here), we calculate the inserted key values are 3, 5, 7, 203, 809. The built tree is shown in Fig 7.

From input data we can calculate the max is 7. When the query interval is (0, 4), finding the node whose time period is [0, 3], then query from the leaf node to right and return all nodes which are $key \% 10^{\alpha} \ge start$ until $809 > 4 \times 10^2 + 4 + 7$. The return key values of leaf nodes are 3, 5, 7, 203, the corresponding time interval is (0, 3), (0, 5), (0, 7), (2, 3), we compare these intervals with search interval (0, 4) and get that the return intervals are all intersecting intervals with (0, 4).

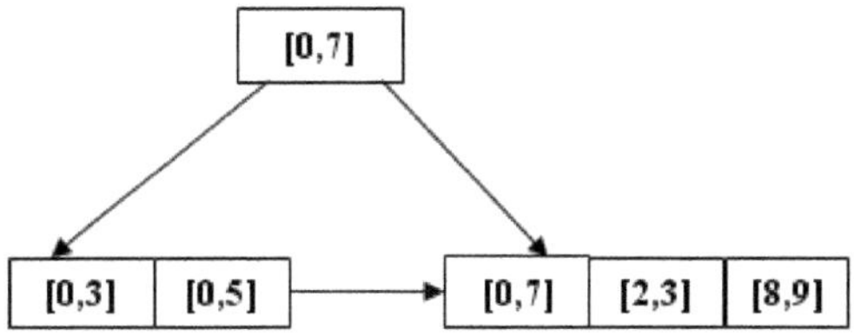

Fig. 7. The index tree built in the example

3.4 NOW Support in BT+-tree

NOW is the current time, which is a temporal variable, it changes over time. NOW records the information which changes over time, its effective value depends on the current time. When we extract Web temporal information, NOW is a problem that can not be ignored, because much time information has been expressed with the NOW in the Web page.

When query process involves NOW, the current time must be clear, because the value of NOW is dependent on the current time. With time changes the same query will get different results at different time even if any tuple doesn't change. We regard NOW as current system time once we query, such as the query (20080110, NOW), we first get the current system time 20080620, then the interval becomes (20080110, 20080620).

From the analysis above, if we need consider the case when build the index, MAP21 algorithm builds a new B+-tree which is defined OET-tree to deal with the situation. But the establishment of two B +-tree index trees does not only affect the query performance but also affect a corresponding increase in index size. Our approach is to assign a special value to NOW, the set of the value depends on the mapping function:

$$\phi(V^k) = \phi(V_s^k, V_e^k) = V_s^k \times 10^\alpha + V_e^k \tag{1}$$

If the time information takes four-digit, according to the requirements of mapping function α should take the integer greater than four-digit. We take six-digit and NOW takes five-digit such as 10000, so when the left end point is 1990, according to the insertion rules, the interval which is inserted into some positions is the far right endpoint in all intervals which are all from the beginning of 1990. After the experiment verification the method can support the insertion, query and so on to the interval contains NOW.

3.5 Duplicated Time

A time interval is mapped into the only key value according to the mapping function, but there is only one value can be inserted into B+-tree when the two time intervals generate the same key value. MAP21 algorithm does not support the duplication of key values. Our project is for Web temporal information index, the num of pages is very large and has more than 100 million, then there are many same intervals, so we need consider how handle the situation and never miss any pages. We propose one method to improve the point, the main idea is to create a public overflow area, and create a linked list of overflow area for the duplicate time interval. When we find the key value of inserted node in the index tree, we put the next time interval into the list of the overflow area the node pointed to. When we query the node, we should return the node and the other nodes which are in the corresponding overflow area. The following example is a test case we use in the experiment. We show it in Fig 8.

```
10
1 1992 1998 www.ustc.edu.cn
2 1994 2001 www.ruc.edu.cn
3 1996 2003 www.baidu.com
4 2002 2004 www.sohu.com
5 1992 1998 www.bbc.com
6 1991 1998 email.ustc.edu.cn
7 1990 1995 www.sina.com
8 1994 1998 www.163.com
9 1992 1998 www.yahoo.com
10 1991 1995 www.qq.com
```

Fig. 8. A test case

When the database file stores according to the above form of organization, the generated BT+-tree index is shown in Fig. 9.

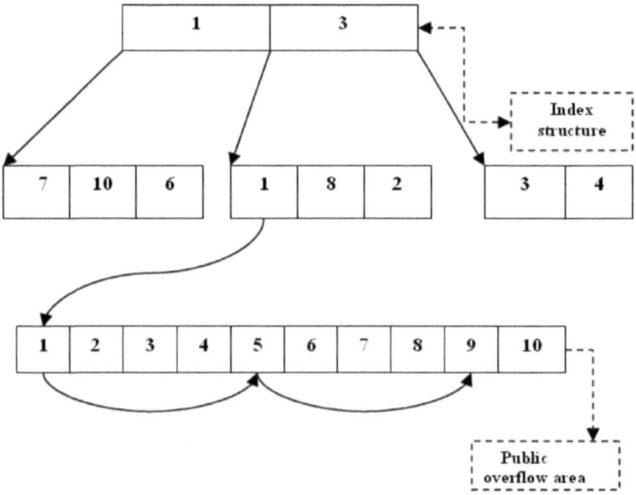

Fig. 9. The generated index structure

As shown in Fig 9, Block 1, 5, and 9 store repeated data. When query node 1, we find block 1 is in the public overflow area and block 1 stores the next block number which has the same interval, then continue to find the next one, so we can return all of the same interval.

4 Performance Evaluation

We implement BT-tree and MAP21-tree using Microsoft Visual Studio 6.0. The test machine is running on Windows XP.

The main difference between BT+-tree and MAP21-tree is BT+-tree support duplicated key values. Since Web pages typically contain duplicated temporal words, MAP21-tree can not be directly used in Web search engines. BT+-tree enhances MAP21-tree with duplicated keys support, so BT+-tree is more suitable for Web search than MAP21-tree. This is also the motivation of BT+-tree.

Compared with MAP21-tree, BT+-tree not only supports duplicated key values, but also has comparable performance as MAP21-tree. In order to compare the performance between BT+-tree and MAP21-tree, we restrict our experiment to the data set without duplicated values. According to such dataset, we will show BT+-tree has comparable performance with MAP21-tree.

In the experiment, we manually generate a simulated dataset. In this dataset, each Web page has one primary time period. The time included in the dataset is limited between 1 and 2040 and every time period is unique.

In our experiment, we generate five simulated datasets. They contain 1 thousand records, 10 thousand records, 100 thousand records, 20 thousand records and 50 thousand records. We want to know whether the results may change with the incremental datasets.

Index size contains the corresponding tree structure size and the size of all Web page records leaf nodes point to. As shown in Table 1, under the same conditions, as the number of Web pages increases, the index size of BT+-tree size is smaller than MAP21-tree's. Because we discuss the situation that there is no duplicate key values, and the size of Web page records leaf nodes point to is the same, the main index size difference between BT+-tree and MAP21-tree is the tree number. BT+-tree has only one tree, the MAP21-tree has two trees.

Table 1. Index size(KB)

Number of Web pages	1000	10000	100000	200000	500000
BT+-tree	169	1,594	15,439	30,911	82,757
MAP21-tree	171	1,639	15,474	30,924	82,801

Second, we compare the query time of two trees. The query type is based on Containment Query. When we input the query time interval, return Web pages whose time intervals include the input time interval. Run time includes two main parts. They are the time for retrieving corresponding tree, and the time for reading page lists from disk. The result is shown in Table 2.

Table 2. Runtimes(s)

Number of Web pages	1000	10000	100000	200000	500000
BT+-tree	0.062	0.499	4.649	9.579	27.456
MAP21-tree	0.078	0.515	4.665	10.635	27.665

Table 2 shows that query time of BT+-tree is comparable with MAP21-tree's, and the time is better as the number of Web pages is larger. Because BT+-tree only need access one tree and their corresponding page lists, MAP21-tree need access two trees and the corresponding page lists. The main time cost is generated from the access to the OET-tree. As the number of Web pages increases, the number of NOW-related temporal words becomes larger and larger. This will result in the increasing of the levels of OET-tree.

Table 3. Runtime (ms)

Number of Web pages	100	1000	10000	100000	1000000
BT+-trees Index Search	5.928	25.319	95.688	305.340	869.222
Sequential	0.812	2.182	19.035	299.651	1066.842

We also conduct a simple experiment to compare the search performance of BT-tree with sequential search, and Table 3 shows the result. Here we use 100, 1000, 10000, 100000, and 1000000 Web pages, whose time intervals fall into the period (1000, 2097). The query intervals are randomly selected from (1256, 1678). Table 3 shows searching on BT+-tree is superior to sequential search as the number of Web pages increases.

5 Related Work

In the real world, the existence of any object or event is always associated with time. Traditional database only saves the current state of one application. Temporal database [4] doesn't only contain current data but also contain history data. In the temporal database all data are associated with time, so we regard the data associated with the time as temporal data.

Traditional commercial search engines, such as Google, Bing have noticed the value of temporal information in Web search. They all provide some ways for users to perform a Web search based on time. E.g. Google uses the *daterange* option to express a temporal predicate. However, those commercial search engines only support the crawled dates of Web pages, i.e., users can only query Web pages towards their creation dates in database. There are also some other temporal information retrieval systems which use similar methods as Google to process temporal information of Web pages, such as Goo [5], Infoseek [6], Namazo [7], Chronica [8], and so on. Generally, there is a gap between the crawled time and content time of a Web page. For example, a news reports the World Exposition hold in May 1th, 2010 in Shanghai but the time of the page may publish in May 3th, 2009. So we consider the content time of one page as the important point focused on. It satisfies users' search demand better.

Index is a fast search technology, database management system mainly uses index to resolve the problem of query efficiency. In the current database system B+ tree [12] which is the variant of B tree [14] is widely used. The basic functions of database systems are to store data efficiently, query and update. How to query the corresponding data user demands effectively, reasonable, fast and accurately, the key technology is index technology. Traditional data index technology has been unable to support efficient access and query to temporal information, so we use temporal databases which adds the time attribute. Current temporal index techniques are mainly R tree, R * tree [9], GR tree [10], 4R trees [11] and so on. R tree in all multi-dimensional databases is generally the same type, and it can't apply to the index in temporal database. Rectangle of R tree is fixed and it can not meet with temporal database indexing which contains NOW and UC and other temporal variables. R * tree based on R tree optimizes the re-insertion method of node, which avoids splitting the node, it provides the optimal splitting method if the node must be split. R* tree makes the overlap between two directories smallest, reduce the I/O operation in the query, and improve the query efficiency. GR tree is a temporal index which contains NOW [15], UC and other temporal variables. It extends R tree, and can deal with the temporal data which contains temporal variables.

In this paper, we research the subject that each page corresponds to a one-dimensional time, and the time only involve valid time except transaction time, so we don't use complex series of R-tree. The paper makes reference to the MAP21 [10] algorithm in the temporal database. In the paper, we propose one new index structure named BT+-tree, which can get better efficiency.

6 Conclusions

In this paper, we presented a new tree called BT+-tree to index the temporal information in Web pages. The index is based on the MAP21-tree, but improves it in both space efficiency and queries on duplicated keys. In the future research, we will focus on the compression and update performance of the index structure and conduct comparison experiments between BT+-tree and other temporal index structures.

Acknowledgments

This work is supported by the National High Technology Research and Development Program ("863" Program) of China (Grant No. 2009AA12Z204), the National Natural

Science Foundation of China under the grant no. 60776801 and 70803001, the Open Projects Program of National Laboratory of Pattern Recognition (no.20090029), the Key Laboratory of Advanced Information Science and Network Technology of Beijing (no. xdxx1005).

References

1. Alonso, O., Gertz, M., Yates, R.B.: On the value of temporal information in information retrieval. In: Proc. Of SIGIR 2007, pp. 35–41 (2007)
2. Jensen, C. S.: Temporal Database Management, PhD Thesis, `http://www.cs.auc.dk/~csj/Thesis/`
3. Nascimento, M., Dunham, M.: Indexing Valid Time Databases via B+-Trees. IEEE Transactions on Knowledge and Engineering 11(6), 929–947 (1999)
4. SnodgrassandIAhn, R.T.: Temporal Databases. Computer 19(9), 35–42 (1986)
5. Goo, `http://www.goo.ne.jp/`
6. Infoseek, `http://www.infoseek.co.jp/`
7. Namazu, `http://www.namazu.org/`
8. Deniz, E., Chris, F., Terence, J.P.: Chronica: a temporal Web search engine. In: Proc. Of ICWE 2006, pp. 119–120 (2006)
9. Nunes, S., Ribeiro, C., David, G.: Use of Temporal Expressions in Web Search. In: Macdonald, C., Ounis, I., Plachouras, V., Ruthven, I., White, R.W. (eds.) ECIR 2008. LNCS, vol. 4956, pp. 580–584. Springer, Heidelberg (2008)
10. Bliujute, R., Jensen, C.S., Saltenis, S., et al.: R-tree based indexing of NOW-relative bitemporal data. In: Proc. of the 24th VLDB Conf.
11. Bliujute, R., Jensen, C.S., Saltenis, S., et al.: Light-weight indexing of bitemporal data. In: Proc. of the 12th International Conf. on Scientific and Statistical Database Management, Berlin, pp. 125–138 (2000)
12. Goh, C.H., et al.: Indexing temporal data using existing B+-trees. Data and Knowledged Engineering 18, 147–165 (1996)
13. Nascimento, M.A., Dunham, M.H.: Indexing Valid Time Databases via B+-Trees. IEEE Trans. Knowl. Data Eng. 11(6), 929–947 (1999)
14. Ang, C.-H., Tan, K.-P.: The interval B-tree. Information Processing Letters 53(2), 85–89 (1995)
15. Clifford, J., Dyreson, C., Isakowitz, T., Jensen, C.S., Snodgrass, R.T.: On the semantics of "NOW" in temporal databases. Technical Report R-94-2047, Dept. of Mathematics and Computer Science, Aalborg University (November 1994)

Rough Set Approximations on Granular Structures and Feature Characterizations

Tong-Jun Li[1] and Yan-Ling Jing[2]

[1] School of Mathematics, Physics and Information Science,
Zhejiang Ocean University, Zhoushan, Zhejiang 316000, China
ltj722@163.com

[2] Library, Zhejiang Ocean University, Zhoushan, Zhejiang 316000, China
jingyl@zjou.edu.cn

Abstract. This paper focuses on generalization of rough set model. First a generalized definition of rough set approximations is proposed on general granular structure, so as to rough set models on some special granular structures are its special cases, these special granular structures include covering, knowledge space, topology space and Pawlak approximation space. Some basic notions are understood as counterparts in topology, and basic properties of new rough set approximations are investigated. Furthermore, with the defined lower and upper approximation operators, we can characterize the features of granular structures, and obtain positive and negative certainty decision rules.

Keywords: Rough sets; Granular structures; Knowledge space; Topology spaces.

1 Introduction

The theory of rough sets proposed by Pawlak in 1982 [1] is a mathematical approach to deal with intelligent systems characterized by insufficient and incomplete information, and has been found very successful in many domains.

In the theory of rough sets, a pair of lower and upper approximation operators is constructed based on an equivalence relation on the universe of discourse. Equivalence classes of equivalence relation form a partition of the universe, partition is a basic concept in Pawlak rough set model. In order to relax the restriction of partition, many authors try to replace the partition with coverings [2], neighborhood systems [3] and abstract approximation space [4], etc.

Granular computing is an emerging field of study focusing on structured thinking, structured problem solving and structured information processing with multiple levels of granularity [5,6]. In the viewpoint of granular computing, knowledge of universe can be described by granular structure. Xu et al construct a generalized approximation space with two subsystems [7], as there may be no connection between two subsystems, so the corresponding lower approximation and upper approximation are not dual to each other. The dual subsystem of a subsystem consists of complement of subsets of the subsystem. For many special

Y. Zhang et al. (Eds.): DTA/BSBT 2010, CCIS 118, pp. 79–88, 2010.

subsystems their dual subsystems are also meaningful, for example, dual subsystem of topology is the set of all closure set of the corresponding topological space, dual subsystem of closure system is knowledge space, etc. The objective of this paper is to construct one kind of approximation spaces with general subsystems and their dual subsystem, such that the approximation operators established on them are generalization of the existing approximation operators defined on some special approximation spaces, meanwhile positive and negative decision rules can be easily made by means of new operator.

In this paper, we choice a subsystem and its dual subsystem to construct a generalized approximation space, which is the generalization of some special knowledge systems, such as Pawlak approximation space, knowledge space [8,9] and topology space [10] etc. In this paper, one type of generalized rough approximations established on granular structures are proposed, the rest of the paper is organized as follows. In Section 2, some basic knowledge about granular structure, knowledge space and topology space are reviewed. Section 3 constructs one kind of generalized rough set model with granular structure, and investigates the properties of the model. In Section 4, axiom sets characterizing different rough approximations are determined. Section 5 concludes the paper.

2 Preliminaries

In granular computing theory, a granule can be considered as a subset of objects of the universe, which is often considered to be definable in some sense. The family of definable granules is called *granular structure*. In some practical applications, granular structures usually satisfy some additional requirements.

2.1 Set Systems and Set Operators

Let U be a nonempty set called a universe of discourse. A *set system* on U is a family $\mathcal{S}$ of subsets of U [10]. *Dual set system* of the set system $\mathcal{S}$ on U, denoted as $\mathcal{S}^c$, consists of complements of all elements of $\mathcal{S}$, that is, $\mathcal{S}^c = \{A \subseteq U| \sim A \in \mathcal{S}\}$, where $\sim A$ denotes the complement of subset A. A set system $\mathcal{C}$ is called a *covering* of U if the condition $\bigcup_{A\in\ \mathcal{C}} A = U$ can be satisfied.

Formal context in formal concept analysis (FCA) is a familiar framework of knowledge, A formal context can be viewed as a set system. A formal context is a triplet (U, A, I), where U is called object set, A attribute set, and I a binary relation between U and A.

Example 1. Table 1 shows formal context (U, A, I), where the object set is $U = \{1, 2, 3, 4, 5, 6\}$, the attribute set is $A = \{a, b, c, d, e, f\}$, and it can be read off that object 1 has attributes a, c, d and e. Thus the object sets of all attributes forms a set system $\mathcal{C}_I$, that is, $\mathcal{C}_I = \{a^*, b^*, c^*, d^*, e^*, f^*\}$, where $a^* = \{1, 3, 4\}$, $b^* = \{3, 5, 6\}$, $c^* = \{1, 4\}$, $d^* = \{1, 3, 4\}$, $e^* = \{1, 2, 4, 6\}$, $f^* = \{3, 5, 6\}$. Moreover, we can find that the set system $\mathcal{C}_I$ is also a covering of U.

Table 1. An formal context (U, A, I)

U	a	b	c	d	e	f
1	1	0	1	1	1	0
2	0	0	0	0	1	0
3	1	1	0	1	0	1
4	1	0	1	1	1	0
5	0	1	0	0	0	1
6	0	1	0	0	1	1

A *set operator* φ on U means a mapping from $\mathcal{P}(U)$ to $\mathcal{P}(U)$, where $\mathcal{P}(U)$ denotes the set of all subsets of U. Two operator φ and ϕ on U are called to be *dual* to each other if they satisfy

$$\varphi(A) = \sim \phi(\sim A) \text{ or } \phi(A) = \sim \varphi(\sim A), \forall A \subseteq U.$$

2.2 Closure Systems and Closure Operators

A *closure system* $\mathcal{C}$ on the universe U is a set system on U which contains U and is closed under set intersections, that is, $U \in \mathcal{C}$ and $\mathcal{D} \subseteq \mathcal{C} \Rightarrow \cap\mathcal{D} \in \mathcal{C}$.

It should be noted that a closure system $\mathcal{C}$ on U is a complete lattice under the set inclusion relation " $\subseteq$". For any $A_1, A_2 \in \mathcal{C}$, their least upper bound or supremum is $A_1 \vee A_2 = \cap\{A \in \mathcal{C} | A_1, A_2 \subseteq A\}$, and their greatest lower bound or infimum is $A_1 \wedge A_2 = A_1 \cap A_2$.

A *closure operator* on U is a mapping φ on $\mathcal{P}(U)$ which satisfies the monotony $(A \subseteq B \Rightarrow \varphi(A) \subseteq \varphi(B))$, the extensity $(A \subseteq \varphi(A))$ and idempotenc $(\varphi(\varphi(A)) = \varphi(A))$.

There is a one-to-one correspondence between the set of all closure systems on U and the set of all closure operators on $\mathcal{P}(U)$. For any closure operator φ on $\mathcal{P}(U)$, the corresponding closure system $\mathcal{C}_\varphi$ is the set of all fixed points of φ, where a fixed point of φ is a subset $A \subseteq U$ with $\varphi(A) = A$, which is also called a closure set of φ. On the other hand, the closure operator $\varphi_\mathcal{C}$ corresponding to a closure system $\mathcal{C}$ is defined by $\varphi_\mathcal{C}(A) = \cap\{B \in \mathcal{C} | A \subseteq B\}, \forall A \subseteq U$, that is, $\varphi_\mathcal{C}(A)$ is the least closure set in $\mathcal{C}$ including A.

The dual set system of a closure system is called *knowledge space* [8,9], *dual closure system* [10] or *pro-topology* [11,12], that is, it contains the empty set and is closed under set union. In this paper, we use the terminology "knowledge space".

The dual operator ϕ of a closure operator is called *interior operator*, therefore it satisfies the monotony $(A \subseteq B \Rightarrow \phi(A) \subseteq \phi(B)$), the contraction $(\phi(A) \subseteq A)$ and idempotence $(\phi(\phi(A)) = \phi(A))$.

Similarly there is also an one-to-one correspondence between the set of all knowledge spaces and the set of all interior operators.

Significantly, the set of all closure systems and the set of all knowledge spaces on U are closure systems. For any set system $\mathcal{D} \subseteq \mathcal{P}(U)$, the set systems $CS(\mathcal{D}) = \{\cap\mathcal{E}|\mathcal{E} \subseteq \mathcal{D}\}$ and $KS(\mathcal{D}) = \{\cup\mathcal{E}|\mathcal{E} \subseteq \mathcal{D}\}$ are the least closure system and knowledge space containing $\mathcal{D}$ respectively. $CS(\mathcal{D})$ and $KS(\mathcal{D})$ are called the closure system and knowledge space generated by $\mathcal{D}$, and $\mathcal{D}$ is also called the basis of $CS(\mathcal{D})$ and $KS(\mathcal{D})$.

2.3 Topologies and Interior Operators

A set system $\mathcal{T}$ on a universe U is called a *topology* on if the following conditions can be satisfied: (1) $\emptyset, U \in \mathcal{T}$, (2) $O_1, O_2 \in \mathcal{T} \Rightarrow O_1 \cap O_2 \in \mathcal{T}$, (3) $O_i, i \in I \Rightarrow \bigcup_{i\in I} O_i \in \mathcal{T}$, . Then $(U, \mathcal{T})$ is called a *topology space*, the elements in $\mathcal{T}$ are called *open sets* on U.

A mapping int:$\mathcal{P}(U) \to \mathcal{P}(U)$ is an *interior operator* on topology space $(U, \mathcal{T})$ if the following conditions can be satisfied: (1) $\text{int}(U)=U$, (2) $\text{int}(A) \subseteq A$, (3) $\text{int}(A \cap B)=\text{int}(A) \cap \text{int}(B)$, (4) $\text{int}(\text{int}(A))=\text{int}(A)$.

Evidently topologies on a universe are also closure systems. For a topology $\mathcal{T}$ on U, the elements of its dual set system $\mathcal{T}^c$ are called *closed sets*, and the following conditions hold: (1) $\emptyset, U \in \mathcal{T}^c$, (2) $C_1, C_2 \in \mathcal{T}^c \Rightarrow C_1 \cup C_2 \in \mathcal{T}^c$, (3) $C_i \in \mathcal{T}^c, i \in I \Rightarrow \bigcap_{i\in I} C_i \in \mathcal{T}^c$.

Correspondingly, *closure operator* $cl : \mathcal{P}(U) \to \mathcal{P}(U)$ with respect to the dual set system $\mathcal{T}^c$ of $\mathcal{T}$ satisfies (1) $cl(\emptyset) = \emptyset$, (2) $A \subseteq cl(A)$, (3) $cl(A \cup B) = cl(A) \cup cl(B)$, (4) $cl(cl(A)) = cl(A)$.

Topologies and interior operators correspond to each other. Given a topology $\mathcal{T}$ on U, an interior operator can be defined as $\text{int}_{\mathcal{T}}(A) = \cup\{O \in \mathcal{T}|O \subseteq A\}$, $\forall A \subseteq U$. Conversely, for any interior operator int on U, $\mathcal{T}_{\text{int}} = \{O \subseteq U|\text{int}(O) = O)\}$ is the topology induced by int, and it is easy to check that $\mathcal{T}_{\text{int}}$ is a topology on U.

2.4 Partitions

Partition of a universe U is a special set system. If a set system $\mathcal{P}$ on U is a covering and its elements are pairwise disjoint, then $\mathcal{P}$ is called a *partition*. In Pawlak rough set model, partition of the universe is primitive knowledge, based on which a pair of upper and lower approximation operators are constructed.

3 Rough Set Model Based on Granular Structure

In this section we propose a rough set model on general granular structure, which is one kind of extension of Pawlak rough set model. So we first review Pawlak rough sets as follows.

3.1 Pawlak Rough Sets

Let U be a universe, R an equivalence relation on U. For any $x \in U$, the successor neighborhood of x is $R(x) = \{y \in U|(x, y) \in R\}$, which is also the equivalence class, also denoted as $[x]_R$. The set of all equivalence classes forms a partition of U, denoted as U/R. Then the pair (U, R) is called Pawlak approximation space, and the equivalence classes and the union of some classes are understood as known knowledge, also called definable sets. For any subset $X \subseteq U$, it can be approximated by two definable sets as follows:

$$\underline{R}(X) = \cup\{A \in U/R | A \subseteq X\}, \quad \overline{R}(X) = \cup\{A \in U/R | A \cap X \neq \emptyset\}.$$

Then $\underline{R}(X)$ and $\overline{R}(X)$ are referred to as the lower and upper approximations of X in (U, R).

The lower and upper approximations are dual to each other, that is,

$$\underline{R}(X) = \sim \overline{R}(\sim X), \ \overline{R}(X) = \sim \underline{R}(\sim X), \forall A \subseteq U.$$

And the following properties hold: $\forall A, B \subseteq U$,

(L1) $\underline{R}(U) = U$,
(L2) $A \subseteq B \Rightarrow \underline{R}(A) \subseteq \underline{R}(B)$,
(L3) $\underline{R}(A \cap B) = \underline{R}(A) \cap \underline{R}(B)$,
(L4) $\underline{R}(A) \subseteq A$,
(L5) $\overline{R}(\underline{R}(A)) \subseteq A$,
(L6) $\underline{R}(\underline{R}(A)) = \underline{R}(A)$,

(U1) $\overline{R}(\emptyset) = \emptyset$;
(U2) $A \subseteq B \Rightarrow \overline{R}(A) \subseteq \overline{R}(B)$;
(U3) $\overline{R}(A \cup B) = \overline{R}(A) \cup \overline{R}(B)$;
(U4) $A \subseteq \overline{R}(A)$;
(U5) $A \subseteq \underline{R}(\overline{R}(A))$;
(U4) $\overline{R}(\overline{R}(A)) = \overline{R}(A)$.

3.2 Rough Set Approximations on Granular Structure

Definition 1. *Let U be a universe, $\mathcal{S}$ a set system on U. Then the triplet $(U, \mathcal{S}, \mathcal{S}^c)$ is called a generalized approximation space. For any $X \subseteq U$, its lower and upper approximations in $(U, \mathcal{S}, \mathcal{S}^c)$ are defined respectively by*

$$\underline{\mathcal{S}}(X) = \cup\{A \in \mathcal{S} | A \subseteq X\}, \quad \overline{\mathcal{S}}(X) = \cap\{A \in \mathcal{S}^c | X \subseteq A\}.$$

$\underline{\mathcal{S}}(X)$, $\sim \overline{\mathcal{S}}(X)$ and $\overline{\mathcal{S}}(X) - \underline{\mathcal{S}}(X)$ are called the positive domain, negative domain and boundary domain of X in $(U, \mathcal{S}, \mathcal{S}^c)$, respectively.

Example 2. (Continued from Example 1) The dual set system of $\mathcal{C}_I$ is $\mathcal{C}_I^c = \{\sim a^*, \sim b^*, \sim c^*, \sim d^*, \sim e^*, \sim f^*\}$, then $(U, \mathcal{C}_I, \mathcal{C}_I^c)$ is a generalized approximation space.

Let $X = \{1, 3, 4, 5\}$, $Y = \{2, 5\}$. By Definition 1 we have $\underline{\mathcal{C}_I}(X) = \{1, 3, 4\}$, $\overline{\mathcal{C}_I}(Y) = \{2, 5, 6\}$.

It should be pointed out that the notions of the lower and upper approximation proposed in Definition 1 are generalization of Pawlak rough approximations. Taking $\mathcal{S} = U/R$, it is not difficult to check that $\underline{\mathcal{S}}(X) = \underline{R}(X)$ and $\overline{\mathcal{S}}(X) = \overline{R}(X)$ for all $X \subseteq U$.

The lower and upper approximations defined in Definition 1 can be interpreted by topology terminology.

Definition 2. *Let $\mathcal{S}$ be a set system on U, for any $X \subseteq U$ and $x \in U$, then*

(1) *x is called an interior point of X in $(U, \mathcal{S}, \mathcal{S}^c)$ if there exists a $C \in \mathcal{S}$ such that $x \in C$ and $C \subseteq X$;*

(2) *x is called an outer point of X in $(U, \mathcal{S}, \mathcal{S}^c)$ if there exists a $C \in \mathcal{S}$ such that $x \in C$ and $C \cap X = \emptyset$;*

(3) *x is called an accumulation point of X in $(U, \mathcal{S}, \mathcal{S}^c)$ if for any $C \in \mathcal{S}$, $x \in C$ implies $C \cap X - \{x\} \neq \emptyset$;*

(4) *x is called a boundary point of X in $(U, \mathcal{S}, \mathcal{S}^c)$ if for any $C \in \mathcal{S}$, $x \in C$ implies $C \cap X \neq \emptyset$ and $C \cap (\sim X) \neq \emptyset$.*

Then the following propositions hold.

Proposition 1. *Let $(U, \mathcal{S}, \mathcal{S}^c)$ be a generalized approximation space, $X \subseteq U$. Then*

(1) *The positive domain $\underline{\mathcal{S}}(X)$ of X in $(U, \mathcal{S}, \mathcal{S}^c)$ consists of all interior points of X in $(U, \mathcal{S}, \mathcal{S}^c)$.*

(2) *The negative domain $\sim \overline{\mathcal{S}}(X)$ of X in $(U, \mathcal{S}, \mathcal{S}^c)$ consists of all outer points of X in $(U, \mathcal{S}, \mathcal{S}^c)$.*

(3) *The boundary domain $\overline{\mathcal{S}}(X) - \underline{\mathcal{S}}(X)$ of X in $(U, \mathcal{S}, \mathcal{S}^c)$ consists of all boundary points of X in $(U, \mathcal{S}, \mathcal{S}^c)$.*

Proof. (1) It directly follows from Definition 1 and 2.

(2) As $\overline{\mathcal{S}}(X) = \cap\{\sim C | C \in \mathcal{S}, X \subseteq \sim C\}$, $\sim \overline{\mathcal{S}}(X) = \cup\{C \in \mathcal{S} | C \subseteq \sim X\}$, so

$$x \in \sim \overline{\mathcal{S}}(X) \Leftrightarrow \exists C(x \in C, C \subseteq \sim X) \Leftrightarrow \exists C(x \in C, C \cap X = \emptyset).$$

The proof has been completed.

(3) For any $x \in \overline{\mathcal{S}}(X) - \underline{\mathcal{S}}(X)$, by $x \in \overline{\mathcal{S}}(X)$ we have that for any $C \in \mathcal{S}$ if $X \subseteq \sim C$, i.e. $C \cap X = \emptyset$, then $x \notin C$. By $x \notin \underline{\mathcal{S}}(X)$ we have that for any $C \in \mathcal{S}$ if $x \in C$ then $C \nsubseteq X$, i.e. $C \cap (\sim X) \neq \emptyset$. Thus for any $C \in \mathcal{S}$ if $x \in C$ then $C \cap X \neq \emptyset$ and $C \cap (\sim X) \neq \emptyset$, it follows that x is a boundary point of X in $(U, \mathcal{S}, \mathcal{S}^c)$. Similarly we can prove that any boundary point of X in $(U, \mathcal{S}, \mathcal{S}^c)$ must belong to the boundary domain of X in $(U, \mathcal{S}, \mathcal{S}^c)$.

If we denote the set of all accumulation points of X in $(U, \mathcal{S}, \mathcal{S}^c)$ as X' then the following result holds.

Proposition 2. *Let $(U, \mathcal{S}, \mathcal{S}^c)$ be a generalized approximation space, $X \subseteq U$. Then $\overline{\mathcal{S}}(X) = X \cup X'$.*

Proof. Assume that $x \in \overline{\mathcal{S}}(X)$. It directly follows from Definition 1 that $X \subseteq \overline{\mathcal{S}}(X)$. For any $C \in \mathcal{S}$ if $X \subseteq \sim C$, i.e. $C \cap X = \emptyset$, then $x \notin C$, equivalently, if $x \in C$ then $C \cap X \neq \emptyset$. If $x \notin X$, we have $C \cap (X - \{x\}) \neq \emptyset$, then $x \in X'$. Thus we have $\overline{\mathcal{S}}(X) \subseteq X \cup X'$.

Suppose that $x \in X \cup X'$. For any $C \in \mathcal{S}$, if $x \in C$ and $x \in X$ then evidently $C \cap (X - \{x\}) \neq \emptyset$. If $x \in C$ and $x \in X'$ then $C \cap (X - \{x\}) \neq \emptyset$, which implies $C \cap X \neq \emptyset$. Thus if $C \cap X = \emptyset$ then $x \notin C$, equivalently if $X \subseteq \sim C$ then $x \in \sim C$. By Definition 1 we have $x \in \overline{\mathcal{S}}(X)$. Thus we can conclude $X \cup X' \subseteq \overline{\mathcal{S}}(X)$.

According to the above proof we obtain $\overline{\mathcal{S}}(X) = X \cup X'$.

From Proposition 1 and 2 we can see that in Definition 1 the generalized approximation space is an extension of topology space, and the lower and upper approximations in a generalized approximation space are extensions of interior and closure in topology space. The following properties of new operators can be easily proved.

Theorem 1. *Let $(U, \mathcal{S}, \mathcal{S}^c)$ be a generalized approximation space, the lower and upper approximations satisfy the following properties: $\forall X, Y \in \mathcal{P}(U)$,*

(L0) $\underline{\mathcal{S}}(X) = \sim \overline{\mathcal{S}}(\sim X)$, (U0) $\overline{\mathcal{S}}(X) = \sim \underline{\mathcal{S}}(\sim X)$;
(L1′) $\underline{\mathcal{S}}(\emptyset) = \emptyset$, (U1′) $\overline{\mathcal{S}}(U) = U$;
(L2) $A \subseteq B \Rightarrow \underline{\mathcal{S}}(A) \subseteq \underline{\mathcal{S}}(B)$, (U2) $A \subseteq B \Rightarrow \overline{\mathcal{S}}(A) \subseteq \overline{\mathcal{S}}(B)$;
(L4) $\underline{\mathcal{S}}(A) \subseteq A$, (U4) $A \subseteq \overline{\mathcal{S}}(A)$;
(L6) $\underline{\mathcal{S}}(\underline{\mathcal{S}}(A)) = \underline{\mathcal{S}}(A)$, (U6) $\overline{\mathcal{S}}(\overline{\mathcal{S}}(A)) = \overline{\mathcal{S}}(A)$.

For any $X \subseteq U$, based on Definition 1, (L4) and (U4) of Theorem 2 we can gain two types of decision rule with respect to X, one is depicted with positive logical description, and another with negative logical description.

Example 3. (Continued from Example 2) As $\underline{\mathcal{C}_I}(X) = a^* \cup c^* \cup d^*$ and $\overline{\mathcal{C}_I}(Y) = \sim a^* \cap \sim c^* \cap \sim d^*$, so two certainty rules can be obtained. One is a positive rule: if an object has one of properties a, c and d, then it must belong to X, we represent it by

$$a^*, c^*, d^* \rightarrow X.$$

Another is a negative rule: if an object belongs to Y, then it dose not has properties a, c and d, which is denoted as

$$Y \rightarrow \sim a^* \text{ or } \sim c^* \text{ or } \sim d^*.$$

The properties (L0) and (U0) in Theorem 1 shows that the approximation operators $\underline{\mathcal{S}}$ and $\overline{\mathcal{S}}$ are dual to each other. Therefore the properties (L1′) and (U1′), (L2) and (U2), (L4) and (U4), (L6) and (U6) can be viewed as dual properties. It should be pointed out that in a generalized approximation space properties (L1), (U1), (L3), (U3), (L5) and (U5) may not hold.

Example 4. Let $U = \{1, 2, 3, 4\}$, $C_1 = \{1\}$, $C_2 = \{1, 2\}$, $C_3 = \{2, 3\}$ and $\mathcal{S} = \{C_1, C_2, C_3\}$. Then $(U, \mathcal{S}, \mathcal{S}^c)$ is a generalized approximation space. Taking $X = \{1, 3\}$, $Y = \{2, 4\}$, $X_1 = \{1, 2\}$, $X_2 = \{2, 3\}$, $Y_1 = \{1\}$ and $Y_2 = \{3, 4\}$. By Definition 1 we have $\underline{\mathcal{S}}(U) = \{1, 2, 3\}$ and $\overline{\mathcal{S}}(\emptyset) = \{4\}$, so

$$\underline{\mathcal{S}}(U) \neq U \text{ and } \overline{\mathcal{S}}(\emptyset) \neq \emptyset.$$

We also figure out that $\overline{\mathcal{S}}(\underline{\mathcal{S}}(X)) = \{1, 4\}$ and $\underline{\mathcal{S}}(\overline{\mathcal{S}}(Y)) = \{2, 3\}$. Thus

$$\overline{\mathcal{S}}(\underline{\mathcal{S}}(X)) \not\subseteq X, Y \not\subseteq \underline{\mathcal{S}}(\overline{\mathcal{S}}(Y)).$$

Moreover we compute that $\underline{\mathcal{S}}(X_1) = \{1,2\}$, $\underline{\mathcal{S}}(X_2) = \{2,3\}$, $\underline{\mathcal{S}}(X_1 \cap X_2) = \{2\}$; $\overline{\mathcal{S}}(Y_1) = \{1,4\}$, $\overline{\mathcal{S}}(Y_2) = \{3,4\}$, $\overline{\mathcal{S}}(Y_1 \cup Y_2) = U$. Thus

$$\underline{\mathcal{S}}(X_1 \cap X_2) \neq \underline{\mathcal{S}}(X_1) \cap \underline{\mathcal{S}}(X_2), \overline{\mathcal{S}}(Y_1 \cup Y_2) \neq \overline{\mathcal{S}}(Y_1) \cup \overline{\mathcal{S}}(Y_2).$$

It should be point out that although the properties (L1), (U1), (L3), (U3), (L5) and (U5) of $\underline{\mathcal{S}}$ and $\overline{\mathcal{S}}$ on generalized approximation space $(U, \mathcal{S}, \mathcal{S}^c)$ may be not satisfied, some special granular structures can be depicted by them

4 Approximation Operator Characterizations of Granular Structures

Rough approximations based on different granular structures satisfy different properties, conversely granularity with different structures can be characterized by the lower and upper approximation operators.

Theorem 2. *Let $(U, \mathcal{S}, \mathcal{S}^c)$ be a generalized approximation space. Then $\mathcal{S}$ is a covering of U if and only if* (L1) $\underline{\mathcal{S}}(U) = U$ *or* (U1) $\overline{\mathcal{S}}(\emptyset) = \emptyset$.

Proof. It directly follows from Definition 1 and the duality of $\underline{\mathcal{S}}$ and $\overline{\mathcal{S}}$.

The following theorem can also be proved easily.

Theorem 3. *Let $(U, \mathcal{S}, \mathcal{S}^c)$ be a generalized approximation space. Then $\mathcal{S}$ is a knowledge space on U, or $\mathcal{S}^c$ is a closure system on U, if and only if $\underline{\mathcal{S}}(\mathcal{S}) \subseteq \mathcal{S}$ or $\overline{\mathcal{S}}(\mathcal{S}^c) \subseteq \mathcal{S}^c$.*

Before we propose operator characterization of topology, we first review the concept of topological base. Let $\mathcal{T}$ be a topology on U, $\mathcal{B} \subseteq \mathcal{T}$. If $KS(\mathcal{B}) = \mathcal{T}$ then $\mathcal{B}$ is called a topological base, or a base of $\mathcal{T}$. Then the following lemma is useful.

Lemma 1. *$\mathcal{B}$ is a base of topology $\mathcal{T}$ on U if and only if for any $G \in \mathcal{T}$, $x \in G$, there exists a $C \in \mathcal{B}$ such that $x \in C \subseteq G$.*

Theorem 4. *Let $(U, \mathcal{S}, \mathcal{S}^c)$ be a generalized approximation space. Then $\mathcal{S}$ is a topological base on U if and only if* (L3) $\underline{\mathcal{S}}(X_1 \cap X_2) = \underline{\mathcal{S}}(X_1) \cap \underline{\mathcal{S}}(X_2)$ *or* (U3) $\overline{\mathcal{S}}(X_1 \cup X_2) = \overline{\mathcal{S}}(X_1) \cup \overline{\mathcal{S}}(X_2)$ *for all $X_1, X_2 \in \mathcal{P}(U)$.*

Proof. ($\Rightarrow$) Assume that $\mathcal{S}$ is a topological base on U. Then $KS(\mathcal{S})$ is the topology generated by $\mathcal{S}$. For any $x \in \underline{\mathcal{S}}(X_1) \cap \underline{\mathcal{S}}(X_2)$, there are $C_1, C_2 \in \mathcal{S}$ such that $C_1 \subseteq X_1$, $C_2 \subseteq X_2$ and $x \in C_1 \cap C_2$. As $KS(\mathcal{S})$ is a topology, and $\mathcal{S}$ is its a base, so $C_1 \cap C_2 \in KS(\mathcal{S})$. By Lemma 1 there exists a $C \in \mathcal{S}$ such that $x \in C \subseteq C_1 \cap C_2 \subseteq X_1 \cap X_2$. Thus $x \in \underline{\mathcal{S}}(X_1 \cap X_2)$. We can conclude $\underline{\mathcal{S}}(X_1) \cap \underline{\mathcal{S}}(X_2) \subseteq \underline{\mathcal{S}}(X_1 \cap X_2)$. On the other hand, $\underline{\mathcal{S}}(X_1 \cap X_2) \subseteq \underline{\mathcal{S}}(X_1) \cap \underline{\mathcal{S}}(X_2)$ follows from (L2) of Theorem 1. Finally we gain $\underline{\mathcal{S}}(X_1 \cap X_2) = \underline{\mathcal{S}}(X_1) \cap \underline{\mathcal{S}}(X_2)$. By the duality of $\underline{\mathcal{S}}$ and $\overline{\mathcal{S}}$ we know $\overline{\mathcal{S}}(X_1 \cup X_2) = \overline{\mathcal{S}}(X_1) \cup \overline{\mathcal{S}}(X_2)$.

($\Leftarrow$) By the duality of $\underline{\mathcal{S}}$ and $\overline{\mathcal{S}}$ we only suppose (L3) holds. For any $G \in KS(\mathcal{S})$, $x \in G$, there exist $\{C_i, i \in I\} \subseteq \mathcal{S}$ such that $G = \bigcup_{i\in I} C_i$. It is easy to prove that $\underline{\mathcal{S}}(G) = G = \bigcup_{i\in I} C_i$. Thus there is a $C \in \{C_i, i \in I\}$ with $x \in C \subseteq G$. By virtue of Lemma 1 we know that $\mathcal{S}$ is a base of the topology $KS(\mathcal{S})$ on U.

Theorem 5. *Let $(U, \mathcal{S}, \mathcal{S}^c)$ be a generalized approximation space. Then there is a set system $\mathcal{D} \subseteq \mathcal{S}$ such that $\mathcal{D}$ is a partition of U and $\underline{\mathcal{D}} = \underline{\mathcal{S}}$, $\overline{\mathcal{D}} = \overline{\mathcal{S}}$ if and only if $\underline{\mathcal{S}}$ satisfies* (L1), (L3) *and* (L5), *or $\overline{\mathcal{S}}$ satisfies* (U1), (U3) *and* (U5).

Proof. ($\Rightarrow$) It is obvious.

($\Leftarrow$) Assume that $\underline{\mathcal{S}}$ satisfies (L1), (L3) and (L5), or $\overline{\mathcal{S}}$ satisfies (U1), (U3) and (U5). Denote $\mathcal{D} = \{\overline{\mathcal{S}}(x) | x \in U\}$, where $\overline{\mathcal{S}}(x) = \overline{\mathcal{S}}(\{x\})$. In the following we will prove that $\mathcal{D}$ just be the partition of U. The whole proof contains five parts:

(1) $y \in \overline{\mathcal{S}}(x) \Rightarrow x \in \overline{\mathcal{S}}(y), \forall x, y \in U$.

By (U5) we have $y \in \underline{\mathcal{S}}(\overline{\mathcal{S}}(y))$. As $\sim \overline{\mathcal{S}}(\sim \overline{\mathcal{S}}(y)) = \underline{\mathcal{S}}(\overline{\mathcal{S}}(y))$, $\overline{\mathcal{S}}(\sim \overline{\mathcal{S}}(y)) = \bigcup_{x \notin \overline{\mathcal{S}}(y)} \overline{\mathcal{S}}(x)$ follows from (U3), we have that if $x \notin \overline{\mathcal{S}}(y)$ then $y \notin \overline{\mathcal{S}}(x)$, i.e. $y \in \overline{\mathcal{S}}(x)$ implies $x \in \overline{\mathcal{S}}(y)$.

(2) $x \in \overline{\mathcal{S}}(y), y \in \overline{\mathcal{S}}(z) \Rightarrow x \in \overline{\mathcal{S}}(z), \forall x, y, z \in U$

By (U6) we have $\overline{\mathcal{S}}(z) = \overline{\mathcal{S}}(\overline{\mathcal{S}}(z)) = \bigcup_{y \in \overline{\mathcal{S}}(z)} \overline{\mathcal{S}}(y)$. So if $y \in \overline{\mathcal{S}}(z)$ then $\overline{\mathcal{S}}(y) \subseteq \overline{\mathcal{S}}(z)$. Thus $x \in \overline{\mathcal{S}}(z)$ follows from $x \in \overline{\mathcal{S}}(y)$ and $y \in \overline{\mathcal{S}}(z)$.

(3) $\mathcal{D}$ is a partition of U.

By (U1$'$) and (U3) we know that $\mathcal{D}$ is a covering of U.

Assume $\overline{\mathcal{S}}(x) \cap \overline{\mathcal{S}}(y) \neq \emptyset$, and take $z \in \overline{\mathcal{S}}(x) \cap \overline{\mathcal{S}}(y)$. If $w \in \overline{\mathcal{S}}(x)$, by (1) we have $x \in \overline{\mathcal{S}}(z)$, according to (2) we gain $w \in \overline{\mathcal{S}}(z)$. As $z \in \overline{\mathcal{S}}(y)$ so $w \in \overline{\mathcal{S}}(y)$ follows from (2). Thus $\overline{\mathcal{S}}(x) \subseteq \overline{\mathcal{S}}(y)$. Similarly we can prove $\overline{\mathcal{S}}(y) \subseteq \overline{\mathcal{S}}(x)$. Therefore we can conclude $\overline{\mathcal{S}}(x) = \overline{\mathcal{S}}(y)$.

Consequently we obtain that $\mathcal{D}$ is a partition of U.

(4) $\mathcal{D} \subseteq \mathcal{S}$.

Let $\overline{\mathcal{S}}(x) \in \mathcal{D}$, by Definition 1 we have that if $y \in \overline{\mathcal{S}}(x)$ then for any $C \in \mathcal{S}$ if $x \notin C$ then $y \notin C$, i.e. if $y \in C$ then $x \in C$. As $x \in \overline{\mathcal{S}}(y)$ follows from $y \in \overline{\mathcal{S}}(x)$, so we can conclude that $y \in \overline{\mathcal{S}}(x)$ is equivalent to that for any $C \in \mathcal{S}$, $x \in C$ and $y \in C$ are equivalent. Thus we can conclude $\overline{\mathcal{S}}(x) = \bigcap_{x \in C, C \in \mathcal{S}} C$. Furthermore, by (L3) we have $\underline{\mathcal{S}}(\overline{\mathcal{S}}(x)) = \underline{\mathcal{S}}(\bigcap_{x \in C, C \in \mathcal{S}} C) = \bigcap_{x \in C, C \in \mathcal{S}} \underline{\mathcal{S}}(C) = \bigcap_{x \in C, C \in \mathcal{S}} C$. Thus there exists a $C \in \mathcal{S}$ such that $x \in C \subseteq \overline{\mathcal{S}}(x) = \bigcap_{x \in C, C \in \mathcal{S}} C$, so $\overline{\mathcal{S}}(x) = C \in \mathcal{S}$. By the arbitrary of x we have $\mathcal{D} \subseteq \mathcal{S}$.

(5) $\underline{\mathcal{D}} = \underline{\mathcal{S}}$, $\overline{\mathcal{D}} = \overline{\mathcal{S}}$.

For any $C \in \mathcal{S}$, $x \in C$, as $\overline{\mathcal{S}}(x) \in \mathcal{D}$ is the minimum set containing x in $\mathcal{S}$, we know that C can be represented as an union of some elements of $\mathcal{D}$. Again considering $\mathcal{D} \subseteq \mathcal{S}$ and Definition 1 it is easy to get $\underline{\mathcal{D}}(X) = \underline{\mathcal{S}}(X)$ for all $X \subseteq U$, i.e. $\underline{\mathcal{D}} = \underline{\mathcal{S}}$. According to the duality of the lower and upper approximation operators it is not difficult to obtain $\overline{\mathcal{D}} = \overline{\mathcal{S}}$.

5 Conclusions

The generalization of rough set model on various knowledge structures is an interesting work. In this paper, generalization in two aspects has been done. The generalization of approximation space is based on a general granular structure and its dual set system, and do not assumed them to be closed under any set operations, on the other hand, the lower and upper approximations on generalized approximation space are defined by means of subsystem-based formalism. It is shown that the new model is extension of the rough set models on covering approximation space, knowledge space, topology space and Pawlak approximation space. In our model, the notions of positive domain, negative domain and boundary domain can be well interpreted by the language of topology. In addition, the properties of the lower and upper approximations are examined in detail, axiom sets characterizing some special granular structures are obtained. With the lower and upper approximations proposed in this paper, positive and negative decision rules may be extracted from information systems.

Acknowledgments. This work is supported by a grant from the National Natural Science Foundation of China (No. 11071284 and 61075120).

References

1. Pawlak, Z.: Rough Sets. International Journal of Computer and Information Sciences 11, 341–356 (1982)
2. Bonikowski, Z., Bryniariski, E., Skardowska, V.W.: Extension and Intensions in the Rough Set Theory. Information Sciences 107, 149–167 (1998)
3. Yao, Y.Y.: Relational Interpretations of Neighborhood Operators and Rough Set Approximation Operators. Information Sciences 111, 239–259 (1998)
4. Cattaneo, G.: Abstract Approximation Spaces for Rough Theories. In: Polkowski, L., Skowron, A. (eds.) Rough Sets in Data Mining and Knowledge Discovery, pp. 59–98. Physica, Heidelberg (1998)
5. Bargiela, A., Pedrycz, W.: Granular Computing: An Introduction. Kluwer Academic Publishers, Boston (2002)
6. Polkowski, L., Semeniuk-Polkowska, M.: On Foundations and Applications of the Paradigm of Granular Rough Computing. International Journal of Cognitive Informatics and Natural Intelligence 2, 80–94 (2008)
7. Xu, F.F., Yao, Y.Y., Miao, D.Q.: Rough Set Approximations in Formal Concept Analysis and Knowledge Spaces. In: An, A., et al. (eds.) ISMIS LNCS (LNAI), vol. 4994, pp. 319–328. Springer, Heidelberg (2008)
8. Doignon, J.P., Falmagne, J.C.: Spaces for the Assessment of Knowledge. International Journal of Man-Machine Studies 23, 175–196 (1985)
9. Doignon, J.P., Falmagne, J.C.: Knowledge Spaces. Springer, Heidelberg (1999)
10. Monjardet, B.: The Presence of Lattice Theory in Discrete Problems of Mathematical Social Sciences. Why. Mathematical Social Sciences 46, 103–144 (2003)
11. Danilov, V.I.: Knowledge Spaces from a Topological Point of View. Journal of Mathematical Psychology 53, 510–517 (2009)
12. Kortelainen, J.: On the Relationship between Modified Sets, Topolo-gical Spaces and Rough Sets. Fuzzy Sets and Systems 61, 91–95 (1994)

Recursive Query Facilities in Relational Databases: A Survey

Piotr Przymus[1], Aleksandra Boniewicz[1], Marta Burzańska[1], and Krzysztof Stencel[1,2]

[1] Faculty of Mathematics and Computer Science, Nicolaus Copernicus University, Toruń, Poland
[2] Institute of Informatics, University of Warsaw, Warsaw, Poland

Abstract. The relational model is the basis for most modern databases, while SQL is the most commonly used query language. However, there are data structures and computational problems that cannot be expressed using SQL-92 queries. Among them are those concerned with the bill-of-material and corporate hierarchies. A newer standard, called the SQL-99, introduced recursive queries which can be used to solve such tasks. Yet, only recently recursive queries have been implemented in most of the leading relational databases. In this paper we have reviewed and compared implementations of the recursive queries defined by SQL:1999 through SQL:2008 and offered by leading vendors of DBMSs. Our comparison concerns features, syntax and performance.

1 Introduction

Recursion is fundamental for common programming languages. People working with databases quickly realized that there are problems, such as bill-of-material or corporate hierarchy problem that cannot be efficiently solved using early SQL standards. Because there was a need for working with hierarchical data, several DBMS vendors have introduced proprietary ways of providing such functionality for their users. One of them was Oracle, which provided hierarchical queries using *Start With ... Connect By* construction in their DBMS version 5.0. The first DBMS to provide recursive Common Table Expressions was IBM's DB2 v.7 in 1997. The research on SQL's recursive CTEs have been greatly influenced by Datalog ([1], [2]). But it was ANSI SQL-99 standard that officially introduced recursive queries and views into SQL language. Throughout the next 10 years there has been little happening in the field of recursive SQL queries, especially in comparison to other relational database research. But with each passing year the problem of recursion became more popular with database vendors. Nowadays most major DBMSs support recursive CTE. In the 2003 Sybase released SQL Anywhere 9 - their first DBMS equipped with recursive CTEs. Two years later Microsoft included them within MS SQL Server 2005. Among other popular databases Firebird supports relational CTEs since 2008 (Firebird 2.1), PostgreSQL and Oracle since 2009 and their databases PostgreSQL 8.4 and Oracle 11gR2. The modern research on recursive queries focuses on optimization. The

Y. Zhang et al. (Eds.): DTA/BSBT 2010, CCIS 118, pp. 89–99, 2010.

papers [3,4,5] present propositions of new optimization techniques specially designed for recursive SQL-99 queries. However even though the topic of recursive queries in SQL is not something new, there is still no complete survey on the subject. The paper [6] provides a basic comparison of features and performance of recursive CTEs implemented by MS SQL 2005 and PostgreSQL 8.4 databases. Unfortunately this paper lacks a more general approach.

The latest public versions of the selected DBMSs have been tested: IBM DB2 Express-C 9.5, RDBMS X, Sybase SQL Anywhere 11, PostgreSQL 8.4, Firebird 2.1.3 and Microsoft SQL Server 2008. This paper is organized as follows: §2 discusses the general construction of recursive CTE in SQL; §3 compares syntax and features of the recursive queries; §4 presents the results of performance testing on those databases under two operating systems: MS Windows 7 and Ubuntu 9.10; § 5 concludes. Detailed data from the experiments is presented in the Appendix.

2 Recursive Common Table Expressions

In SQL a recursive query specifies a temporary result set known as a recursive Common Table Expression (CTE). It is usually defined within the scope of a SELECT query, but may also be used with INSERT, UPDATE, MERGE or DELETE statements. The recursive clause itself may be defined explicitly as a CTE or it may be used in a CREATE VIEW statement as a part of its definition. Each recursive CTE definition consist of a CTE header, a seed query and a recursive query. The general syntax of a recursive SQL query is presented below:

$$\textbf{WITH RECURSIVE}\ \mathit{cte}_0\ \ (A_{01},\cdots,A_{0n})\ \textbf{AS}$$
$$(\mathit{seed\_query}\ \textbf{UNION ALL}\ \mathit{recursive\_query}\ \text{additional\_clauses}),$$
$$[\text{RECURSIVE}]\ \mathit{cte}_1(A_{11},\cdots,A_{1n})\ \ldots\ \mathit{outer\ query\ with\ cte_i}\ (i{\geq}0)$$

In the example given above the CTE header is formed out of the *WITH RECURSIVE* keywords, the CTE's name and a list of columns. Both seed query and a recursive query may consist of multiple SELECT statements, all of them should be connected using *UNION ALL* operator. The seed query forms the initial set of tuples which will be used by the recursive query. The SQL standard disallows non-linear recursion thus the recursive query is restricted to exactly one reference to the CTE. The CTE definition may contain additional clauses that provide additional functionality. For example in order to specify the search order one may use the SEARCH clause with DEPTH FIRST or BREADTH FIRST keywords. The CYCLE clause may be used to detect cycles that could result in loops.

There are subtle differences in the implementations of the recursive queries provided by various vendors. The next section discusses those issues in detail.

3 Features Comparison

This section's aim is to compare syntax differences, available features and limitations between databases. The Table 1 contains compact information about

Table 1. Features Comparison

	Features	MS SQL	SQL Anywhere	PostgreSQL	DB2	RDBMS X	Firebird
Multiple queries in	Initial step	Y	Y	Y	Y	N*	Y
	Recursive step	Y	N	N	Y	N	Y
Other (then UNION ALL) set operators	Between initial queries	Y*	Y*	Y*	Y	-*	Y*
	Between recursive and initial queries	N	N	Y*	N	N	N
	Between recursive queries	N	-	-	N	-	N
Referencing CTE	Recursive CTE to other recursive CTE	Y	Y*	Y	Y	Y	Y
	Recursive CTE to non-recursive CTE	Y	Y	Y	Y	Y	Y
	Non-recursive CTE to recursive CTE	Y	Y	Y	Y	Y	Y
	Mutual recursive CTE	N	N	N	N	N	N
Using Recursive CTE in	SELECT	Y*	Y*	Y	Y	Y	Y
	INSERT	Y	Y*	N	Y	Y	N
	UPDATE	Y	N	N	Y*	Y	N
	DELETE	Y	N	N	N	Y	N
	CREATE VIEW	Y	Y*	N	N	Y	N
	OTHER	Y*	N	N	N	Y*	N
Various	Protection from infinite loops	Y*	Y*	N	N	Y	Y*
	Search clause	N	N	N	N	Y	N
	Cycle clause	N	N	N	N	Y	N
Recursive query part limitations	Group by	N	N	N	N	N	N
	Having	N	N	N	N	N	N
	Order by	N	N	N	N	Y	Y
	Distinct	N	N	Y	N	N	N
	Non aggregate functions	Y*	Y	Y	N	Y*	Y
	Agregate functions	N	N	N	N	N	N
	Recursive in subquery	N	N	N	N	N	N
	Left outer join	N	Y*	Y	N	Y*	N
	Right outer join	N	Y*	Y	N	Y*	N
	Full outer join	N	N*	N	N	N	N
	Limiting results	N	Y	N	Y	Y	Y
Initial subquery part support for	Group by, Order by, Distinct, Having, Aggregate and Non aggregate Functions	Y	Y	Y	Y	Y	Y
Syntax	obligatory RECURSIVE keyword	N	Y	Y	N	N	Y
	obligatory column list after WITH	?	Y	N	Y	Y	N

features and information about support in specified databases ("Y" - yes, "N" - no or "-" not applicable, "*" - additional information is in Table 2). The comparison has been made based on the product documentation, supported error codes and results of list of control queries. **Impact of recursive CTE features on RDBMS functionality**: Multiple queries in the seed part of the CTE are supported in almost every tested database, only RDBMS X does not support this feature. Multiple queries in the recursive part are more troublesome. Only half of the database vendors decided to implement this feature. The lack of this feature

Table 2. Features Comparison-Details

Database	Feature	Description
MS SQL Server 2008	Protection from infinite loops	Query hint MAXRECURSION from 0 to 32767
	Other (then UNION ALL) set operators in CTE between initial queries	UNION, EXCEPT, or INTERSECT
	Recursive subquery limitations - Non aggregate functions	Functions not allowed: functions with parameters, functions with side effects
	Using Recursive CTE in OTHER	MERGE
SQL Anywhere	Protection from infinite loops	Database option max_recursive_iterations(default 100). Additionally two connection type options appear MAX_TEMP_SPACE prevents the server from utilizing an unlimited amount of space for the TEMP file. TEMP_SPACE_LIMIT_CHECK limits temporary space usage on a connection-by-connection basis.
	Other (then UNION ALL) set operators in CTE between initial queries	UNION, EXCEPT, or INTERSECT
	LEFT,RIGHT OUTER JOIN	the reference to the recursive table cannot appear on the null-supplying side of an outer join
	Referencing CTE recursive CTE to other recursive CTE	Works, but documentation states differently
Postgre SQL	Other (then UNION ALL) set operators in CTE between initial queries	UNION, EXCEPT, or INTERSECT
	Other (then UNION ALL) set operators in CTE between recursive and initial queries	UNION
	LEFT,RIGHT OUTER JOIN	the reference to the recursive table cannot appear on the null-supplying side of an outer join
DB2	Using Recursive CTE in UPDATE	Non official way (but suggested by a DB2 developer) `WITH v AS (..)SELECT COUNT(*)FROM OLD TABLE( UPDATE ..)`
RDBMS X	Recursive query part analytic functions are permitted	limitations on analytic functions are permitted
	Multiple queries in initial step	Feature described in documentation but not implemented
	Other (then UNION ALL) set operators in CTE between initial queries	Feature described in documentation but not implemented. Documentation mentions UNION, INTERSECT and MINUS.
	LEFT,RIGHT OUTER JOIN	the reference to the recursive table cannot appear on the null-supplying side of an outer join
	Using Recursive CTE in OTHER	CREATE TABLE
Firebird	Other (then UNION ALL) set operators in CTE between initial queries	UNION
	Protection from infinite loops	Hardcoded recursion depth 1024

probably could be overcome by a user with other language constructions, but this would lead to overcomplicated solutions. Therefore, native support would be appreciated since no obvious difficulties are involved.

All databases supporting multiple initial queries are allowing other than UNION ALL set operators in the initial queries part. On the other hand, there is a lack of other set operators in the recursive part. Additionally only PostgreSQL supports other than UNION ALL operators (in this case UNION) between initial and recursive part. This construction may be used as some form of protection against infinite loops, but it will work only in some cases (for example when cycle is a result of duplicate rows).

Usage of CTE in various SQL statements slightly differs for each SQL dialect. Lack of support for some statements (especially UPDATE or DELETE) may be burdensome and may lead to strange, non-standard solutions (see row 4 of the Table 2). Differences are probably engendered by general implementation of CTE and not related strictly to recursive extension.

Protection from infinite loops is a major feature. Data with cycles may cause unconstrained reachability queries. There is a serious threat for stability of databases without this protection. From the authors' experience such cases may end with depletion of system resources. Only four databases assure such protection.

Cases of support for DISTINCT (PostgreSQL) and ORDER BY (RDBMS X and Firebird) clauses are worth mentioning and further investigation. As OUTER JOIN may lead to various problems, there are different policies for using them. However, after restricting OUTER JOIN construction there is no technical reason to forbid them.

4 Performance Tests

Tests have been conducted on two identical computers equipped with processor Intel(R) Core(TM)2 Quad CPU Q9400, 2.66GHz, 3072 KB cache size, 4 GB RAM and 320 GB hard-drive. Used operating systems: Windows 7 Enterprise 64bit, and Ubuntu 9.10 Server Edition. We tested following databases: MS SQL Server 2008 Express, PostgreSQL 8.4, DB2 Express-C 9.7, Firebird 2.1.3, Sybase SQL Anywhere 11, RDBMS X. RDBMS X on Windows does not support SQL-99 compliant recursive queries and MS SQL Server is only available on MS Windows, so they were tested only on Linux and Windows respectively.

Database schema consists of the following relations:

- CITIES(cid, city) contains 200 records,
- TRAINS(departure, arrival, railline, tid, price) contains 800 records,
- FLIGHTS(departure, arrival, carrier, fid, price) contains 800 records.

Each query was executed 6 times. The average duration of the query was measured. The tests were conducted for the data containing cycles and without cycles, on both with and without indexes. Because some databases automatically place indexes on key columns, by "no indexes" we mean that no indexes have been placed manually - a default situation after simple CREATE TABLE command. Indexes were placed on: the cid column from relation CITIES, departure and arrival columns from TRAINS and FLIGHTS.

We have identified two kinds of problem being solved by recursion: hierarchy (like in corporate hierarchy or bill-of-material) and arbitrary graph (like in transport connections). Therefore, we have prepared two sets of data: the one without cycles and the one with cycles. In our opinion those cover most of the sorts of problems the recursive queries have been designed to solve.

The cyclic data are tested with the parameter $I = 1 \cdots 9$ which limits the recursion depth. For $I = 9$ it was enough to fully exhaust the system resources and in many cases to crash of the database system.

4.1 Test Suite 1

Query Q1 displays all the cities reachable by plane starting from Toronto, the number of connections is limited by the parameter I. This query has been tested on: Windows - DB2, Firebird, MS SQL, PostgreSQL, SQL Anywhere and Ubuntu - DB2, Firebird, PostgreSQL, RDBMS X and SQL Anywhere.

```
WITH destinations (origin, departure, arrival, connections) AS
  (SELECT a.departure, a.departure, a.arrival, 1
   FROM flights a, cities c
   WHERE a.departure = c.cid AND c.city = 'Toronto'
  UNION ALL
   SELECT r.origin, b.departure, b.arrival, r.connections + 1
   FROM destinations r, flights b
   WHERE r.arrival = b.departure AND r.flight_count < I )
SELECT count(*) FROM destinations
```

Listing 1.1. Query Q1

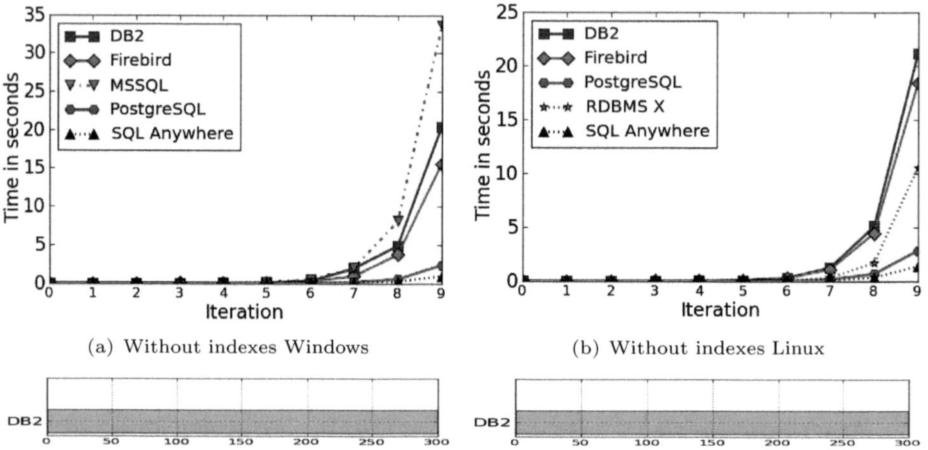

(a) Without indexes Windows

(b) Without indexes Linux

(c) Overall percentual speed increase (if any) with indexes Windows

(d) Overall percentual speed increase (if any) with indexes Linux

Fig. 1. Result of the query Q1 for data with cycles

For data without cycles query Q1 completed instantly for $I = 0 \cdots 9$ with and without indexes.

SQL Anywhere had the best performance on both OSs. Only for the DB2, placing indexes made a huge speed improvement, for data without indexes full table scans were used, while after adding indexes, light index scans and immediate tuple fetching occurred.

4.2 Test Suite 2

Query Q2 has been tested on the same set of databases as Q1. Q2 displays all the cities reachable by plane starting from Toronto or Warsaw, the number of connections is limited by the parameter I.

```
WITH destinations (departure, arrival, connections, cost) AS
 (SELECT a.departure, a.arrival, 0, price
  FROM flights a, cities c
  WHERE a.departure = c.cid
   AND c.city = 'Toronto' OR c.city = 'Warsaw'
 UNION ALL
  SELECT r.departure, b.arrival, r.connections + 1, r.cost + b.price
  FROM destinations r, flights b
  WHERE r.arrival = b.departure AND r.connections < I)
SELECT count(*) FROM destinations
```

Listing 1.2. Query Q2

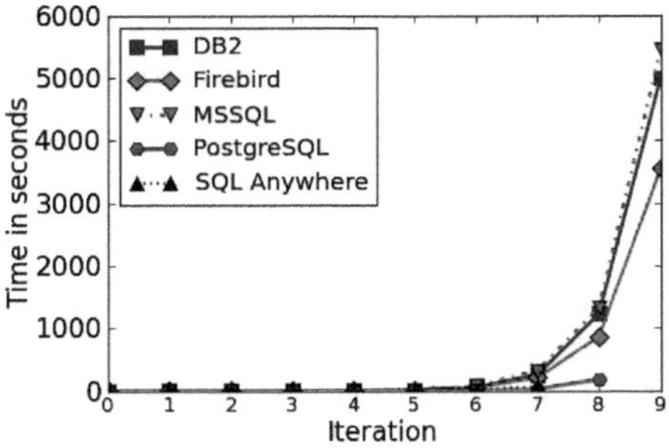

(a) Without indexes Windows

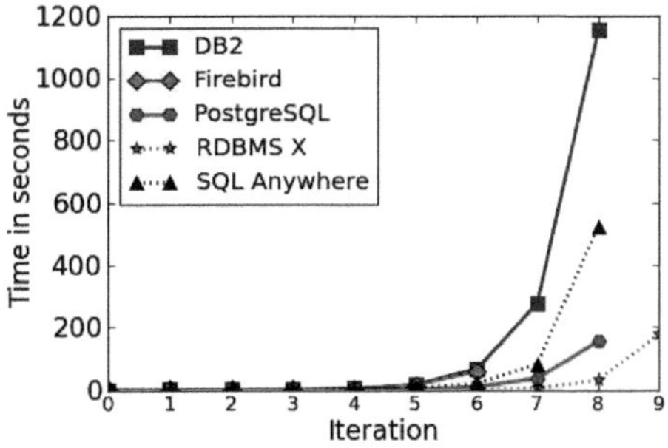

(b) Without indexes Linux

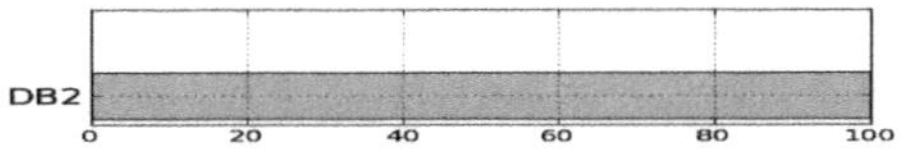

(c) Overall percentual speed increase (if any) with indexes Windows

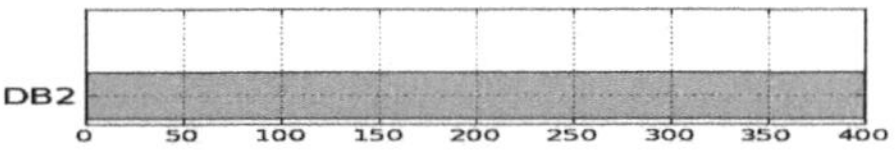

(d) Overall percentual speed increase (if any) with indexes Linux

Fig. 2. Result of the query Q2 for data with cycles

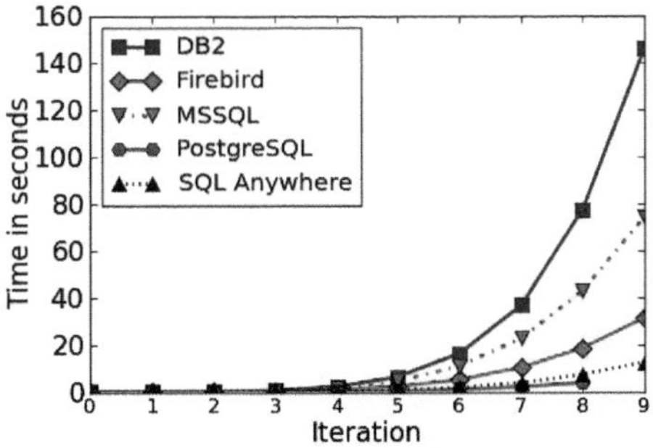

(a) Without indexes Windows

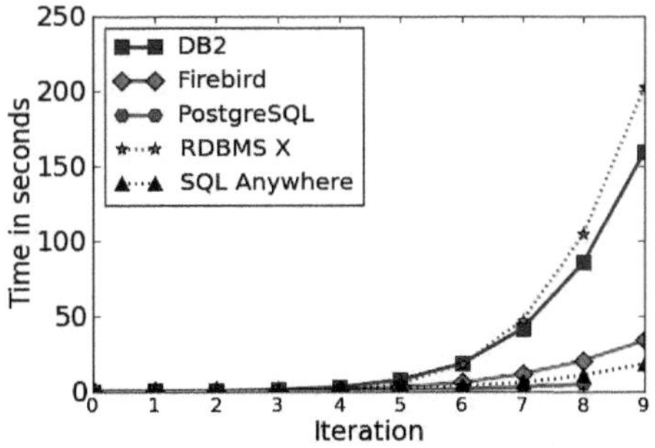

(b) Without indexes Linux

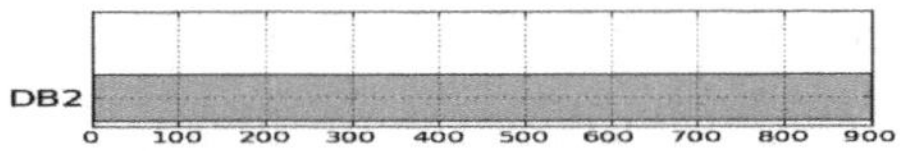

(c) Overall percentual speed increase (if any) with indexes Windows

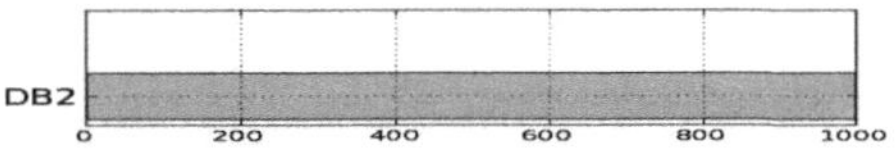

(d) Overall percentual speed increase (if any) with indexes Linux

Fig. 3. Result of the query Q2 for data without cycles

For data with cycles the test results indicate that for the parameter $I = 0 \cdots 8$ PostgreSQL has the best execution times on Windows system, while RDBMS X has the best execution times on Ubuntu system.

On Ubuntu for $I = 9$ and data with or without indexes only RDBMS X completed successfully. While on Windows system only two databases returned errors for data without indexes(SQL Anywhere, PostgreSQL), and after adding indexes only one(PostgreSQL). All occurring errors where results of memory allocation error, despite of 4 GB RAM and 10 GB free hard drive space.

For $I = 0 \cdots 8$ and data without cycles PostgreSQL was leading database, unfortunately for $I = 9$ it was the only database that returned memory allocation error.

Only DB2 got huge performance bost from indexes, the same methods of optimization as in Q1 was used.

4.3 Test Suite 3

Query Q3 has been tested on databases Firebird, MS SQL and DB2 on Windows and Firebird and DB2 on Ubuntu (other databases do not support this query's syntax). This query displays all the cities reachable by planes and trains from Toronto, the number of connections is limited by the parameter I.

```
WITH destinations(departure, arrival, connections, flights, trains, cost
    ) AS
 (SELECT f.departure, f.arrival, 0, 1, 0, price
  FROM flights f, cities c
  WHERE f.departure = c.cid AND c.city = 'Toronto'
 UNION ALL
  SELECT t.departure, t.arrival, 0, 0, 1, price
  FROM trains t, cities c
  WHERE t.departure = c.cid AND c.city = 'Toronto'
 UNION ALL
  SELECT r.departure, b.arrival, r.connections+1 , r.flights+1, r.trains
      , r.cost + b.price
  FROM  destinations r, flights b
  WHERE r.arrival = b.departure AND r.connections<I
 UNION ALL
  SELECT r.departure, v.arrival, r.connections+1 , r.flights, r.trains +
      1, r.cost + v.price
  FROM destinations r, trains v
  WHERE r.arrival = v.departure AND r.connections<I )
SELECT count(*) FROM destinations
```

Listing 1.3. Query Q3

For $I = 6 \cdots 8$ DB2 gained best performance, but for $I = 9$ ended with memory allocation error on both OSs. On Ubuntu system for $I = 7$ Firebird returned memory allocation error. On Windows system for $I = 9$ with no indexes Firebird was stooped after 30 hours, after adding indexes, computation ended in reasonable time.

This time MS SQL clearly benefited from indexes. Analysis of the execution plan, indicates that for data without manually created indexes the clustered index scans were used, while with manually created indexes, index seek using user indexes.

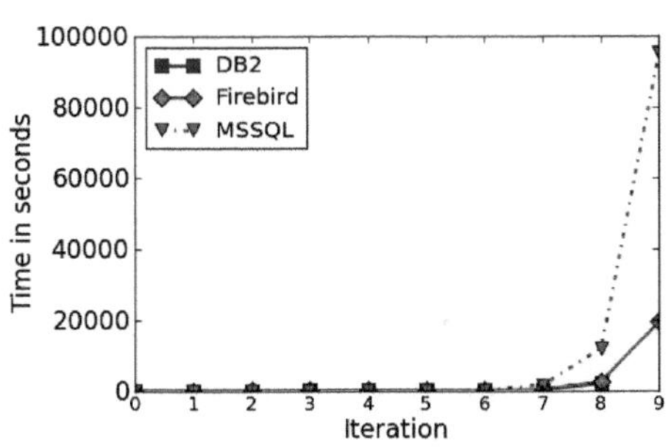

(a) Without indexes Windows

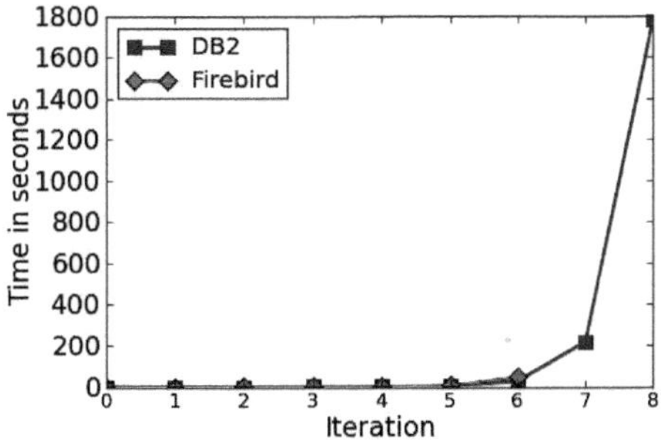

(b) Without indexes Linux

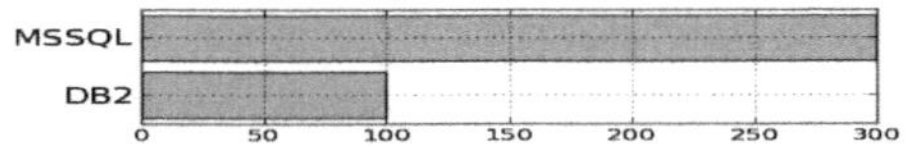

(c) Overall percentual speed increase (if any) with indexes Windows

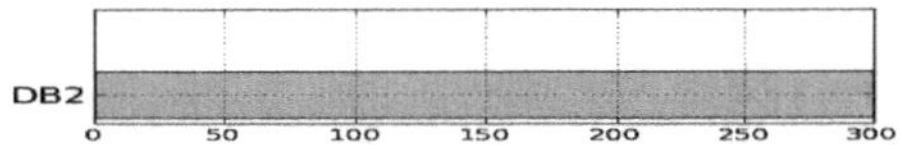

(d) Overall percentual speed increase (if any) with indexes Linux

Fig. 4. Result of the query Q3 for data with cycles

For data without cycles for all of the test cases the query Q3 completed instantly (0s with respect to measurement errors).

4.4 Test Suite 4

Query Q4 is similar to Q2 and has been tested on the same set of databases. This query displays all the cities reachable by plane starting from Toronto or Warsaw without placing limits on the number of connections. This query has been tested only for data without cycles.

```
WITH destinations (departure, arrival, connections, cost) AS
 (SELECT a.departure, a.arrival, 0, price
  FROM flights a, cities c
  WHERE a.departure = c.cid AND c.city = 'Toronto' OR c.city = 'Warsaw'
 UNION ALL
  SELECT r.departure, b.arrival, r.connections+1, r.cost + b.price
  FROM destinations r, flights b WHERE r.arrival = b.departure)
SELECT count(*) FROM destinations
```

Listing 1.4. Query Q4

On Windows best performance was gained by PostgreSQL, on Ubuntu SQL Anywhere showed best performance without indexes, and DB2 was the leading competitor for data with indexes. On Ubuntu OS Firebird ended exceeding limit level of depth, and PostgreSQL returned memory allocation error. Both errors occurred for data with and without indexes.

Only DB2 benefited from manually created indexes (with a huge performance boost), additional indexes resulted in replacement of expensive full table scans with operations of index scans and row fetching.

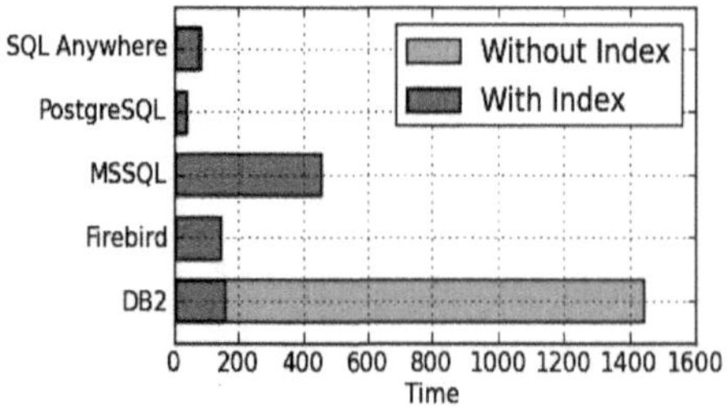

(a) Windows

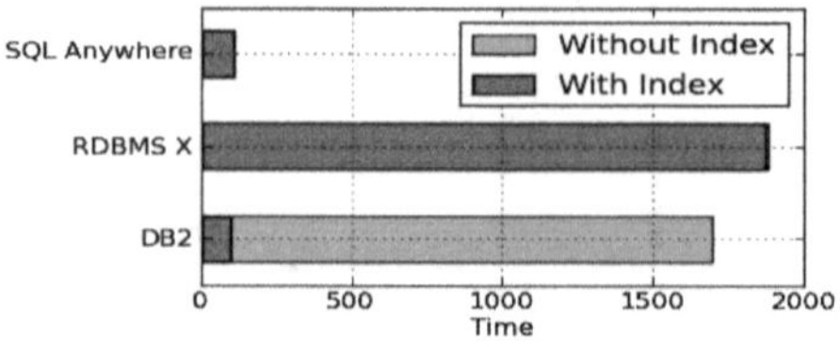

(b) Ubuntu

Fig. 5. Result of the query Q4 for data without cycles

4.5 Test Suite 5

For the pure purpose of testing the data independent query labeled Q5 has been executed on all of the tested databases. Its purpose was to calculate the sum of the numbers starting from 0 to 10 multiplied by $I = 0 \cdots 10$. Each time it was completed instantly (app. 0s).

```
WITH RECURSIVE t(n) AS ( VALUES (1) UNION ALL SELECT n+1 FROM t WHERE n
    < 100*I) SELECT sum(n) FROM t;
```

Listing 1.5. Query Q5

5 Conclusions

In this paper we have investigated the implementation of recursive SQL Common Table Expressions in most of the popular DBMSs. We have cataloged and compared common features of the syntax of such queries choosing SQL-99 standard as a base for comparison. Furthermore we performed performance tests on MS Windows and Linux systems, including cases of data with and without cycles and data independent queries. Also, we have checked whether placement of indexes would make a significant difference. Last, but not least, the schema, data and queries presented in this paper may be a starting point for development of a benchmark for recursive queries.

References

1. Bancilhon, F., Maier, D., Sagiv, Y., Ullman, J.D.: Magic sets and other strange ways to implement logic programs. In: Proceedings of the Fifth ACM SIGACT-SIGMOD Symposium on Principles of Database Systems, PODS 1986, pp. 1–15. ACM, New York (1986)
2. Letuchy, D.: Recursive queries in sql. In: Proceedings of SYRCoDIS (2005)
3. Ordonez, C.: Carlos Ordonez. Optimization of linear recursive queries in sql. IEEE Trans. on Knowl. and Data Eng. 22(2), 264–277 (2010)
4. Burzańska, M., Stencel, K., Wiśniewski, P.: Pushing Predicates into Recursive SQL Common Table Expressions. In: Grundspenkis, J., Morzy, T., Vossen, G. (eds.) ADBIS 2009. LNCS, vol. 5739, pp. 194–205. Springer, Heidelberg (2009)

5. Ghazal, A., Crolotte, A., Seid, D.Y.: Recursive sql query optimization with k-iteration lookahead. In: Bressan, S., Küng, J., Wagner, R. (eds.) DEXA 2006. LNCS, vol. 4080, pp. 348–357. Springer, Heidelberg (2006)
6. Stuparu, D., Petrescu, M.: Common Table Expression: Different Database Systems Approach. Journal of Communication and Computer 6(3), 9–15 (2009)
7. Melton, J., Simon, A.R.: SQL: 1999: understanding relational language components. Morgan Kaufmann Pub., San Francisco (2002)

Semantic Constraint-Based XML Updating

Md. Sumon Shahriar[1] and Jixue Liu[2]

[1] Tasmanian ICT Centre, CSIRO, Hobart, Australia
[2] University of South Australia, Australia
mdsumon.shahriar@csiro.au, jixue.liu@unisa.edu.au

Abstract. In this paper, we propose a novel semantic constraint based XML updating. In XML updating, we consider pair-wise-close semantics for ordered XML documents. Further, how semantic constraints can be incorporated with integrity constraints for XML updating and XML view updating are discussed.

1 Introduction

In traditional database management systems, updating is an important operation. In recent years, XML [1] is widely used in many data centric applications such as data exchange [2] and data integration [3]. In the database approach of XML has necessitated the update operations in XML [4,5] in the past. Most update operations in XML are based on un-ordered semantics of XML data. However, updating XML using semantic constraints using ordered approach is little investigated to the best of our knowledge.

XML updating can be accomplished in two ways. First, updating can be done on the base XML document(s). Second, updating can be done on the materialized view derived from the base XML documents [6,7,8,9,10]. Without loss of generality, we say XML updating to mean the update operations on the base XML document and we say XML view updating to mean the update operations on XML materialized view. We present some motivating examples of XML updating.

Updating value of a node: Consider the DTD in Fig.1. The DTD gives the salary information a person having first name (*fname*) and last name (*lname*).

```
<!ELEMENT db((fname, lname), salary)+ >
<!ELEMENT fname(#PCDATA) >
<!ELEMENT lname(#PCDATA) >
<!ELEMENT salary(#PCDATA) >
```

Fig. 1. An XML DTD

The XML document (in tree structure) in Fig.2 conforms to the DTD in Fig.1. We want to update the salary of the person having first name **Sam** and last name **Andrew** to **80K**. First, we can update the node v_3 or the node v_6 labeled by salary considering the first name and the last name value pair for the

Y. Zhang et al. (Eds.): DTA/BSBT 2010, CCIS 118, pp. 100–109, 2010.

nodes v_1 and v_5. Second, we can update the node v_9 labeled by salary considering the first name and the last name value pair for the nodes v_7 and v_8. We show the problem using question(?) marks in Fig.3. Clearly, the first option is not semantically correct because the salary node v_3 is for the person having the first name **Sam** and the last name **Kim** and the salary node v_6 is for the person having the first name **Kim** and the last name **Andrew**.

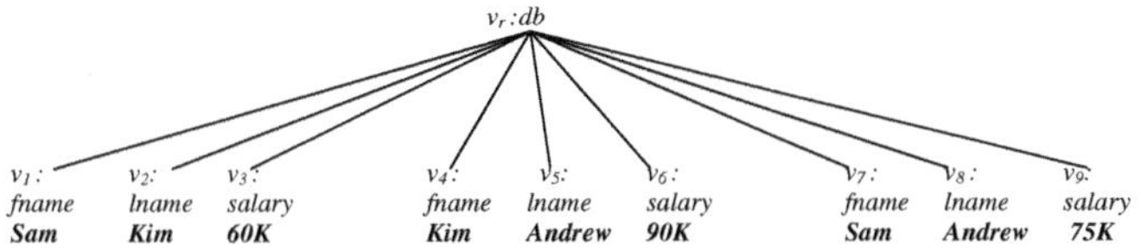

Fig. 2. An XML tree

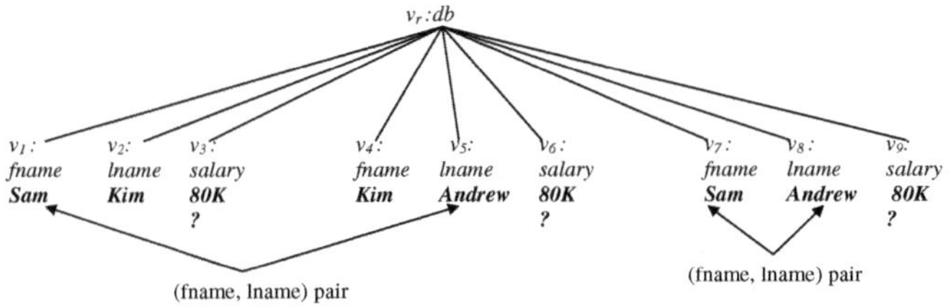

Fig. 3. An XML tree (updating salary of Sam Andrew to 80K)

Adding a node: Consider the DTD in Fig.4. The DTD gives the information of name(first name *fname* and last name *lname*) and telephone numbers (*tell*).

```
<!ELEMENT info((fname, lname),tell+)+ >
<!ELEMENT fname(#PCDATA) >
<!ELEMENT lname(#PCDATA) >
<!ELEMENT tell(#PCDATA) >
```

Fig. 4. An XML DTD

The XML document (in tree structure) in Fig.5 conforms to the DTD in Fig.4.

We want to add one more telephone number **8334** to the person having first name **John** and last name **Simpson**. Now we have a problem in adding a node labeled *tell* with value **8334** in Fig.6. We can add a node v_7 because the nodes v_1 and v_5 have the value pair **(John,Simpson)** or we can add a node v_{11} because the nodes v_8, v_9 have the same value pair. We see clearly that the first value pair for nodes v_1 and v_5 is not semantically correct because the v_1 having value **John** is the first name of the person with the last name **Smith** of node v_2. The second value pair for the nodes v_8 and v_9 is semantically correct.

Deleting a node: Consider the DTD in Fig.7. It gives the enrollment information of students (with student id *sid*) in courses (with course id *cid*). The XML tree in Fig.8 conforms to the DTD in Fig.7.

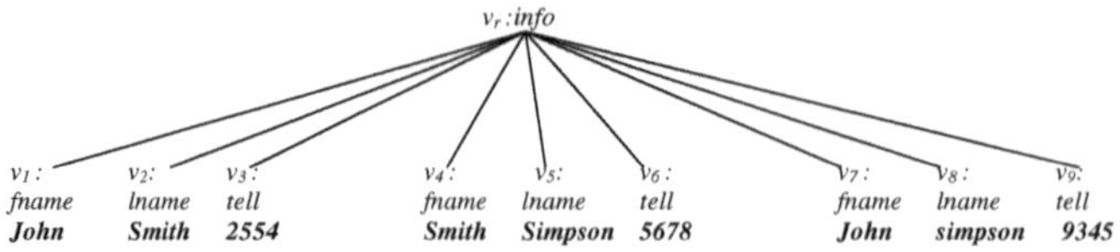

Fig. 5. An XML tree

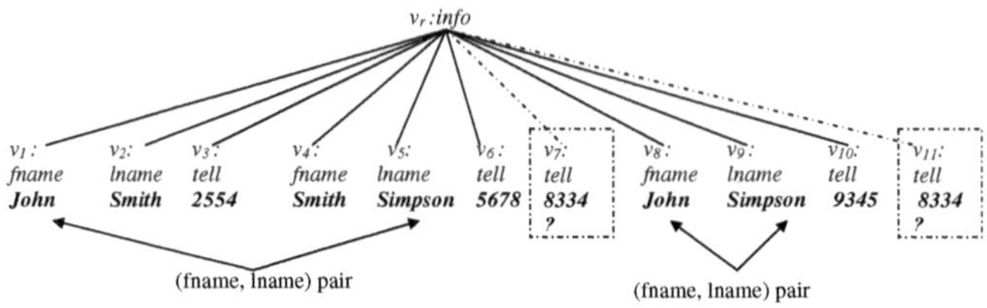

Fig. 6. An XML tree (adding a telephone number 8334 for John Simpson)

Now we want to delete a course having id **CS04** for the student having id **S001**. Like previous examples, we have the problem of deleting either the node v_3 or the node v_5(shown in Fig.9). Surely, the deletion of node v_3 is semantically correct.

We aim to contribute the followings. (a) We propose semantic constraints for XML updating using close value pair concept for ordered XML documents. (b) The close value pair concept can be incorporated with XML integrity constraints. (c) We also study the XML view updating using semantics constraints.

2 Basic Definitions

In this section, we present basic definitions needed throughout the paper.

Definition 1 (DTD). *An XML DTD is defined as* $D = (EN, G, \beta, \rho)$ *where*

(a) EN *is a set of element names.*

(b) G *is the set of type constructs and* $g \in G$ *is defined as one of the followings:*

(i) $g = Str$ *where* Str *means* $\#PCDATA$*;*

(ii) $g = e$ *where* $e \in EN$*;*

(iii) $g = \epsilon$ *means* $EMPTY$ *type;*

(iv) $g = g_1{\times}g_1$ *or* $g_1|g_1$ *is called conjunctive or disjunctive* **sequence** *respectively where* $g_1 = g$ *is recursively defined,* $g_1 \neq Str \wedge g_1 \neq \epsilon$*;*

```
<!ELEMENT enroll((sid, cid+))+ >
<!ELEMENT sid(#PCDATA) >
<!ELEMENT cid(#PCDATA) >
```

Fig. 7. An XML DTD

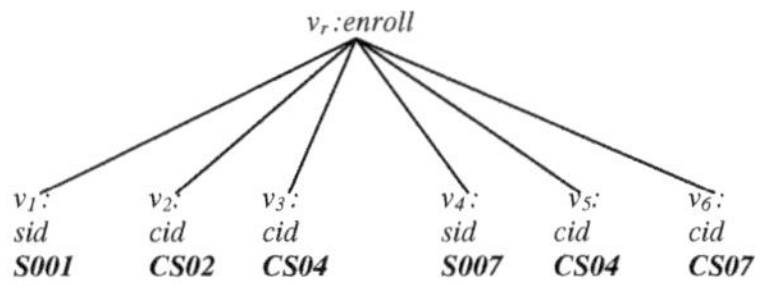

Fig. 8. An XML tree

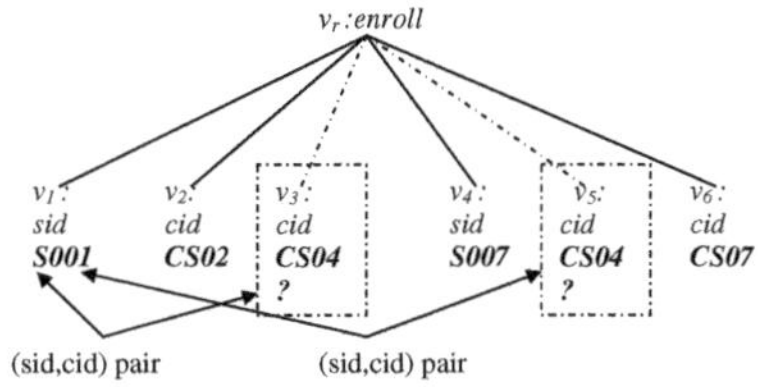

Fig. 9. An XML tree(deleting a course CS04 for the student id S001)

$g = g_2^c \wedge (g_2 = e \wedge e \in EN$ *or* $g_2 = [g_\times \cdots {}_\times g]$ *or* $g_2 = [g|\cdots|g])$, *called a* **component** *where* $c \in \{?, 1, +, *\}$ *is the multiplicity of* g_2, *'[ ]' is the component constructor;*

(c) $\beta(e) = [g]^c$ *is the function defining the type of an element e where* $e \in EN$ *and* $g \in G$. *We term* $\beta(e)$ *as* **element definition***.*

(d) ρ *is the root of the DTD and that can only be used as* $\beta(\rho)$. □

Example 1. Consider the DTD in Fig.1. The DTD is defined as $D = \{EN, G, \beta, \rho\}$ where $EN = \{db, fname, lname, salary\}$, $G = \{[[fname_\times lname]_\times salary]^+\}$, $\beta(db) = [[fname_\times lname]_\times salary]^+$, $\beta(fname) = Str$, $\beta(lname) = Str$, $\beta(salary) = Str$ and $\rho = db$.

We now define paths on the DTD.

Definition 2 (Path). *Given a* $D = (EN, G, \beta, \rho)$, *a simple path* $\wp$ *on* D *is a sequence* $e_1/\cdots/e_m$, *where* $\forall i(e_i \in EN)$ *and* $\forall e_w \in [e_2, \cdots, e_m]$ (e_w *is an element in* $\beta(e_{w-1})$). *A simple path* $\wp$ *is a complete path if* $e_1 = \rho$. *A path* $\wp$ *is empty if* $m = 0$, *denoted by* $\wp = \epsilon$. *When we say path, we mean simple path. We use function* $last(\wp)$ *to return* e_m, $beg(\wp) = e_1$ *and* $par(e_w) = e_{w-1}$, *the parent of* e_w. *We use* $len(\wp)$ *to return* m.

Paths satisfying this definition are said ***valid*** *on D.* □

Example 2. Consider the DTD in Fig.1. In Figure, $db/fname$ is simple path. $db/fname$ is also a complete path because db is the root of the DTD. $last(db/fname)$ is $fname$ and $beg(db/fname)$ is db. $len(db/fname)$ is 2.

Definition 3 (XML Tree). *The XML tree T representing an XML document is defined to be one of the followings:*

(i) $T = \phi$*(empty).*

(ii) A labeled node $T = (v : e : txt)$.

(iii) If $T_1, \cdots, T_f$ are trees, then $T = (v : e\ T_1 T_2 \cdots T_f)$ is also a tree. $T_1 \cdots T_f$ are called subtrees.

□

The symbol v is the node identifier which can be omitted when the context is clear.

Example 3. Consider the document in Fig.2. The document is represented as $T_{v_r} = (v_r : db(T_{v_1} T_{v_2} T_{v_3} T_{v_4} T_{v_5} T_{v_6} T_{v_7} T_{v_8} T_{v_9}))$. Then, $T_{v_1} = (v_1 : fname : Sam)$, $T_{v_2} = (v_2 : lname : Kim)$, $T_{v_3} = (v_3 : salary : 60K)$, $T_{v_4} = (v_4 : fname : Kim)$, $T_{v_5} = (v_5 : lname : Andrew)$, $T_{v_6} = (v_6 : salary : 90K)$, $T_{v_7} = (v_7 : fname : Sam)$, $T_{v_8} = (v_8 : lname : Andrew)$, $T_{v_9} = (v_9 : fname : 75K)$.

Definition 4 (Hedge and Conformation). *Let $(v : eT_1 \cdots T_n)$ be a tree where $T_1 \cdots T_n$ are subtrees of a node v with label e. A hedge H is a maximal sequence of consecutive subtrees from $T_1 \cdots T_n$ that conforms to a component g under the context node v, denoted by $H \Subset g$ or H^g.*

When there are multiple hedges for g, we use H_j^g to denote one of them and H^{g*} to denote all of them.

Example 4. Consider the DTD in Fig.1 and the tree in Fig.2. Let the component be $g_1 = [[fname_{\times} lname]_{\times} salary]$ be a component. Then $H_1^{g_1} = T_{v_1} T_{v_2} T_{v_3}$, $H_2^{g_1} = T_{v_4} T_{v_5} T_{v_6}$ and $H_3^{g_1} = T_{v_7} T_{v_8} T_{v_9}$ are the hedges conforming to g_1. Let $g_2 = [[fname_{\times} lname]_{\times} salary]^+$ be the component. Then the hedge $H_1^{g_2} = T_{v_1} T_{v_2} T_{v_3} T_{v_4} T_{v_5} T_{v_6} T_{v_7} T_{v_8} T_{v_9}$ is the only hedge conforming to g_2.

We define the tree conformation using the hedge.

Definition 5 (Tree Conformation). *Given a DTD $D = (EN, G, \beta, \rho)$ and XML Tree T, T conforms to D denoted by $T \Subset D$ if $T{=}(\rho\ H^{\beta(\rho)})$.* □

Example 5. Consider the DTD in Fig.1 and the tree representation of the XML document in Fig.2. The root $\rho = db$ and the the element definition $\beta(\rho) = [[fname_{\times} lname]_{\times} salary]^+$. The hedge $H^{[[fname_{\times} lname]_{\times} salary]^+} = T_{v_1} T_{v_2} T_{v_3} T_{v_4} T_{v_5} T_{v_6} T_{v_7} T_{v_8} T_{v_9}$ conforms to the component $[[fname_{\times} lname]_{\times} salary]^+$. We then get $T_r = (db H^{[[fname_{\times} lname]_{\times} salary]^+})$. Thus the tree in Fig.2 conforms to the DTD in Fig.1.

Definition 6 (Tree Equivalence). *Two trees T_a and T_b are value equivalent, denoted by $T_a =_v T_b$, if*

(1) $T_a = (v_1 : e : txt1)$ and $T_b = (v_2 : e : txt1)$, or

(2) $T_a = (v_1 : e\ T_1 \cdots T_m)$ and $T_b = (v_2 : e\ T_1^{'} \cdots T_n^{'})$ and $m = n$ and for $i = 1, \cdots, m (T_i =_v T_i^{'})$. □

$T_x \equiv T_y$ if T_x and T_y refer to the same tree. We note that, if $T_x \equiv T_y$, then $T_x =_v T_y$.

Example 6. Consider two trees $T_a = (v_1 : fname : Sam)$ and $T_b = (v_7 : fname : Sam)$. Then $T_a =_v T_b$.

Definition 7 (Hedge Equivalence). *Two hedges H_x and H_y are value equivalent, denoted by $H_x =_v H_y$, if*

(1) Both H_x and H_y are empty, or

(2) $H_x = T_1 \cdots T_m$ and $H_y = T_1' \cdots T_n'$ and $m = n$ and for $i = 1, \cdots, m (T_i =_v T_i')$. □

We define minimal structure of a DTD.

Definition 8 (Minimal Structure). *Given an element definition $\beta(e)$ and two elements e_1 and e_2 in $\beta(e)$ of a DTD, the minimal structure g of e_1 and e_2 in $\beta(e)$ is the innermost component belonging to the component $\beta(e)$ that contains both e_1 and e_2.* □

Example 7. Let *db* be an element and $[[fname_\times lname]_\times salary]^+$ be the element definition of *db* in the DTD in Fig.1. Now consider two elements *fname* and *salary*. Then the minimal structure for the elements *fname* and *salary* is $[[fname_\times lname]_\times salary]$. But the minimal structure of *fname* and *lname* is $[fname_\times lname]$.

Definition 9 (Minimal Hedge). *A hedge that conforms to a minimal structure g is a minimal hedge.* □

We now define prefixed trees. The reason for defining prefixed trees is to get the trees for a path.

Definition 10 (Prefixed Trees). *We use T^e to denote a tree rooted at a node labeled by the element name e. Given path $e_1/\cdots/e_m$, we use $(v_1 : e_1).\cdots.(v_{m-1} : e_{m-1}).T^{e_m}$ to mean the tree T^{e_m} with its ancestor nodes in sequence, called the prefixed tree or the prefixed format of T^{e_m}.*

Given path $\wp = e_1/\cdots/e_m$, $T^\wp$ denotes $(v_1 : e_1).\cdots.(v_{m-1} : e_{m-1}).T^{e_m}$ and is called a tree of the path $\wp$. When the context is clear, we use T^{e_m} to mean $T^\wp$.

Given path $\wp = e_1/\cdots/e_k/\cdots/e_m$, $prec(T^\wp) = (v_1 : e_1).\cdots.(v_k : e_k).\cdots.(v_m : e_m)$ means the precision of the tree $T^\wp$ for path $\wp$.

$\langle T^\wp \rangle$ is the set of all $T^\wp$. $\langle T^\wp \rangle = \{T_1^\wp, \cdots, T_f^\wp\}$. $|\langle T^\wp \rangle|$ returns the number of $T^\wp$ in $\langle T^\wp \rangle$.

Example 8. Consider an element *salary*. Then T^{salary} means one of the trees T_{v_3}, T_{v_6} and T_{v_9} in Fig.2. Let $\wp = db/salary$. The prefixed trees of T^{salary} are $(v_r : db).T_{v_3}^{emp}$, $(v_r : db).T_{v_6}^{emp}$ and $(v_r : db).T_{v_9}^{emp}$. The precisions of the tree $T^\wp$ are $prec(T^\wp) = (v_r : db)$. Then $\langle T^\wp \rangle = \langle T^{emp} \rangle = \{T_{v_3}, T_{v_6}, T_{v_9}\}$ and $|\langle T^{emp} \rangle| = 3$.

3 XML Updating

In the section, we study updating on XML. First, we define updating language.

3.1 Updating Language

We now define the language for updating XML.

Update a node: We define the language of updating value of a node as: update Q set $val(Q)$ where P_1 ==$val(P_1)$ $\cdots$ and/or $\cdots$ P_n==$val(P_n)$.

We now give an example.

Example 9. Consider the DTD in Fig.1 and the conforming document in Fig.2. We want to update salary $75K$ of *Sam Andrew* to $80K$. The language for this updating is: "update $db/salary$ set $val(db/salary)$==$80K$ where $db/fname$==Sam and $db/lname$==$Andrew$".

Add a node: We define the language of adding a node as: add Q where P_1==$val(P_1)$ $\cdots$ and/or $\cdots$ P_n==$val(P_n)$.

We now give an example.

Example 10. Consider the DTD in Fig.4 and the conforming document in Fig.5. We want to add a telephone number 8334 for *John* and *Simpson*. The language for this updating is: "add $info/tell/8334$ where $info/fname$==$John$ and $info/lname$==$Simpson$".

Delete a node: We define the language of deleting a node as: delete Q where P_1==$val(P_1)$ $\cdots$ and/or $\cdots$ P_n==$val(P_n)$.

We now give an example.

Example 11. Consider the DTD in Fig.7 and the conforming document in Fig.8. We want to delete the course $CS04$ for the student id $S001$. The language for this updating is: "delete $enroll/cid$ where $enroll/cid$=$CS04$ and $enroll/sid$=$S001$".

We first define the following definition.

Definition 11 (Close Value Pair). *Given a set of complete paths* $\{P_1, ..., P_l\}$. *A close value pair is a sequence of pair-wise-close subtrees, denoted by* $F[P] = (T^{P_1} \cdots T^{P_l})$, *where 'pair-wise-close' is defined next.*
Let $\wp_i = e_1/\cdots/e_k/e_{k+1}/\cdots/e_m \in P_i$ *and* $\wp_j = e'_1/\cdots/e'_k/e'_{k+1}/\cdots/e'_n \in P_j$ *be two paths for any* P_i *and* P_j. *Let* $prec(T^{P_i}) = (v_1 : e_1).\cdots.(v_k : e_k).(v_{k+1} : e_{k+1}).\cdots.(v_m : e_m)$ *and* $prec(T^{P_j}) = (v'_1 : e'_1).\cdots.(v'_k : e'_k).(v'_{k+1} : e'_{k+1}).\cdots.(v'_n : e'_n)$. *Then* T^{P_i} *and* T^{P_j} *are pair-wise-close if,* $e_1 = e'_1, \cdots, e_k = e'_k, e_{k+1} \neq e'_{k+1}$, *then* $v_k = v'_k$, $(v_{k+1} : e_{k+1})$ *and* $(v'_{k+1} : e'_{k+1})$ *are two nodes in the same minimal hedge of* e_{k+1} *and* e'_{k+1} *in* $\beta(e_k)$. □

Example 12. Consider the DTD in Fig.1 and the document in Fig.2. Consider the $db/fname, db/lname$ and $db/salary$ are the complete paths. We see that db is common to all of these complete paths. Then minimal structure for the elements $fname$, $lname$ and $salary$ is $g_m = [[fname_\times lname]_\times salary]$. Then the minimal hedges are $H_1^{g_m} = T_{v_1}T_{v_2}T_{v_3}$, $H_2^{g_m} = T_{v_4}T_{v_5}T_{v_6}$ and $H_3^{g_m} = T_{v_7}T_{v_8}T_{v_9}$. So $(Sam, Kim, 60K)$ is the close pair value for the hedge $H_1^{g_m}$, $(Kim, Andrew, 90K)$ is the close pair value for the hedge $H_2^{g_m}$ and $(Sam, Andrew, 75K)$ is the close pair value for the hedge $H_3^{g_m}$. However, the pair $(Sam, Andrew, 90K)$ is not the close pair value. There are two possibilities for not being the close pair value. First, *Sam* is taken from hedge H_1 and $Andrew, 90K$ are taken from H_2. Second, $Sam, Andrew$ are taken from hedge H_3 and $90K$ is taken from H_2.

We now present the algorithm for updating XML.

Algorithm 1. UPDATE_XML_DOCUMENT(T, D, Q, P)

Input: An XML document T, DTD D, update parameters Q,P;
Output: Updated document T;
Calculate minimal structure g_m for Q,P;
Calculate minimal hedges for g_m;
Generate close value pairs for minimal hedges;
Search and locate the close value pair according to value pair for Q,P;
Update the located close value pair;

Example 13. Consider the example 11. In the example, $Q = enroll/cid$, $P_1 = enroll/cid$ and $P_2 = enroll/sid$. So the minimal structure is $g_{m1} = [sid_{\times}cid^{+}]$ in the DTD in Fig.7 and the minimal hedges are $H_1^{g_{m1}} = T_{v_1}T_{v_2}T_{v_3}$ and $H_2^{g_{m1}} = T_{v_4}T_{v_5}T_{v_6}$. The close value pairs are $(T_{v_1}T_{v_2}) = (S001, CS02)$ and $(T_{v_1}T_{v_3}) = (S001, CS04)$ for minimal hedge $H_1^{g_{m1}}$ and $(T_{v_4}T_{v_5}) = (S007, CS04)$ and $(T_{v_4}T_{v_6}) = (S007, CS07)$ for minimal hedge $H_2^{g_{m1}}$. The close value pair for Q, P is $(S001, CS04)$. So we only delete the node T_{v_3} in Fig.8.

4 XML Integrity Constraints with Semantics for Updating

We first define the close value pair equivalence.

Definition 12 (Close Value Pair Equivalence). *Two close value pairs $F_1[P] = (T_1^{\wp_1} \cdots T_1^{\wp_l})$ and $F_2[P] = (T_2^{\wp_1} \cdots T_2^{\wp_k})$ are value equivalent, denoted by $F_1[P] =_v F_2[P]$ if $l = k$ and for each $i = 1, \cdots, k$ $(T_1^{\wp_i} =_v T_2^{\wp_i})$.* □

Example 14. Consider the XML document in Fig.10. The close value pairs $(T_{v_1}T_{v_2})$ and $(T_{v_4}T_{v_5})$ are value equivalent because $T_{v_1} =_v T_{v_4}$ and $T_{v_2} =_v T_{v_5}$.

In the XML document, there may be duplicate close value pairs meaning that any two close value pairs are equivalent. In that case, updating is done for duplicate close value pairs. If there is an XML key (Q, P) defined, meaning that the close value pairs are distinct, then we only update only one close value pair.

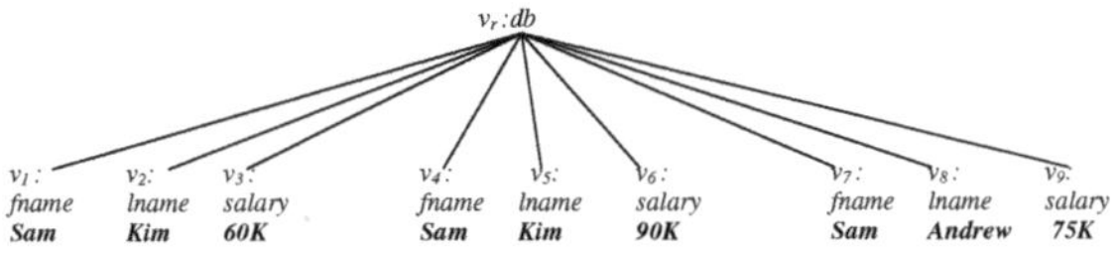

Fig. 10. An XML tree

<!ELEMENT db(person+) >
<!ELEMENT person((fname, lname), tell) >
<!ELEMENT tell(home, office) >
<!ELEMENT fname(#PCDATA) >
<!ELEMENT lname(#PCDATA) >
<!ELEMENT home(#PCDATA) >
<!ELEMENT office(#PCDATA) >

Fig. 11. An XML DTD

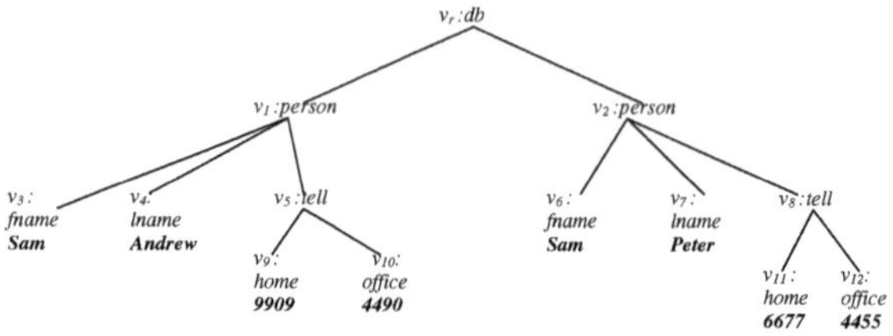

Fig. 12. An XML tree

5 Updating on XML Views

We now study how XML view can be updated. We first give a motivating example. Consider the DTD in Fig.11 and the conforming document in Fig.12. In the document, the first name and last name of persons and their home and office telephone numbers are stored. Now, we transform the DTD and the document (using either XQuery [11] or transformation operators) to the DTD in Fig.13 and the document in Fig.14 as view.

Now we want to update the contact telephone number 6677 to 6999 of the person having first name *Sam* and last name *Peter* in the view document in Fig.14. The *fname*, *lname* and *contact* in the view document maps to the *fname*, *lname* and *home* in the base document in Fig.12.

<!ELEMENT db((fname, lname), contact) >
<!ELEMENT fname(#PCDATA) >
<!ELEMENT lname(#PCDATA) >
<!ELEMENT contact(#PCDATA) >

Fig. 13. An XML transformed DTD for view

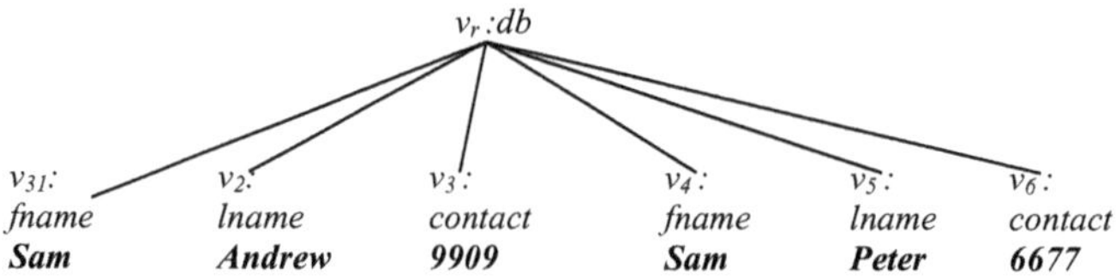

Fig. 14. An XML transformed document for view

However, we need to locate the correct person to update the contact at the view document as well as home telephone number in base document. In this case, we need the close value pair mechanism to locate the correct place for updating.

6 Conclusions

We proposed a novel method for XML updating using close value pair in the ordered XML documents. In updating, updating in base XML document and in XML materialized view are considered. We addressed how the semantic constraint expressed using close value pair can be incorporated with XML integrity constraints.

References

1. Bray, T., Paoli, J., Sperberg-McQueen, C. M.: Extensible Markup Language (XML) 1.0. World Wide Web Consortium (W3C) (February 1998), `http://www.w3.org/TR/REC-xml`
2. Arenas, M., Libkin, L.: XML Data Exchange: Consistency and Query Answering. In: PODS, pp. 13–24 (2005)
3. Bertino, E., Ferrari, E.: XML and Data Integration. IEEE Internet Computing, pp. 75–76 (2001)
4. Damiani, E., Fansi, M., Gabillon, A., Marrara, S.: Securely Updating XML. In: Apolloni, B., Howlett, R.J., Jain, L. (eds.) KES 2007, Part III. LNCS (LNAI), vol. 4694, pp. 1098–1106. Springer, Heidelberg (2007)
5. Wang, L., Rundensteiner, E., Mani, M., Jiang, M.: HUX: Handling Updates in XML. In: VLDB, pp. 1235–1238 (2006)
6. Braganholo, V.P., Davidson, S.B., Heuser, C.A.: From XML view updates to relational view updates: old solutions to a new problem. In: VLDB 2004, pp. 276–287 (2004)
7. Cong, G.: Query and Update Through XML Views. In: Bhalla, S. (ed.) DNIS 2007. LNCS, vol. 4777, pp. 81–95. Springer, Heidelberg (2007)
8. Leonidas, F.: Propagating updates through XML views using lineage tracing. In: ICDE, pp. 309–320 (2010)
9. Jiang, M., Wang, L., Mani, M., Rundensteiner, E.: Updating views over recursive xml. In: Proc. of ICDT Worshop on Emerging Research Opportunities in Web Data Management (2007)
10. Wan, L., Rundensteiner, E., Mani, M.: Updating XML views published over relational databases: Towards the existence of a correct update mapping. Data & Knowledge Engineering 58, 263–298 (2006)
11. XQuery 1.0: An XML Query Language. W3C Working Draft (May 2003), `http://www.w3.org/TR/xquery/`

Coupling Ontology with Rule-Based Theorem Proving for Knowledge Representation and Reasoning

Xiuqin Zhong*, Hongguang Fu, and Yan Jiang

School of Computer Science and Engineering,
University of Electronic Science and Technology of China,
Chengdu, 610054, China
zhongxiuqin2009@gmail.com, fu_hongguang@hotmail.com, yancqit@gmail.com

Abstract. Theorem proving is an important topic in artificial intelligence. Several methods have already been proposed in this field, especially in geometry theorem proving. Since they belong to algebraic elimination method or artificial intelligence method, it is difficult to use them to express domain knowledge clearly or represent hierarchy of systems. Therefore, we build a loosely coupled system to combine ontology and rule-based reasoning. Firstly, we construct elementary geometry ontology with OWL DL by creating classes, properties and constraints for searching or reasoning in domains. Then we design bidirectional reasoning based on rules and reasoning strategies such as all-connection method, numerical test method, rule classification methods and so on for complex reasoning. The system greatly improves sharing and reusability of domain knowledge and simultaneously implements readable proofs for geometry theorem proving efficiently.

Keywords: Knowledge representation, Reasoning, Ontology, Rule-based, Theorem Proving.

1 Introduction

It is well known that theorem proving is an important topic in artificial intelligence (AI) and knowledge engineering. Geometry Theorem Proving is firstly proposed by Hibert; and Tarski proved the possibility of elementary geometry mechanization by algebraic method [1]in 1951; In 1970's, Wen-Tsun Wu presented Method Wu[2] which is the first effective method to prove a large quantity of theorems systematically (such as Pythagoras' theorem, the Simson theorem, Feuerbach theorem, and so on). And then J. Z. Zhang put forward a geometric invariant method (i.e. the area method[3]) with readable proofs. Later, many related methods appeared such as resultant elimination method[4],

* Xiuqin Zhong is with School of Computer Science and Engineering, University of Electronic Science and Technology of China, 610054, Sichuan, P.R.C. (phone: 86-028-61830223; fax: 86-028-61830670; e-mail: zhongxiuqin2009@gmail.com).

Y. Zhang et al. (Eds.): DTA/BSBT 2010, CCIS 118, pp. 110–119, 2010.

Grobner basis[5], Clifford's algebraic method, the exemplification method and so on. These methods are algebraic elimination methods, which are hardly comprehensible and not intuitive enough for students to understand.

While the logic method[6,7,8,9,10] proposed by Grant. has always been the dream of mankind, but it is limited by search space. However, it is difficult for algebraic method and AI method to express domain knowledge clearly (without ontology supporting), and impossible to consider the hierarchy of system. So it's necessary to introduce ontology in our systems.

What's more, Ontology has been a popular research area for knowledge engineering, natural language processing and knowledge representation since 1990. Ontology is an explicit specification of a conceptualization. It primarily contains concepts, relations, instances, rules and methods, all of which have unified, normal and certain representations, which facilitates communication between humans and machines. Then the methodology of ontology modeling is built step by step.

In recent years, a lot of work has been done on ontology, such as CYC, Enterprise, NKI, Biobike, high performance knowledge base and so on. Recently, literature[11] briefly described three layers (domain layer, inference layer and task layer) of knowledge base for elementary geometry, and literature[12] was mainly focused on knowledge acquisition.

We all know that description logic is an excellent carrier of ontology. Represent knowledge with OWL DL can not only classify domain concepts, and describe the semantics of their own by constraints, but also express complex dependencies of concepts, which can avoid the deep link and cascade query of database. Furthermore, it can reason and derive new facts on ontology.

In this paper, we build ontology to represent concepts and properties (i.e. relations) of elementary geometry, and simultaneously establish rules by lisp to execute theorem proving.

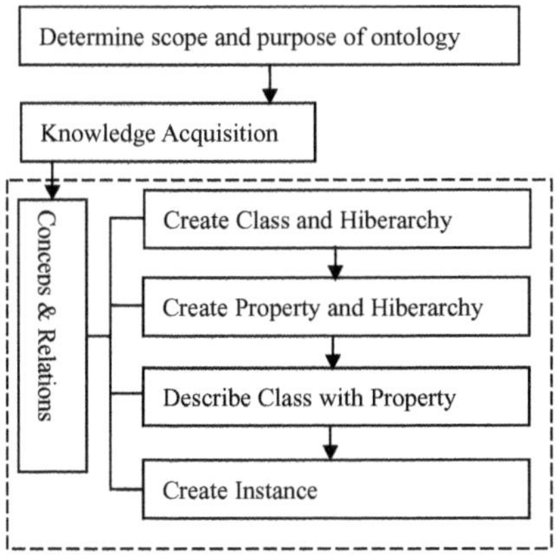

Fig. 1. The flow chat of EGOM

2 Ontology of Elementary Geometry

The flow chat of Elementary Geometry Ontology Modeling (EGOM) can be illustrated as figure 1. Determine the scope and the purpose of EGOM is easy for our special experience in this filed. What's more, knowledge acquisition of elementary geometry has already been implemented in [12], and in the following we'll discuss how to build EGOM in details.

2.1 Creating Classes and Properties

In the following, we will utilize Protégé 3.4.4 to create classes, properties and their hierarchy.

Firstly, analyze the basic concepts of elementary geometry such as point, line, segment, ray, angle, triangle, quadrilateral, polygon, circle, arc, sector, and so on, and then classify them into classes, and then add rdfs:subClassOf to construct the class hierarchy. It's shown in the left part of figure 2.

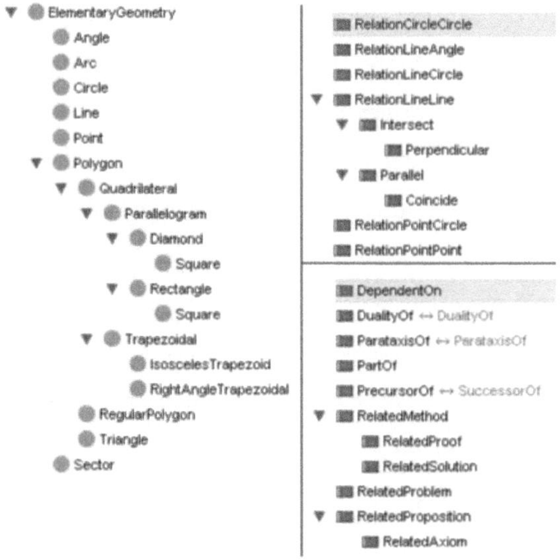

Fig. 2. Ontology built by Protégé 3.4.4

Secondly, classify the relations according to the basic concepts, such as RelationPointPoint (coincidence, collinear); RelationPointLine (PointOnLine); RelationPointCircle (PointInsideCircle, PointOnCircle and PointOutsideCircle); RelationLineLine (parallel, perpendicular, intersection, coincide); RelationQuantity (GreaterThan, LessThan, Equal, ProportionOf, etc.) and so on. And then add rdfs:subPropertyOfto construct the property hierarchy in this field. It's displayed in the upper right corner of figure 2.

In succession, extract many derived concepts based on the above basic concepts and relations, such as: MiddlePoint (which contains concepts of point and line, and relation of PointOnLine and RelationQuantity), TangentPoint, CommonTangentPoint, CentreOfGravity, Tangent, CommonTangent, Altitude, RightTriangle, IsoscelesTriangle, EquilateralTriangle, Parallelogram, Rectangle, Diamond, Square and so on, and then add them to the class hierarchy or the property hierarchy respectively.

By this time, we obtain all concepts and relations in this filed. Thirdly, it's necessary for us to specify the relationship (such as PrecursorOf, SuccessorOf, DependantOn, ParataxisOf, RelatedMethod, RelatedProposition, Hasdefinition and so on) based on above all, which is revealed in the lower right corner of figure 4. And it's also required to add proper constraints such as Owl:TransitiveProperty, Owl:SymmetricProperty, owl:inverseOf and so on for each relation. For instance, if PrecursorOf has constraint Owl:TransitiveProperty, then we can get PrecursorOf (A,B)∩PrecursorOf(B,C)→PrecursorOf(A,C); and if PrecursorOf has constraint owl:inverseOf with SuccessorOf, then we can get PrecursorOf(A,B)→SuccessorOf (B,A). We can still describe class with constraints, for example, we describe triangle (hasVertex 3, hasSide 3, hasArea positiveRealNumber).

2.2 Individuals and Reasoning

Up to now we build the TBox in description logic, and represent it with OWL DL. Finally, we should build individuals so as to create ABox in description logic, and import instances into EGOM. As is shown in figure 3, we present some part of class hierarchy and property hierarchy (TBox) over the horizontal dotted line, and some part of individuals (ABox) under the horizontal dotted line. And in the following we list out part of source codes of TBox in the figure.

```
<owl:Class rdf:ID="EquilateralTriangle">
<rdfs:subClassOf>
<owl:Class rdf:ID="Triangle"/>
</rdfs:subClassOf>
</owl:Class>
<owl:Class rdf:ID="Polygon"/>
<owl:Class rdf:about="#Triangle">
<rdfs:subClassOf rdf:resource="#Polygon"/>
</owl:Class>
<owl:Class rdf:ID="Circle"/>
<owl:ObjectProperty rdf:ID="CircleInscribedTriangle">
<owl:inverseOf>
<owl:ObjectProperty rdf:ID="TriangleExCircle"/>
</owl:inverseOf>
<rdfs:range rdf:resource="#Triangle"/>
<rdfs:subPropertyOf>
<owl:ObjectProperty rdf:ID="Tangent"/>
</rdfs:subPropertyOf>
```

<rdfs:domain rdf:resource="#Circle"/>
</owl:ObjectProperty>
<owl:ObjectProperty rdf:about="#TriangleExCircle">
<rdfs:range rdf:resource="#Circle"/>
<owl:inverseOf rdf:resource="#CircleInscribedTriangle"/>
<rdfs:subPropertyOf rdf:resource="#Tangent"/>
<rdfs:domain rdf:resource="#Triangle"/>
</owl:ObjectProperty>

Solid arrow means the known relations or restraints. If the relation from ABC to C1 is supposed to TriangleExCircle, and the relation from C2 and C1 is set to sameAs, then we can derive many potential relations such as ABC is Type of Triangle, and the relation from C2 to ABC is CircleInscribedTriangle, and others that are displayed in dashed arrows by ontology reasoning.

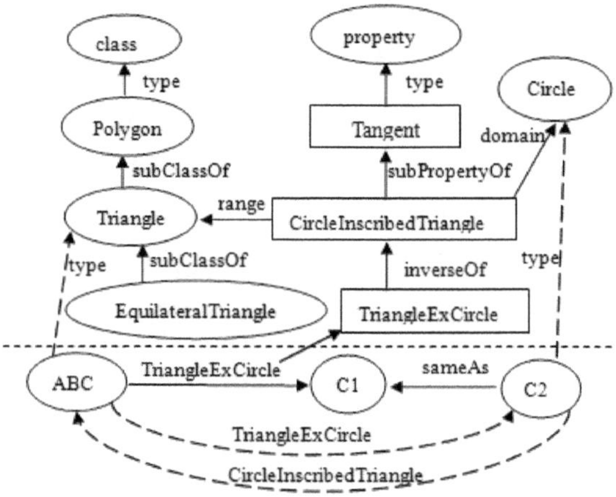

Fig. 3. Knowledge representation and reasoning on ontology

Therefore, we can construct our EGOM by creating class hierarchy, property hierarchy, and restraints of them, so that the semantic web is formed, and we can use it to execute a straight forward query, meta relation query and graph query by reasoning.

But it's not efficient for description logic to represent rule-based knowledge, such as "If ... Then ...". It's also impossible for description logic to provide compositional reasoning of relations, for example, if we know hasFather (Tom, Jack) and hasSpouse (Jack, Rose), but we can not derive hasMother (Tom, Rose) by description logic reasoning. In fact, most information systems involve reasoning of complex relations, so it's necessary to combine rules with ontology. [13] did a research on adopting geo-ontology to represent semantic and spatio-temporal relationships and applying SWRL to express spatio-temporal rules, and [14] integrated ontologies and rules in the same layer of web Semantics. In the following, we build a rule-based theorem proving modeling on EGOM.

3 Rule-Based Theorem Proving

3.1 Rule Base

The rule database contains basic rules in this field such as axioms, lemmas, definitions, theorems (including propertyTheorems and determinationTheorems), some algebraic rules and so on. Each rule has five properties: usable or not, id number, name, property (Does it belong to an axiom, definition, theorem or algebraic rule) and natural language description. It's given in table 1. Obviously, each rule is corresponding to its formal description in lisp. In order to increase the inference efficiency, all the rules are divided into classes by their relations (RelatedAxiom, RelatedTheorem, RelatedAlgebraicRule, etc) with object classes (point, line, triangle, quadrangle, circle, etc). Thus we can combine EGOM with rules. And according to these basic rules, we can still generate some combination rules[15] for increasing inference efficiency.

Table 1. Rule list in elemantary geometry

Use	Id	Name	Property	NaturalLanguageDescription
Y	0	CompactAxiom	GeoAxiom	Completely coincident figures are equal.
	...	...	...	...
Y	11	Midpoint	Definition	If point C divides segment AB into two equal
Y	12	PropertyMidP	Theorem	If point C is midpoint of segment AB, segment
	...	...	...	...
Y	227	AlgebricRule26	AlgebricRule	a+b=c+d,a=c ⇒ b=d .

3.2 Bidirectional Reasoning

We utilize bidirectional reasoning which combines forward reasoning and backward reasoning. As for forward reasoning, we will start with conditions, and obtain many temporary facts by calling propertyTheorems of concepts in conditions, and then match the result with temporary facts. if successful, proving process is finished, otherwise it's continued. While, as for backward reasoning, we will start with result, and obtain many temporary sub results by calling definition or determinationTheorems of concepts in result, and then match the conditions with temporary sub results. Usually, bidirectional reasoning is carried out by forward reasoning and backward reasoning alternately, which is different from literature[11].

In the following we'll illustrate an example to demonstrate it. Let ABCD be a random quadrangle, and points E, F, G, H are middle points of segments AB, BC, CD and DA respectively. Prove: quadrangle EFGH is a parallelogram. The bidirectional reasoning process can be shown in figure 4.

For such node as "11; (px4 E F G H); 71; (9,10);" which is the result of this example, we can divide the result into two sub-results "9; (// (:line E F) (:line G H)); 24; (5 6); (11)" and "10; (// (:line E H) (:line G F)); 24; (7 8); (11)" by backward reasoning. Expression "11;(px4 E F G H);71;(9,10);" represents that

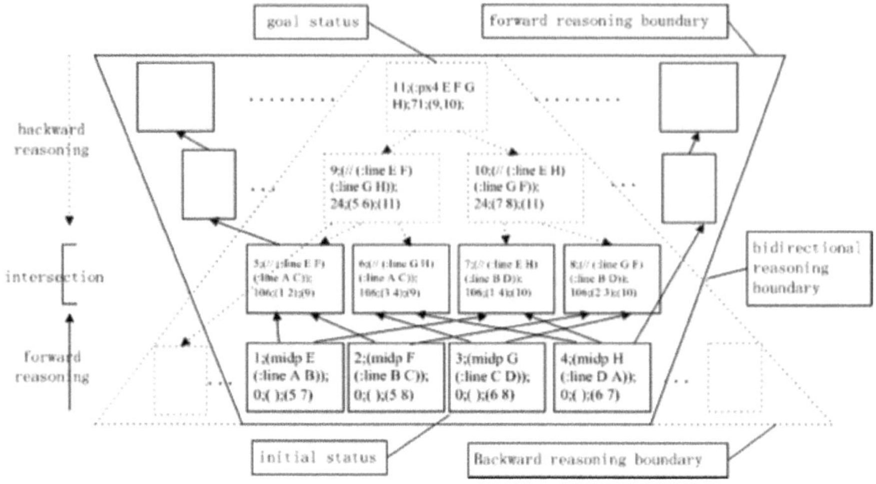

Fig. 4. Bidirectional reasoning process

in step 11, quadrangle EFGH is a parallelogram which is obtained by theorem 71 (definition of parallelogram), step 9 and step 10. While for such node as "5; (// (:line E F) (:line A C)); 106; (1 2); (9)", which can be inferred from "1; (midp E (:line A B) 0; (); (5 7)" and "2; (midp F (:line B C) 0; (); (5 8)" by forward reasoning. Expression "1; (midp E (:line A B)); 0; (); (5 7)" shows that in step 1, point E is the midpoint of line AB, and symbols "0" and "()" mean that the expression is known condition, and "(5 7)" means that step 1 will be used in steps 5 and 6. All the other nodes are similar. Therefore, we can get the rule-based theorem proving by bidirectional reasoning, and it can reduce the search space greatly.

3.3 Reasoning Strategies

In order to increase reasoning efficiency, we should take strategies into consideration. We can image that the problem will be much more difficult when there are too many input points in conditions, and that all the points are connected with each other. In order to prevent exponential explosion, a threshold of all-connection points is set, the default is 7 (i.e. if the number of input points is less than 7, then all the points will be connected with each other by adding auxiliary line). We can modulate the threshold according to the time or space limitation.

What's more, numerical test is needed, too. Numerical test is a pre-treatment method for reasoning. On one hand, numerical computation is much faster than symbolic computation based on rules. On the other hand, combinatorial explosion will occur when the number of points is too large, and numerical test can deal with this problem properly. In addition, considering the problem of collinear points, we substitute triangles with points. Now we take similar triangles for example.

1. Create a temporary library of similar triangles.
2. If the number of triangles exceeds threshold, then we find all similar triangles by calculating their corresponding coordinates of points.
3. Search the pair of triangles to be solved in the temporary library. authors of the paper.
4. If they are found, we can continue proving the result. Otherwise, give it up.

Note that, we can adopt numerical test only because our dynamic graphing with geometric constraints provides us with relatively accurate numerical input.

Therefore, the reasoning strategies based on rules can be listed as follows:

1. Set up all-connection condition.
2. Filter rules which are not related to the problem by RelatedAxiom, RelatedTheorem, RelatedAlgebraicRule, etc.
3. Numerical test will be used when the number of triangles is grearer than the threshold.
4. Call the combination rules if necessary.

Now, we can proceed bidirectional reasoning based on rules and reasoning strategies. And the readable proof of the previous example is given in figure 5.

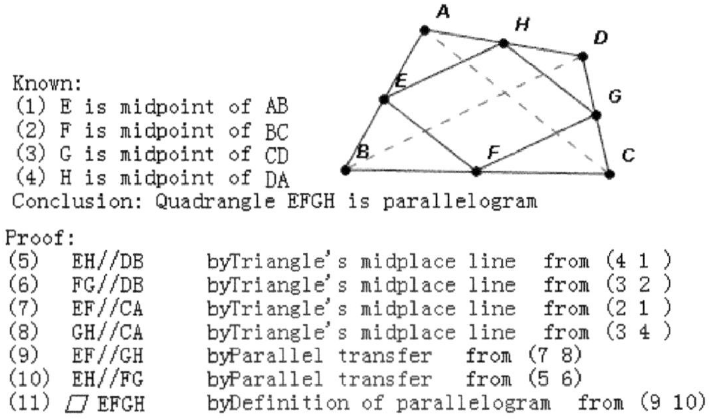
A
H
D
E
G
B
F
C
Known:
(1) E is midpoint of AB
(2) F is midpoint of BC
(3) G is midpoint of CD
(4) H is midpoint of DA
Conclusion: Quadrangle EFGH is parallelogram
Proof:
(5) EH//DB byTriangle's midplace line from (4 1)
(6) FG//DB byTriangle's midplace line from (3 2)
(7) EF//CA byTriangle's midplace line from (2 1)
(8) GH//CA byTriangle's midplace line from (3 4)
(9) EF//GH byParallel transfer from (7 8)
(10) EH//FG byParallel transfer from (5 6)
(11) ▱ EFGH byDefinition of parallelogram from (9 10)

Fig. 5. Reasoning process for example

4 Loosely Coupling Ontology with Rule-Based Theorem Proving

In this paper, we use ontology to represent the elementary geometry knowledge, and use rules to execute theorem proving. That is to say, we couple ontology with rule-based theorem proving loosely as is shown in figure 6. Therefore, it can not only explicitly express domain knowledge hierarchy in elementary geometry as is shown in figure 2, and execute ontology reasoning as is shown in figure 3, but also efficiently implement theorem proving as is presented in figure 5. What's more, it can greatly enhance efficiency of knowledge search, knowledge reuse and knowledge sharing. The structure of knowledge representation and reasoning is shown as figure 6.

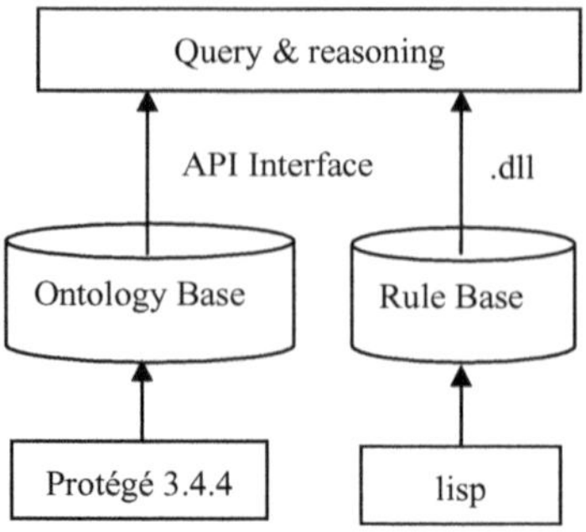

Fig. 6. Structure of knowledge representation and reasoning

5 Conclusions and Discussion

In a word, with the development of EGOM and rule-based reasoning, it is possible to implement a huge, virtual and distributed knowledge base accessible on internet, and to utilize it as an intelligent internet agent which is transparent to both human and machines. What's more, this method provides a new way for problem solving of geometric theorem proving, and it can be extended to problem solving systems of domain knowledge such as subject knowledge base for education and so on.

However, it still remains several challenging topics. For example: Building EGOM Semi-automatically or automatically, applying EGOM to semantic annotation, tightly coupling ontology with rules, etc.

Acknowledgements

The authors wish to express gratitude to Jon L. White and Sheng-Chuan Wu for their kind help and valuable suggestions in our work.

This work was partially supported by National Natural Science Foundation of China (Grant No. 61073099) and by the Fundamental Research Funds for the Central Universities (Grant No. ZYGX2009J059).

References

1. Tarski, A.: Decision Method for Elementary Algebra and Geometry, 2nd edn. University of California Press, Berkeley (1951)
2. Wen-Tsun, W.: Basic Principles of Mechanical Theorem Proving in Elementary Geometries. Journal of Automated Reasoning 2(3), 221–252 (1986)
3. Chou, S.C., Gao, X.S., Zhang, J.Z.: Machine Proofs in Geometry. World Scientific, Singapore (1994)
4. Wang, D.M.: Elimination procedures for mechanical theorem proving in geometry. Annals of Mathematics and Artificial Intelligence 13(1-2), 1–24 (1995)

5. Kapur, D.: Using Grobner bases to reason about geometry problems. Journal of Symbolic Computation 2(4), 399–408 (1986)
6. Coelho, H., Pereira, L.M.: Automated reasoning in geometry theorem proving with Prolog. Journal of Automated Reasoning 2(4), 329–390 (1986)
7. Chou, S.C., Gao, X.S., Zhang, J.Z.: A deductive database approach to automated geometry theorem proving and discovering. Journal of Automated Reasoning 25(3), 219–246 (2000)
8. Matsuda, N., Vanlehn, K.: GRAMY: A Geometry Theorem Prover Capable of Construction. Journal of Automated Reasoning 32(1), 3–33 (2004)
9. Zhang, J.Z., Gao, X.S., Chou, S.C.: Geometry Information Search System by Forward Reasoning. Chinese Journal of Computers 19(10), 721–724 (1996)
10. Zhang, J.Z., Li, C.Z.: Automatic Reasoning and Intellectual Platform of CAI Software. Journal of Guangzhou University 15(2) (2001)
11. Wu, W.Y., Zeng, Z.B., Fu, H.G.: Designing Knowledge Base for the Elementary Geometry Based on Ontology. Computer Applications 22(3), 10–14 (2002)
12. Zhong, X.Q., Fu, H.G., She, L., Huang, B.: Geometry Knowledge Acquisition and Representation on ontology. Chinese Journal of Computers 33(1), 167–174 (2010)
13. Huang, Y.Q., Deng, G.Y.: Research on Representation of Geographic Spatio-temporal Information and Spatio-temporal Reasoning Rules Based on Geo-ontology and SWRL. Environmental Science and Information Application Technology 3, 381–384 (2009)
14. Rosati, R.: On the decidability and complexity of integrating ontologies and rules. Journal of Web Semantics 3(1), 61–73 (2005)
15. Fu, H.G., Zhong, X.Q., Zeng, Z.B.: Automated and Readable Simplification of Trigonometric Expressions. Mathematical and Computer Modeling 44, 1169–1177 (2006)

On Defining Functional Dependencies in XML Schema*

Haitao Chen, Husheng Liao, and Zengqi Gao

College of Computer Sciences, Beijing University of Technology, Beijing, China
chenheyuzhi@yahoo.com.cn, liaohs@bjut.edu.cn, zengqigao@gmail.com

Abstract. The concept of functional dependency plays a foundational role in relational databases where it is used in integrity enforcement and database design. Similarly, functional dependencies in XML (XFDs) will play a centric role in providing richer data semantic information and normalizing XML data. In this paper, we present a new approach for defining XFDs on XML Schema. While showing how to extend XML Schema, we analyze the expressive power of our XFDs. We focus on supporting complex value (e.g. list, set) in our proposal. The satisfaction of XFDs in an XML document is defined in terms of the value equality. The checking algorithm for our proposal is given. Finally, we discuss the advantages of our XFDs over other previous work.

1 Introduction

The Constraints for semi-structured data and XML have been widely studied [1,2]. The concept of functional dependency plays a foundational role in relational databases where it is used in integrity enforcement and database design. Similarly, functional dependencies in XML (XFDs) will play a centric role in providing richer data semantic information and normalizing XML data. Although XFDs have been investigated more intensively than other data semantic constraints for XML [3,4], there seems to be no consensus on how to define such dependencies. XML Schema provides the mechanism for specifying the identity constraints. However, XFDs can not be defined in XML Schema.

Due to the structural difference between XML data and relational data, it is not a trivial task to extend functional dependency from relational database to XML. Hierarchical structure makes the information items related to XFDs may appear in different branches or at different levels of XML tree. An appropriate and reasonable approach towards the definition of XFDs should provide users with the mechanism to express these differences. And semi-structured XML document would produce more null values. In relational databases, the interpretation of null values as *value existent but unknown* has been assumed for most of approaches [5]. However, the flexibility of XML sometimes makes it more reasonable to consider it as *nonexistent* at all. Therefore, we should distinguish

* This research is supported by Beijing Municipal Natural Science Foundation (4082003).

Y. Zhang et al. (Eds.): DTA/BSBT 2010, CCIS 118, pp. 120–131, 2010.

between different types of null values. Another important problem is how to define *value equality* for complex elements. If v_1 and v_2 are complex elements, most of previous methods say they are *value equal* if (1) their sequence of child elements are pairwise value equal; (2) for every attribute α of v_1, there is an attribute β of v_2 such that α and β are *value equal* and vice versa [3]. However, the *Choice* and *occurs* component in XML Schema make it unreasonable to only consider the sequence of child elements are pairwise value equal.

In this paper, we present a new approach for defining XFDs in XML Schema. In summary, the main contributions are:

1. We extend XML Schema with the ability to specify XFDs by adding some new components to XML Schema.
2. We extend the *field* component in XML Schema to support complex value (e.g. set, list).
3. We distinguish the difference between *existent but unknown* null values and *nonexistent* null values. We believe it is necessary, because they have different semantics and there are more null values in XML.
4. A novel concept *match tree* is introduced for judging the value equality of complex type elements.
5. The checking algorithm for our proposal is given.

This paper is organized as follows. Section 2 provides preliminary terminology and notations. Section 3 presents an example and section 4 formally defines XFDs on XML Schema and the satisfaction in an XML document. The checking algorithm for our proposal is given in section 5. Section 6 gives an overview of related work and discusses the advantages of our proposal. Section 7 concludes the paper and states some future work.

2 Terminology and Notations

In this section, we introduce the basic definitions. Similar concepts can be found from the early literature [3,6,7,8,9]. the concepts of some existing components (*selector* and *field*) in XML Schema can be found from [10], which is omitted for saving space. And some new components for defining XFDs on XML Schema are previewed informally.

2.1 Basic Definitions and Notations

An **XML document tree** is denoted by D, which is a labeled tree. For a node v in XML document tree, $lab(v)$ is the label of node v and $val(v)$ is the value of node v. An **XML Schema files** is denoted by S, which includes definitions of attributes, elements, types and constraints. There are two kinds of type: simple type and complex type. We focus on complex type. The content model has three model groups: *All*, *Choice* and *Sequence*. We do not consider *All* model group in our approach since it can be treated separately. *Sequence* model group needs the element information items must appear in specified order, whereas *Choice* model group demands the element information item must match one element in the content. *Sequence* and *Choice* can be nested with each other. *Occurs*, including *minOccurs* and *maxOccurs*, represents the occurrences of element, wildcard or model group. The combination of model group and occurs makes the elements nested together, and then, in schema level, an element with complex type can be seen as a tree whose root is the element itself and children are the set of attributes and elements specified by the content model, and the like.

Given a set of XML Schema files $X = \{S_1, ..., S_n\}$, we can define various (schema) paths, which is omitted in this paper and can be found from [3]. Typically, in XML Schema, paths are described by *xpath* component.

2.2 New Components for Defining XFDs

In general, a functional dependency expresses a set of information items functionally determines another set of information items, written $X \rightarrow Y$ where $X = \{x_1, \ldots, x_n\}, Y = \{y_1, \ldots, y_n\}$. We call X the **determinant** and Y the **dependency**. For defining XFDs on XML Schema, we add a new component *functional*, which can be declared in an element. All the information items participating in a functional dependency are called **participant**. We use a new component *participant* to express it, which is specified in a *functional* component. Like identity constraints, in a *participant* component we use *selector* to select the elements to which the constraint applies. D*eterminant* and *dependency* components are introduced to represent the above corresponding concepts, which include a set of *field* components respectively and appear in the *participant* component. The *participant* component can contain another *participant* component to show the rest of participants that should have a new context element specified by a new *selector* component, which caters the hierarchical structure of XML.

3 An Example

For saving space, we express the XML Schema file in graphical notations. As shown in Fig. 1, the XML Schema file models a set of courses. The course has its number (*cno*) and name. In particular, the course also includes a set of classes. The class records its number (*clno*), teacher who teaches the class and several reference books that are specified by the teacher. The teacher has his number (*tno*), name, optional birthday and the telephone information. We assume a teacher must provide one telephone number, phone or cell phone. The reference book has name, publisher and year, which must occur in order.

A fragment of XML document instance is shown in Fig. 2, in which some data are omitted for saving space, such as the name of teacher. We consider the following example constraints: (A) tno $\rightarrow$ telephone, which means the number of teacher functionally determines his phone information. (B) cno, tno $\rightarrow$ reference-books, which means the number of course and the number of teacher functionally determines the specified reference books. The reference books are treated as a whole, i.e. it is an element with complex type. (C) cno, tno → referencebook, which means the number of course and the number of teacher functionally determines the set of the specified reference books. (D) tno → birthday, which means the number of teacher functionally determines his birthday.

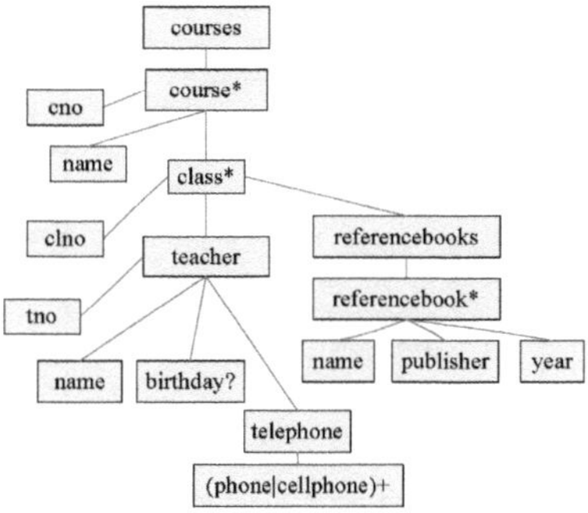

Fig. 1. XML Schema

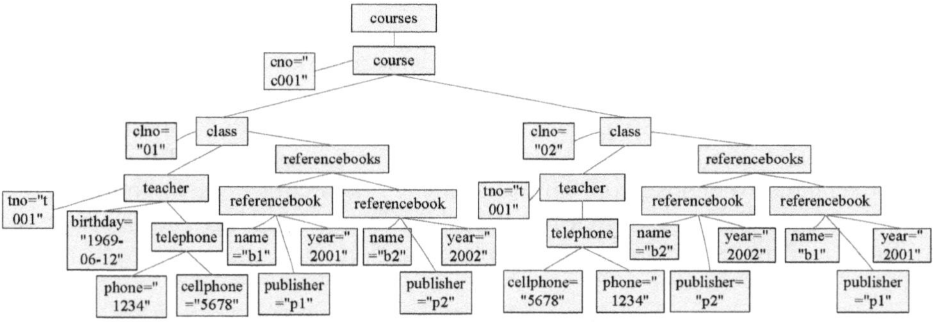

Fig. 2. XML document instance

By constraint (A), we show the information item in XFDs may be complex type. And the content model of *telephone* is *Choice*, it can occur more than once. Thus, it makes the two options both appeared. However, they are not in the same order. They do have the same values for the same labeled nodes. We think the two *telephone* elements should be regarded as having equal values. For constraints (B), they have similar situation to constraint (A). We use it to show there is the same problem with *Sequence* group. For constraints (C), we consider the set of reference books, by which we show the information item may be a set(or list) of nodes. By constraints (D), we can find there is no *birthday* information in the right "teacher". We care about how to deal with null values, which will be discussed in the following section.

We give three fragments of XML Schema to illustrate how to specify the first three example constraints in our approach, which is shown in Fig. 3.

```
Declared in courses
<functional>
  <participant>
    <selector xpath="//teacher"/>
    <determinant>
      <field xpath="tno"/>
    </determinant>
    <dependency>
      <field xpath="telephone"/>
    <dependency>
  </participant>
</functional>
```

(a)

```
Declared in courses
<functional>
  <participant>
    <selector xpath="course"/>
    <determinant>
      <field xpath="cno"/>
    </determinant>
    <participant>
      <selector xpath="class"/>
      <determinant>
        <field xpath="teacher/tno"/>
      </determinant>
      <dependency>
        <field
          xpath="referencebooks"/>
      </dependency>
    </participant>
  </participant>
</functional>
```

(b)

```
Declared in courses
<functional>
  <participant>
    <selector xpath="course"/>
    <determinant>
      <field xpath="cno"/>
    </determinant>
    <participant>
      <selector xpath="class"/>
      <determinant>
        <field xpath="teacher/tno"/>
      </determinant>
      <dependency>
        <field
          xpath="referencebooks/
                 referencebook"/>
      </dependency>
    </participant>
  </participant>
</functional>
```

(c)

Fig. 3. XFDs on XML Schema: (a) XFDs for example constraint (A), (b) XFDs for example constraint (B), (c)XFDs for example constraint (C)

Notably, there is recursive style in our definition, i.e. *participant* component can be nested into another *participant*, which is necessary to adapt to the hierarchical structure of XML data. We need the mechanism to specify the new context, which is the effect of the participant component. We also note that it is a technique to choose appropriate element in which the XFDs are declared. Our approach can express rich function dependency by clever choice of applied element and use of recursive *participant* definition.

4 XFDs in XML Schema

4.1 XFDs

We add *functional* component to XML Schema to define XFDs, which is shown in Fig. 4. and can be declared in an element.

```
<functional Id=ID Name=NCName /> Content: participant </functional>
<participant> Content: selector, determinant? , dependency? , participant? </participant>
<determinant> Content: field+ </determinant> <dependency> Content: field+ <dependency>
<field Id=ID xpath= a subset of XPath expression type=set or list/>
```

Fig. 4. The definition of functional component on XML Schema

If we focus on the paths (specified by *selector* and *field*), the above definition can be abstracted as $p_1[q_{11} \dots q_{1i} \rightarrow r_{11} \dots r_{1j}]$ $(p_2[q_{21} \dots q_{2k} \rightarrow r_{21} \dots r_{2l}]$

$(p_3[q_{31} \cdots q_{3m} \rightarrow r_{31} \cdots r_{3t}](\cdots p_n[q_{n1} \cdots q_{nu} \rightarrow r_{n1} \cdots r_{nv}]\underbrace{)\ldots)}_{n-1}$ where (1) i, j, k, l, m, t, u, v and n are nonnegative integer; (2) for $i \in [1, n], p_i$ is the schema path and specified by *selector*, if $i \in [2, n], p_i$ is relative to p_{i-1}, p_1 is relative to the element in which the XFDs is defined; (3) q_{ij} and r_{ik} are two *schema paths* that is relative to p_i. q_{ij} can be regarded as the abstraction of a *field* in the *determinant* component and r_{ik} can be regarded as the abstraction of a *field* in the *dependency* component where i is the level number; (4) the nodes identified by the upper level *schema path* are the context of the nodes identified by the current level *schema path.*

As illustrated in Sec. 3, we need recursive style definition of *participant* (nested *participant* component) because of the hierarchical structure of XML and the restriction of XPath in XML Schema. The nested *participant* components provide desired flexibility and can capture the information items that occur at different levels of XML document trees. We also extend the existing component *field* since it is not reasonable to limit the node identified by *field* is a single node and whose value is of simple type. In XML document, an information item maybe a set of nodes or a list of nodes if order is important. And the value of the information item can be a complex value if the corresponding node is a complex type element node. Therefore, we allow *field* to identify a set or a list of nodes. One node can be regard as a set having only one node. By this extension, we can efficiently support the functional dependencies (A), (B) and (C) in Sec. 3.

4.2 Value Equality

In order to define and check the satisfaction of our XFDs, we have to give the concept of *value equality* of nodes firstly. Since the XML data is semi-structured, there is a slight difference in treating null values between XML and relational database. And it is also not a trivial task to compare two complex type elements.

Two Types of Null Values. There are two types of null values: *existent but unknown* and *nonexistent*. In our approach, we employ an optimistic strategy, that is, we think an *existent but unknown* null value is equal to a real value when they both occurs in the **dependency**, since there is the possibility. However, nonexistent null value is not equal to a real value, since there is no possibility. Obviously, an *existent but unknown* null value should not be equal to a *nonexistent* null value. For checking XFDs, we analyze the effects of null values further and conclude that

(A) if null values appear in the **determinant**, there is no difference between *existent but unknown* and *nonexistent*, because if the **determinant** is not integral, it is nonsensical to check the XFDs.
(B) If null values appear in the **dependency**, there are only two cases that incur the conflicts. For a *nonexistent* null value, if the corresponding value is not *nonexistent* null value (*existent but unknown* null values and real values), then there is a conflict.

Match Tree. It is crucial to compare two complex type elements when we define and check the satisfaction of XFDs. Crux of the matter is *Occurs* component makes elements repeat, but it is unordered at semantic level. We propose a novel concept *match tree* to capture the semantics (order relationship of children nodes) of different model groups.

We express a complex type as a *particle*, which is the way to implement XSV. A *particle* has three properties: *min occurs*, *max occurs* and *term*. The *term* is either an *element declaration*, a *wildcard*, a *Sequence* or a *Choice*. A *Sequence* has a sequence of *particles* and a *Choice* has a set of *particles*. We define *match tree* by a constructive manner. Given a complex type T, an element e that has type T and a current position at which we are working, a *match tree* of e is a tree that can be constructed by following steps:

(A) If the *term* of the current *particle* is an *element declaration* or a *wildcard* and can be matched by the current sub-element, we return the element as the leaf node of the *match tree* below the current position.

(B) If the *term* of the current *particle* is a *Choice*, we construct an *unordered node* below current position and set this new *unordered node* as the current position. Then by using the left sub-elements, we try to match the *particles* in the set of *Choice* in terms of the *occurs* of the current *particle*. The matched sub-elements are organized under the new unordered node. If there is no match, the new unordered node is removed. We return to the upper level *particle* to match the left sub-elements.

(C) If the *term* of the current *particle* is a *Sequence*, we match the sequence of particles in the *Sequence* and organize them under an ordered node, which is constructed below current position. And the new ordered node is set as the current position. If there is no match, the new ordered node is removed. If the *Sequence particle* occurs more than once, a new unordered node is inserted at the current position and the ordered nodes are organized under this unordered node. When there is no more match, we return to the upper level *particle* to match the left sub-elements.

Informally, for a valid complex type element, a *match tree* records how an element instance matches its type definition. In Fig. 5, we show how the left "teacher" in Fig. 2 matches its type definition by *match tree*. A rectangle represents an *ordered node*, and a diamond represents an *unordered node*. It is easy to find that for a complex type definition, if two valid element instances, v_1 and v_2, are value equal, the match tree of v_1 can be matched by v_2 using a top-down match algorithm. Since the match tree can capture the ordered and unordered information between elements, it is more reasonable to compare element with complex type by this concept.

Definition of Value Equality. By applying the concept of "match tree", we define *value equality* below.

Let v_1 and v_2 be nodes in D. v_1 and v_2 are called value equal, denoted $v_1 =_v v_2$, if $lab(v_1) = lab(v_2)$, and (A) if v_1 and v_2 are both simple type elements or

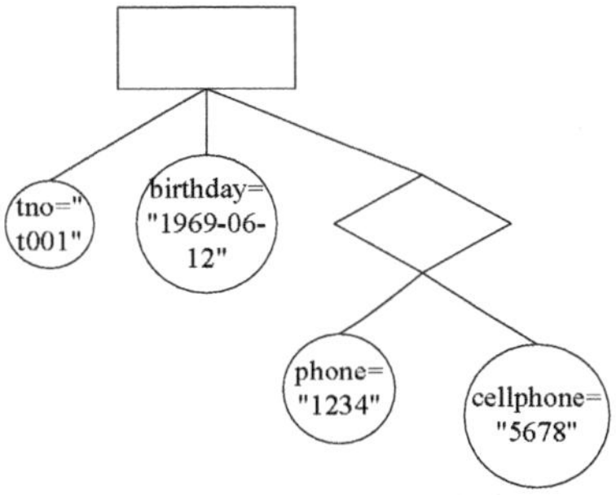

Fig. 5. Match tree for a teacher element

attributes, $val(v_1) = val(v_2)$; (B) if v_1 and v_2 are complex type elements, the match tree of v_1 can be matched by v_2 .

Let l_1 and l_2 be list of nodes. l_1 and l_2 are called value equal, denoted $l_1 =_v l_2$, if the members of two lists are pairwise value equal.

Let s_1 and s_2 be set of nodes. s_1 and s_2 are called value equal, denoted $s_1 =_v s_2$, if for every member m_1 of s_1, there is a member m_2 of s_2 such that $m_1 =_v m_2$ and vice versa.

4.3 The Satisfaction of XFDs in an XML Document

For an XFDs specified in our approach, the values of outer level are the context of the values of the inner level. By unnesting (that is similar to [11]), we can transform the result into a table that is similar to relational databases. Thus, we define the satisfaction of XFDs in an XML document by using tuples in the table, which is easy to describe and understand. The similar idea for generating tuples can be found from [11]. We employ the concept of *value equality* discussed before.

For an XML document D and a specified XFDs fd, we assume we have got the table $T(Q_1, \ldots, Q_n, R_1, \ldots, R_m)$. We say D satisfies fd if $\forall t_i, t_j \in T$ where $t_i = \{q_{i1}, \ldots, q_{in}, r_{i1}, \ldots, r_{im}\}$, $\{q_{j1}, \ldots, q_{jn}, r_{j1}, \ldots, r_{jm}\}$, (1) $\exists k \in [1, n]$, q_{ik} or q_{jk} is null value or $q_{ik} \neq_v q_{jk}$; or (2) $\forall k \in [1, n], r_{ik} =_v r_{jk}$.

5 Checking XFDs

In this section, we first give the unnesting algorithm. Then we show the checking algorithm. Our work is similar to [11]. However, we have a different definition. More importantly, we consider the effects of complex values and null values.

5.1 Generating Table by Unnesting

We consider the constraint (C) in Sec. 3, which is defined on XML Schema by Fig. 3 (c). We use the XML document in Fig. 2. There are three information items: *cno*, *tno* and *referencebook*. Instead of tree structure, we get the table by unnesting, which is shown in Tab. 1. Notably, it is not the same with relational table. For the third column, any tuple has two *referencebook*s and the *referencebook* is not of atomic value.

Table 1. Table for constraint (C)

cno="c001"	tno="t001"	referencebook (name="b1" publisher="p1" year="2001") referencebook (name="b2" publisher="p2" year="2002")
cno="c001"	tno="t001"	referencebook (name="b2" publisher="p2" year="2002") referencebook (name="b1" publisher="p1" year="2001")

The algorithm for generating table reads the XFDs definition and an XML document, and then it generates the involved information table. Let *elt* be the element that is being processed and *par* be the *participant* of XFDs that is declared in that element. We use *par.x* to denote the *x* component of *par*. We use *det* and *dep* to describe the desired *determinant* table and *dependency* table, whose initialized value is empty. The whole table should be (*det*, *dep*). We give the algorithm 1, which is shown in Fig. 6.

```
Algorithm 1
INPUT: par, elt, det, dep
OUTPUT: a table
Let candidates = find all nodes under elt that are located by par.selector
Foreach s in candidates,
   if there is determinant component in par
      append information items that are located by the fields in par.determinant to det
   if there is dependency component in par
      append information items that are located by the fields in par.dependency to dep
   if there is participant component in par
      return recursive call with par = par.particpant, elt = s, del and dep
   else
      return (det, dep)
```

Fig. 6. Algorithm for generating table of information items

We use k to denote the number of recursive invoking. Generally speaking, k is a countable number. We assume the number of the leaves nodes of the checked XML document is n. The *selector* of *participant* can find n nodes in the worst case. In every recursive invoking, the number of *fields* in *determinant* and *dependency* are *constant*. We assume they are $c_1, c_2, \ldots, c_k$ and c_i is the max of them where $i \in [1, k]$. In the worst case, the algorithm 1 can be achieved in $(c_i * n)^k$. In others words, the algorithm 1 can be achieved in polynomial time in context of the number of leaves nodes in the checked XML document tree.

5.2 Checking Algorithm

For Checking XFDs, our basic idea is the same to [11]. The XFDs can be abstracted as "$P \rightarrow Q$", where $P = \{p_1, \ldots, p_m\}$ and $Q = \{q_1, \ldots, q_n\}$, that is, P is the *determinant* and Q is the *dependency*. For each XFD, we create a hash table. For each tuple in the table, we consider its value on P. If there is no such hash key, we added the new hash key to the hash table and set its hash value to the tuple's value on Q. If there is the same hash key, we check whether its hash value is equal to the tuple's value on Q. If they are not equal, there is a conflict. We pay more attention on how to deal with null values, which has been discussed in Sec. 4.2. And the comparison of two nodes is also based on the concept of *value equality* defined before. We show the algorithm for checking XFDs in Fig. 7. For simplicity, we just think the value on P can be used as key value, that is, we do not mention a hash function again. If t is a tuple, we use $\mathrm{t}(P)$ to show the tuple's value on P and $\mathrm{t}(q_i)$ to denote the tuple's value on q_i, $i \in [1, n]$. Let v is a hash value, the $v(q_i)$ means the hash value on q_i.

Algorithm 2
INPUT: a table (a set of tuples)
OUTPUT: true or false

```
Foreach t in table
  if t(P) is not in hash table
    add t(P) to hash table with hash value: t(Q)
  else
    Let v = hash value whose hash key is t(P)
    Foreach i in n
      if t(qi) is nonexistent null value and v(qi) is not nonexistent null value
        return false
      if t(qi) is existent but unknown null value and v(qi) is nonexistent null value
        return false
      if t(qi) ≠ v(qi)
        return false
return true
```

Fig. 7. Algorithm for cheching XFDs

It is easy to understand that the algorithm 2 is linear in the number of tuples that are generated by algorithm 1 since the checking in hash table is constant.

6 Related Work

XFDs have been studied relatively intensively. Most of them are path-based [3,6,9,12,13,14,15]. Reference [6] proposes the novel concept of tree tuple. Intuitively, a tree is regarded as a set of paths that constitute the tuple. XFDs are defined on DTD and the semantics (satisfaction) is given in terms of tuples, which is similar to the corresponding concept in relational world. Reference [9] extends the definition of functional dependencies in incomplete relations to XML, which has the similar expressive power to [6]. The major difference between them is the treatment of null values. Reference [12] proposes a schema language-independent representation for XFDs. For defining XFDs, a *header* is used, which defines the scope of valuation and has the same effect to the *selector* in XML Schema. Local XFDs discussed in [13] capture the constraints that hold in part of an XML, which is similar to [12]. XFDs in [14] is an expression the form $P : Q : LHS \rightarrow RHS$, where P is the context path, Q is the target path, and LHS and RHS are two non-empty sets of paths. The appearance is similar to the *identity constraints* in XML Schema. XFDs in [15] unifies the global and local XFDs. References [12,13,14,15] have defined XFDs by a path to limit the scope of the constraints, which has the same basic idea to the *selector* in our proposal. Unlike the path-based methods, references [16,17] are tree-based, which can capture further kinds of XFDs. They consider the sub-tree as a unit, which can express richer structures. However, if we can use paths neatly, we can have the same effects by paths.

Comparing with previous methods, we discuss some advantages of our approach: (A) Our proposal extends XML Schema with the mechanism for defining XFDs. The restricted XPath in XML Schema is a kind of general path language and has rich expressive power. (B) The information items involved in a XFDs can be a set or a list of nodes and the single node can be with complex type, which make the expressive power of our proposal richer. A novel concept "match tree" is used to compare two element nodes with complex type. (C) The nested *participant* components are more suitable to the hierarchical structure. Meanwhile, it is necessary to remedy the deficiency of the restricted XPath in XML Schema. (D) We distinguish the difference between *existent but unknown* null values and *nonexistent* null values.

7 Conclusions and Future Work

This paper proposes a new approach on defining XFDs in XML Schema. Our proposal, which has been implemented on XSV, is path-based. Specifically, we use the restricted XPath in XML Schema, which can express rich cases. We extend the *field* component to support complex value (e.g. set, list) and complex type element. A novel concept *match tree* is introduced for judging the value equality of complex type elements more reasonably. We distinguish the difference between *existent but unknown* and *nonexistent* null values. The satisfaction of our XFDs is defined based on *value equality* and an algorithm is implemented in our system for checking the satisfaction against an XML document. Our proposal can make data semantic richer and assist to maintain data integrity.

Currently, we are investigating how to reduce the restrictions on XPath. For future work, we are interested in normalization theory for XML. How to use these integrity constraints in related fields also deserves investigation.

References

1. Buneman, P., Fan, W., Simeon, J., Weinstein, S.: Constraints for semistructured data and XML. SIGMOD RECORD 30(1), 47–54 (2001)
2. Fan, W.: XML constraints: Specification, analysis, and applications. In: Proceedings of the Sixteenth International Workshop on Database and Expert Systems Applications, vol. 805–809, pp. 805–809. IEEE COMPUTER SOC., Los Alamitos (2005)
3. Wang, J.: A comparative study of functional dependencies for XML. In: Zhang, Y., Tanaka, K., Yu, J.X., Wang, S., Li, M. (eds.) APWeb 2005. LNCS, vol. 3399, pp. 308–319. Springer, Heidelberg (2005)
4. Lv, T., Yan, P.: A survey study on XML functional dependencies. In: Proceedings of the First International Symposium on Data, Privacy, and E-commerce, pp. 143–145 (2007)
5. Klein, H.: Null values in relational databases and sure information answers. In: Semantics in Databases, pp. 119–138 (2003)
6. Arenas, M., Libkin, L.: A normal form for XML documents. ACM Transactions on Database Systems 29(1), 195–232 (2004)
7. Wang, J., Topor, R.: Removing xml data redundancies using functional and equality-generating dependencies. In: Proceedings of the 16th Australasian Database Conference, ADC 2005, Darlinghurst, Australia, pp. 65–74 (2005)
8. Buneman, P., Davidson, S., Fan, W., Hara, C., Tan, W.: Keys for XML. Computer Networks. The International Journal of Computer and Telecommunications Networking 39(5), 473–487 (2002)
9. Vincent, M., Liu, J., Liu, C.: Strong functional dependencies and their application to normal forms in XML. ACM Transactions on Database Systems 29(3), 445–462 (2004)
10. W3C: Xml schema part 1: Structures. Technical report (2004), `http://www.w3.org/TR/xmlschema-1/`
11. Vincent, M., Liu, J.: Checking functional dependency satisfaction in XML. In: Bressan, S., Ceri, S., Hunt, E., Ives, Z.G., Bellahsène, Z., Rys, M., Unland, R. (eds.) XSym 2005. LNCS, vol. 3671, pp. 4–17. Springer, Heidelberg (2005)
12. Lee, M., Ling, T., Low, W.: Designing functional dependencies for XML. In: Jensen, C.S., Jeffery, K., Pokorný, J., Šaltenis, S., Hwang, J., Böhm, K., Jarke, M. (eds.) EDBT 2002. LNCS, vol. 2287, pp. 124–141. Springer, Heidelberg (2002)
13. Liu, J., Vincent, M., Liu, C.: Local xml functional dependencies. In: Proceedings of the 5th ACM International Workshop on Web Information and Data Management, WIDM 2003, pp. 23–28. ACM, New York (2003)
14. Ahmad, K., Ibrahim, H.: Functional Dependencies and Inference Rules XML. In: Internaltional Symposium of Information Technology, pp. 494–499 (2008)
15. Shahriar, M.S., Liu, J.: On defining functional dependency for xml. In: International Conference on Semantic Computing, vol. 0, pp. 595–600 (2009)
16. Hartmann, S., Link, S.: More functional dependencies for XML. In: Kalinichenko, L.A., Manthey, R., Thalheim, B., Wloka, U. (eds.) ADBIS 2003. LNCS, vol. 2798, pp. 355–369. Springer, Heidelberg (2003)
17. Lv, T., Yan, P.: XML constraint-tree-based functional dependencies. In: Proceedings ICEBE, pp. 224–228 (2006)

The Lower and the Upper Systems of Rules in Tables with Missing Values

Hiroshi Sakai[1], Michinori Nakata[2], and Dominik Ślęzak[3,4]

[1] Mathematical Sciences Section, Department of Basic Sciences, Faculty of Engineering, Kyushu Institute of Technology
Tobata, Kitakyushu 804, Japan
sakai@mns.kyutech.ac.jp

[2] Faculty of Management and Information Science, Josai International University
Gumyo, Togane, Chiba 283, Japan
nakatam@ieee.org

[3] Institute of Mathematics, University of Warsaw
Banacha 2, 02-097 Warsaw, Poland

[4] Infobright Inc., Poland
Krzywickiego 34 pok. 219, 02-078 Warsaw, Poland
slezak@infobright.com

Abstract. A rule in a *Deterministic Information System* (*DIS*) is often defined by an implication τ such that both $support(\tau) \geq \alpha$ and $accuracy(\tau) \geq \beta$ hold for the threshold values α and β. In the previous work, we focused on the information incompleteness in *DISs*, and investigated rule generation in *Non-deterministic Information Systems* (*NISs*). We also proposed *NIS-Apriori* algorithm for this rule generation. In this paper, we consider *DISs* with missing values, which may be known as *Incomplete Information Systems* (*IISs*). A rule in a *DIS* is extended to either a rule in the *lower system* or a rule in the *upper system* in each *DIS* with missing values. *NIS-Apriori* algorithm is applied to generating such rules.

Keywords: Rough sets, Incomplete information, Missing values, Rule generation, Apriori algorithm.

1 Introduction

We followed rule generation in *DISs* [7,11], and investigated rule generation in *NISs*. *NISs* were proposed by Pawlak [7], Orłowska [6] and Lipski [5] in order to handle information incompleteness in *DISs*, like null values, unknown values, missing values. Since the emergence of incomplete information research [2,4,5,6], *NISs* have been playing an important role. We have also focused on the semantic aspect of incomplete information, and proposed *Rough Non-deterministic Information Analysis* (*RNIA*) [8]. This paper continues the framework of rule generation in *NISs* [8,9,10], and we apply this framework to *DISs* with missing values, which may be known as incomplete information systems.

Y. Zhang et al. (Eds.): DTA/BSBT 2010, CCIS 118, pp. 132–141, 2010.

2 Decision Rule Generation and Apriori Algorithm

A *Deterministic Information System* (DIS) is a quadruplet

$(OB, AT, \{VAL_A|\ A \in AT\}, f)$ [7].

Here, OB, AT and VAL_A are finite sets, and we sequentially call every element an *object*, an *attribute* and an *attribute value*, respectively. Furthermore, f is a mapping such that $f : OB \times AT \rightarrow \cup_{A \in AT} VAL_A$. We usually identify a DIS with a standard table.

We define two sets $CON \subseteq AT$ which we call *condition attributes* and $DEC \subseteq AT$ ($CON \cap DEC = \emptyset$) which we call *decision attributes*. An object $x \in OB$ is *consistent* (with any distinct object $y \in OB$), if $f(x, A) = f(y, A)$ for every $A \in CON$ implies $f(x, B) = f(y, B)$ for every $B \in DEC$. We call a pair $[A, val_A]$ ($A \in AT, val_A \in VAL_A$) a *descriptor*. For CON and DEC, we usually consider an implication (from an object x)

$\wedge_{A \in CON}[A, f(x, A)] \Rightarrow \wedge_{B \in DEC}[B, f(x, B)]$.

If x is consistent, we say this implication is *consistent*. A rule (more correctly, a candidate of a rule) is an appropriate implication. We usually employ two criteria, *support*(τ) and *accuracy*(τ) for the appropriateness [1,7].

A rule generation task in a DIS

Find all implications τ satisfying $support(\tau) \geq \alpha$ and $accuracy(\tau) \geq \beta$ for the threshold values α and β ($0 < \alpha, \beta \leq 1$).

Agrawal proposed *Apriori* algorithm [1] for such rule generation, and *Apriori* algorithm is now a representative algorithm for data mining.

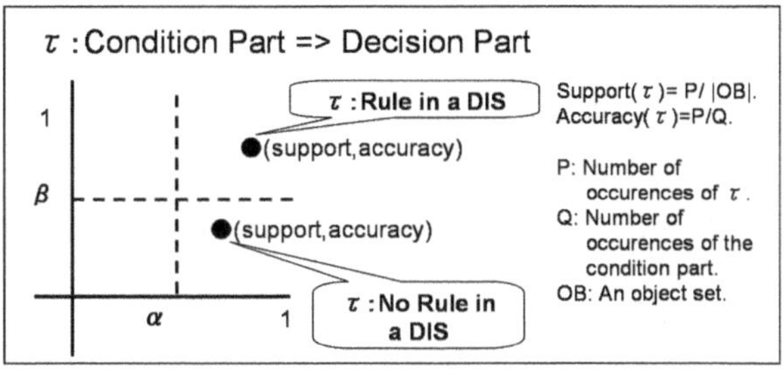

Fig. 1. A pair (*support,accuracy*) corresponding to the implication τ is plotted in a coordinate plane

3 Decision Rule Generation in DISs with Missing Values

A *Deterministic Information System* with *Missing Values* (DIS_{MV}) is also a quadruplet

$(OB, AT, \{VAL_A|A \in AT\}, f_{miss})$ [2,4],

where $f_{miss} : OB \times AT \rightarrow \cup_{A \in AT} VAL_A \cup \{*\}$. Semantically, $*$ is a variable whose value exists in $\cup_{A \in AT} VAL_A$. If $f_{miss}(x, A) \in VAL_A$, we see the information about the pair (x,A) ($x \in OB$, $A \in AT$) is *definite*. If $f_{miss}(x, A) = *$, we see the information about the pair (x,A) ($x \in OB$, $A \in AT$) is *missing*.

For a DIS_{MV}=$(OB, AT, \{VAL_A | A \in AT\}, f_{miss})$, a *derived DIS* (*for ATR from a* DIS_{MV}) is a DIS=$(OB, ATR, \{VAL_A | A \in ATR\}, f)$ satisfying $f(x, A) \in VAL_A$. In a DIS_{MV}, there are finite number of derived DISs due to the information incompleteness. Let $DD(DIS_{MV})$ denote the set of all derived DISs from DIS_{MV}.

Table 1. A Deterministic Information System with Missing Values Φ

OB	$Temperature$	$Headache$	$Nausea$	Flu
1	$high$	$*$	no	yes
2	$*$	yes	yes	yes
3	$*$	no	no	$*$
4	$high$	yes	$*$	$*$
5	$high$	$*$	yes	no
6	$normal$	yes	$*$	yes
7	$normal$	no	yes	no
8	$*$	yes	$*$	yes

Table 1 is an example of DIS_{MV} Φ, which is an artificial data sets. We can pick up implications below from Φ.

(1) In each $\psi \in DD(\Phi)$,

$\tau_1 : [Temperature, high] \Rightarrow [Flu, yes]$ from objects 1.

(2) In a proper selection of $\psi \in DD(\Phi)$,

$\tau_1 : [Temperature, high] \Rightarrow [Flu, yes]$ from objects 1, 2, 3, 4 and 8.

$\tau_2 : [Temperature, high] \wedge [Headache, yes] \Rightarrow [Flu, yes]$
from objects 1, 2, 4 and 8.

$\tau_3 : [Temperature, high] \wedge [Headache, no] \Rightarrow [Flu, yes]$
from objects 1 and 3.

According to this example, we know each τ depends upon the selection of $\psi \in DD(DIS_{MV})$, and τ may not occur in some ψ. Let $DD(DIS_{MV}, \tau)$ denotes a set of all ψ, where τ appears. This condition is equivalent to $support(\tau) > 0$. Due to $DD(DIS_{MV}, \tau)$, we give the definition of rule generation task in DIS_{MV}.

A rule generation task in a DIS$_{MV}$ (A revised definition in [9])
Let us consider the threshold values α and β $(0 < \alpha, \beta \leq 1)$.
(The lower system) Find all implications τ such that $support(\tau) \geq \alpha$ and $accuracy(\tau) \geq \beta$ hold in each $\psi \in DD(DIS_{MV}, \tau)$.
(The upper system) Find all implications τ such that $support(\tau) \geq \alpha$ and $accuracy(\tau) \geq \beta$ hold in some $\psi \in DD(DIS_{MV}, \tau)$.

In a DIS, $DD(DIS, \tau)$ means a singleton set, therefore the lower and the upper systems define the same implications in a DIS. Namely, the above definition is a natural extension from rule generation in DISs. Intuitively, the lower system defines *rules with certainty*, and the upper system defines *rules with possibility*. Especially, if $DD(DIS_{MV}, \tau)$ is equal to $DD(DIS_{MV})$ and τ is a rule in the lower system, this τ is the most reliable.

4 The Minimum and the Maximum Criterion Values, and NIS-Apriori Algorithm

In Table 1, let us consider $\tau_1 : [Temperature, high] \Rightarrow [Flu, yes]$, again. In this case, $support(\tau_1)$ and $accuracy(\tau_1)$ take several values according to the derived $DISs$. Therefore, we handle the minimum and the maximum values.

We define $minsupp(\tau)$ (minimum support) by $\min_{\psi \in DD(DIS_{MV},\tau)} support(\tau)$. Similarly, we define $minacc(\tau)$ (minimum accuracy), $maxsupp(\tau)$ (maximum support) and $maxacc(\tau)$ (maximum accuracy) for each implication τ. These minimum and maximum values depend upon all $\psi \in DD(DIS_{MV}, \tau)$, whose elements usually increase exponentially. However, we have proved the next results, namely the calculation of these criterion values is not exponential order.

Result 1. [9] In a $DIS_{MV}=(OB, AT, \{VAL_A | A \in AT\}, f_{miss})$, we can calculate four criterion values by using the next sets, $inf([A, val_A])$ and $sup([A, val_A])$.

(1) $inf([A, val_A])=\{x \in OB | f_{miss}(x, A)=val_A\}$.
(2) $inf(\wedge_{A \in ATR}[A, val_A])=\cap_{A \in ATR} inf([A, val_A])$.
(3) $sup([A, val_A])=inf([A, val_A]) \cup \{x \in OB | f_{miss}(x, A)=*\}$.
(4) $sup(\wedge_{A \in ATR}[A, val_A])=\cap_{A \in ATR} sup([A, val_A])$.
For $\tau : [CON, \zeta] \Rightarrow [DEC, \eta]$,

$$minsupp(\tau)=|inf([CON,\zeta]) \cap inf([DEC,\eta])|/|OB|,$$
$$minacc(\tau)=\frac{|inf([CON,\zeta]) \cap inf([DEC,\eta])|}{|inf([CON,\zeta])|+|OUTACC|},$$
$$(OUTACC=[sup([CON,\zeta]) - inf([CON,\zeta])] - inf([DEC,\eta])),$$
$$maxsupp(\tau)=|sup([CON,\zeta]) \cap sup([DEC,\eta])|/|OB|,$$
$$maxacc(\tau)=\frac{|inf([CON,\zeta]) \cap sup([DEC,\eta])|+|INACC|}{|inf([CON,\zeta])|+|INACC|}.$$
$$(INACC=[sup([CON,\zeta]) - inf([CON,\zeta])] \cap sup([DEC,\eta])).$$

In each object $y \in OUTACC$, we can pick up $\tau' : [CON, \zeta] \Rightarrow [DEC, \eta']$ ($\eta \neq \eta'$). Such y is counted in the denominator of the $accuracy(\tau)$. On the other hand, each object $y \in INACC$, we can pick up $\tau : [CON, \zeta] \Rightarrow [DEC, \eta]$. Therefore, such y is counted both in the denominator and the numerator of the $accuracy(\tau)$. Due to the inequality $\frac{A}{B} \leq \frac{A+1}{B+1}$ $(0 < A \leq B)$, we proved Result 1.

Result 2. [9] For each implication τ, there is a $\psi_{worst} \in DD(DIS_{MV}, \tau)$, where both $support(\tau)$ and $accuracy(\tau)$ are minimum. Furthermore, there is a derived $\psi_{best} \in DD(DIS_{MV}, \tau)$, where both $support(\tau)$ and $accuracy(\tau)$ are maximum.

Table 2 shows two derived $DISs$, namely ψ_{worst} and ψ_{best}, for τ_1. Generally, either ψ_{worst} nor ψ_{best} may not be unique. Fig.2 shows the distribution of pairs ($support(\tau)$,$accuracy(\tau)$) in each $\psi \in DD(DIS_{MV}, \tau)$.

According to the above two results, we can handle the next definition.

An equivalent rule generation task in a DIS$_{MV}$
Let us consider the threshold values α and β $(0 < \alpha, \beta \leq 1)$.
(The lower system) Find all implications τ such that $minsupp(\tau) \geq \alpha$ and $minacc(\tau) \geq \beta$.

(The upper system) Find all implications τ such that $maxsupp(\tau) \geq \alpha$ and $maxacc(\tau) \geq \beta$.

In the first definition of rule generation task, we needed to examine $support(\tau)$ and $accuracy(\tau)$ in all $\psi \in DD(DIS_{MV}, \tau)$. Since the number of $DD(DIS_{MV}, \tau)$ increases exponentially, so to check all $\psi \in DD(DIS_{MV}, \tau)$ will be hard for large data sets. However we can examine the same results by comparing (*minsupp*, *minacc*) and (*maxsupp*,*maxacc*) with the threshold α and β. Like this, we can extend rule generation in a DIS to rule generation in a DIS_{MV}, and we can apply *NIS-Apriori* [9,10] to this rule generation. This is an adjusted *Apriori* algorithm to NISs, and it can handle not only deterministic information but also incomplete information. *NIS-Apriori* algorithm does not depend upon the number of derived DISs, and the complexity is almost the same as *Apriori* algorithm.

Table 2. Derived ψ_{worst} from Table 1, which causes the minimum support 1/8 and the minimum accuracy 1/4 of τ_1, and derived ψ_{best} which causes the maximum support 5/8 and the maximum accuracy 5/6 of τ_1 to the right

OB	$Temperature$	Flu
1	*high*	*yes*
2	*very_high*	*yes*
3	*high*	*no*
4	*high*	*no*
5	*high*	*no*
6	*normal*	*yes*
7	*normal*	*no*
8	*normal*	*yes*

OB	$Temperature$	Flu
1	*high*	*yes*
2	*high*	*yes*
3	*high*	*yes*
4	*high*	*yes*
5	*high*	*no*
6	*normal*	*yes*
7	*normal*	*no*
8	*high*	*yes*

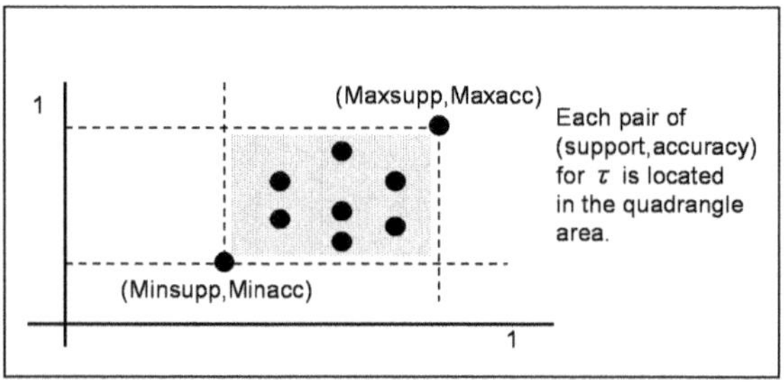

Fig. 2. A distribution of pairs (*support*,*accuracy*) for an implication τ

5 Rule Generation in DIS$_{MV}$

We are now coping with *SQL-NIS-Apriori* on Infobright ICE system [12], and we are discussing on Data Mining in Warehousing and Various Types of Inexact Data [3]. However, we have not completed the implementation in *SQL*.

Therefore, at first, we show the simulation of rule generation in DIS_{MV} by *NIS-Apriori* algorithm. Then, we show the real execution by *NIS-Apriori* in prolog instead of SQL.

Let us consider the next case in this section.

(1) DIS_{MV}: Φ in Table 1, DEC=$\{Flu\}$, α=0.3, β=0.5,

(2) $VAL_{Temperature}$=$\{normal, high, very\_high\}$,
$VAL_{Headache}$=VAL_{Nausea}=VAL_{Flu}=$\{yes, no\}$,

(3) Descriptors:
$[T, n]$, $[T, h]$, $[T, v]$ (T:Temperature, n:normal, h:high, v: very_high),
$[H, y]$, $[H, n]$ (H:Headache, y:yes, n:no), $[N, y]$, $[N, n]$ (N:Nausea),
$[Flu, y]$, $[Flu, n]$ (F:Flu).

5.1 A Simulation of the Lower System by NIS-Apriori Algorithm

Let us show each steps. In both *Apriori* and *NIS-Apriori* algorithms, we do not enumerate implications, but we focus on descriptors.

(STEP 1: Each descriptor and *inf*, *sup* information)
According to Result 1, we first generate inf and sup information in Table 3.

(STEP 2: Selection of the necessary descriptor)
Since α=0.3, an implication τ must occur more than 3 ($\geq 8 \times 0.3$=2.4) times in each $\psi \in DD(\Phi, \tau)$. Therefore, each descriptor which occurs more than 3 times is necessary in the lower system. Namely, we pick up descriptors satisfying $|inf([A, val_A])| \geq 3$ in Table 3. In the upper system we pick up descriptors satisfying $|sup([A, val_A])| \geq 3$ in Table 3.

(STEP 3: Iteration process)
After STEP 2, we obtain four descriptors $[T, h]$, $[H, y]$, $[N, y]$ and $[Flu, y]$. For these necessary descriptors, we generate conjunctions of descriptors, and calculate inf and sup in Result 1 for each conjunction in Table 4.

In Table 4, we know $\tau^* : [H, y] \Rightarrow [Flu, y]$ satisfies $minsupp(\tau^*) \geq 0.3$. Then, we calculate $OUTACC$=$[\{2, 4, 6, 8, 1, 5\} - \{2, 4, 6, 8\}] - \{1, 2, 6, 8\}$=$\{5\}$, and we

Table 3. Both inf and sup for each descriptor in Φ and the necessity in the lower and the upper systems

	$[T, n]$	$[T, h]$	$[T, v]$	$[H, y]$	$[H, n]$
inf	$\{6, 7\}$	$\{1, 4, 5\}$	$\emptyset$	$\{2, 4, 6, 8\}$	$\{3, 7\}$
sup	$\{6, 7, 2, 3, 8\}$	$\{1, 4, 5, 2, 3, 8\}$	$\{2, 3, 8\}$	$\{2, 4, 6, 8, 1, 5\}$	$\{3, 7, 1, 5\}$
$Necessity(Lower)$	*no*	*yes*	*no*	*yes*	*no*
$Necessity(Upper)$	*yes*	*yes*	*yes*	*yes*	*yes*

	$[N, y]$	$[N, n]$	$[Flu, y]$	$[Flu, n]$
inf	$\{2, 5, 7\}$	$\{1, 3\}$	$\{1, 2, 6, 8\}$	$\{5, 7\}$
sup	$\{2, 5, 7, 4, 6, 8\}$	$\{1, 3, 4, 6, 8\}$	$\{1, 2, 6, 8, 3, 4\}$	$\{5, 7, 3, 4\}$
$Necessity(Lower)$	*yes*	*no*	*yes*	*no*
$Necessity(Upper)$	*yes*	*yes*	*yes*	*yes*

Table 4. Both *inf* and *sup* for each conjunction of descriptors and the necessity in the lower system

	$[T,h]\wedge[Flu,y]$	$[H,y]\wedge[Flu,y]$	$[N,y]\wedge[Flu,y]$
inf	$\{1\}$	$\{2,6,8\}$	$\{2\}$
sup	$\{1,2,3,4,8\}$	$\{1,2,4,6,8\}$	$\{2,4,6,8\}$
Necessity(Lower)	*no*	*yes*	*no*

obtain $minacc(\tau^*)=\frac{|\{2,6,8\}|}{|\{2,4,6,8\}|+|\{5\}|}=\frac{3}{5}=0.6 \geq \beta$. Like this, we know τ^* is a rule in the lower system, namely this τ^* satisfies $support(\tau^*) \geq 0.3$ and $accuracy(\tau^*) \geq 0.5$ in each $\psi \in DD(\Phi,\tau^*)$. For conjuction τ' such that $support(\tau') \geq 0.3$ and $minacc(\tau') < \beta$, we generate new conjunctions from such τ', and repeat STEP 3 until there is no necessary conjunction.

5.2 A Simulation of the Upper System by NIS-Apriori Algorithm

In the upper system, we generate τ satisfying $support(\tau) \geq \alpha$ and $accurcay(\tau) \geq \beta$ for some $\psi \in DD(DIS_{MV},\tau)$. Namely, we think the most convenient $\psi \in DD(\Phi,\tau)$ for τ. Each role of the steps is the same as the lower system. In Table 3, $|sup([A,val_A])| \geq 3$ holds for each descriptor, so each of 9 descriptors is necessary. Table 5 shows four conjunctions of 14 conjunctions.

Table 5. Both *inf* and *sup* for some conjunctions of descriptors and the necessity

	$[T,n]\wedge[Flu,y]$	$[T,n]\wedge[Flu,n]$	$[T,h]\wedge[Flu,y]$	$[T,h]\wedge[Flu,n]$
inf	$\{6\}$	$\{7\}$	$\{1\}$	$\{5\}$
sup	$\{2,3,6,8\}$	$\{3,7\}$	$\{1,2,3,4,8\}$	$\{3,4,5\}$
Necessity(Upper)	*yes*	*no*	*yes*	*yes*

In Table 5, we know $\tau^* : [T,n] \Rightarrow [Flu,y]$ satisfies $maxsupp(\tau^*)=4/8 \geq 0.3$. Then, we calculate $INACC=[\{6,7,2,3,8\}-\{6,7\}] \cap \{1,2,6,8,3,4\}=\{2,3,8\}$, and we obtain $maxacc(\tau^*)=\frac{|\{6,7\}\cap\{1,2,6,8,3,4\}|+|\{2,3,5\}|}{|\{6,7\}|+|\{2,3,5\}|}=\frac{1+3}{2+5}=4/7 \geq \beta$. Like this, we know τ^* is a rule in the upper system, namely this τ^* satisfies $support(\tau^*) \geq 0.3$ and $accuracy(\tau^*) \geq 0.5$ in some $\psi \in DD(\Phi,\tau^*)$.

5.3 Real Execution by NIS-Apriori in Prolog

This subsection shows real execution. This program is implemented in prolog on a PC with 3.4GHz Pentium 4 CPU. We at first prepare the data set below:

```
object(8,4). /* 8 objects, 4 attributes */
support(0.3). /* support value */
accuracy(0.5). /* accuracy value */
```

```
decision(4). /* the definition of the decision attribute (4th attribute) */
attrib_values(1,temperature,3,[normal,high,very_high]).
attrib_values(2,headache,2,[yes,no]). /* attribute and attribute values */
attrib_values(3,nausea,2,[yes,no]).
attrib_values(4,flu,2,[yes,no]).

data(1,[high,nil,no,yes]). /* a table data, nil means missing value  */
data(2,[nil,yes,yes,yes]).
data(3,[nil,no,no,nil]).
data(4,[high,yes,nil,nil]).
data(5,[high,nil,yes,no]).
data(6,[normal,yes,nil,yes]).
data(7,[normal,no,yes,no]).
data(8,[nil,yes,nil,yes]).
```

Then, prolog program generates two sets inf, *sup* and etc.. The following is a part of them. The first statement means $inf([temperature, normal])=\{6,7\}$ and $sup([temperature, normal])=\{2,3,6,7,8\}$. Since $|\{6,7\}|/8 < 0.3$ holds, this descriptor does not satisfy $support \geq 0.3$. Therefore, we do not employ this descriptor $[temperature, normal]$ in the lower system, but $|\{2,3,6,7,8\}|/8 > 0.3$ holds, we employ this descriptor in the upper system. Finally, prolog program generates the last four lines, which mean a list of employed descriptors in the lower/upper system and the condition/decision attributes.

```
upper(1,1,[temperature,normal],[6,7],[2,3,6,7,8]).
lower(1,2,[temperature,high],[1,4,5],[1,2,3,4,5,8]).
upper(1,2,[temperature,high],[1,4,5],[1,2,3,4,5,8]).
upper(1,3,[temperature,very_high],[],[2,3,8]).
lowerdesc(1,[2]).
upperdesc(1,[1,2,3]).
     :       :       :
lowerdesc_con_list([[1,2],[2,1],[3,1]]).
lowerdesc_dec_list([[4,1]]).
upperdesc_con_list([[1,1],[1,2],[1,3],[2,1],[2,2],[3,1],[3,2]]).
upperdesc_dec_list([[4,1],[4,2]]).
```

Finally, we apply *NIS-Apriori* algorithm to the generated data set.

```
?- step1.
File Name for Read Open: 'flu_t.pl'.

===== Lower System ================================
  [1] MINSUPP=0.125, MINACC=0.25
  [2] MINSUPP=0.375, MINACC=0.6 [headache,yes]=>[flu,yes]
  [3] MINSUPP=0.125, MINACC=0.25
The Rest Candidates: []
(Lower System Terminated)
```

```
===== Upper System ================================
  [1] MAXSUPP=0.5, MAXACC=0.8 [temperature,normal]=>[flu,yes]
  [2] MAXSUPP=0.25, MAXACC=0.6666666667
  [3] MAXSUPP=0.625, MAXACC=0.8333333333 [temperature,high]=>[flu,yes]
  [4] MAXSUPP=0.375, MAXACC=0.75 [temperature,high]=>[flu,no]
  [5] MAXSUPP=0.375, MAXACC=1.0 [temperature,very_high]=>[flu,yes]
  [6] MAXSUPP=0.125, MAXACC=1.0
  [7] MAXSUPP=0.625, MAXACC=1.0 [headache,yes]=>[flu,yes]
  [8] MAXSUPP=0.25, MAXACC=0.4
  [9] MAXSUPP=0.25, MAXACC=0.6666666667
  [10] MAXSUPP=0.375, MAXACC=1.0 [headache,no]=>[flu,no]
  [11] MAXSUPP=0.5, MAXACC=0.6666666667 [nausea,yes]=>[flu,yes]
  [12] MAXSUPP=0.375, MAXACC=0.75 [nausea,yes]=>[flu,no]
  [13] MAXSUPP=0.625, MAXACC=1.0 [nausea,no]=>[flu,yes]
  [14] MAXSUPP=0.25, MAXACC=0.6666666667
The Rest Candidates: []
(Upper System Terminated)

EXEC_TIME=0.0(sec)
```

In this execution of the lower system, we know that an implication τ : $[headache, yes] \Rightarrow [flu, yes]$ satisfies $support(\tau) \geq 0.3$ and $accuracy(\tau) \geq 0.5$ in each $DD(DIS_{MV}, \tau)$. Namely, this τ will be the most reliable implication. In this execution, no rest candidates exist in both systems, therefore the recursive step is closed after the first step. If we fix value of *accuracy* to 1.0, the program terminated after the third step. We have applied this program to several data sets, and we examined this prolog program works well. This program can handle any data set in the form of the previously shown data set.

6 Concluding Remarks

This paper applied *NIS-Apriori* algorithm to rule generation in *DISs* with missing values. After obtaining *in* and *sup* information, we can similarly obtain rules in the lower system and the upper systems. The rule generation task in DIS_{MV} depends upon the number of $DD(DIS_{MV}, \tau)$, which increases exponentially. However, *NIS-Apriori* calculates rules due to Results 1 and 2. This is not the exponential order. We implemented this functionality on computers in prolog.

Acknowledgment. The first author is supported by the Grant-in-Aid for Scientific Research (C) (No.16500176, No.18500214, No.22500204), Japan Society for the Promotion of Science. The fourth author was partially supported by the grants N N516 368334 and N N516 077837 from the Ministry of Science and Higher Education of the Republic of Poland.

References

1. Agrawal, R., Srikant, R.: Fast Algorithms for Mining Association Rules. In: Proc. 20th Very Large Data Base, pp. 487–499 (1994); Roddick, J.F.: Association mining. ACM Comput. Surv. 38(2) (2006)
2. Grzymala-Busse, J.: Data with Missing Attribute Values: Generalization of Indiscernibility Relation and Rule Induction. Transactions on Rough Sets 1, 78–95 (2004)
3. Infobright.org Forums, `http://www.infobright.org/Forums/viewthread/288/`
4. Kryszkiewicz, M.: Rules in Incomplete Information Systems. Information Sciences 113, 271–292 (1999)
5. Lipski, W.: On Semantic Issues Connected with Incomplete Information Data Base. ACM Trans. DBS. 4, 269–296 (1979)
6. Orłowska, E., Pawlak, Z.: Representation of Nondeterministic Information. Theoretical Computer Science 29, 27–39 (1984)
7. Pawlak, Z.: Rough Sets. Kluwer Academic Publishers, Dordrecht (1991)
8. Sakai, H., Okuma, A.: Basic Algorithms and Tools for Rough Non-deterministic Information Analysis. Transactions on Rough Sets 1, 209–231 (2004)
9. Sakai, H., Ishibashi, R., Nakata, M.: On Rules and Apriori Algorithm in Non-deterministic Information Systems. Transactions on Rough Sets 9, 328–350 (2008)
10. Sakai, H., Nakata, M., Ślęzak, D.: Rule Generation in Lipski's Incomplete Information Databases. In: Szczuka, M. (ed.) RSCTC 2010. LNCS (LNAI), vol. 6086, pp. 376–385. Springer, Heidelberg (2010)
11. Skowron, A., Rauszer, C.: The Discernibility Matrices and Functions in Information Systems. In: Intelligent Decision Support - Handbook of Advances and Applications of the Rough Set Theory, pp. 331–362. Kluwer Academic Publishers, Dordrecht (1992)
12. Ślęzak, D., Sakai, H.: Automatic Extraction of Decision Rules from Non-deterministic Data Systems: Theoretical Foundations and SQL-Based Implementation. In: Proc. of DTA 2009, pp. 151–162. Springer CCIS 64, Heidelberg (2009)

Data Warehouse Discovery Framework: The Foundation

Cas Apanowicz

IT Horizon, Canada
cas.apanowicz@it-horizon.com

Abstract. The cost of building an Enterprise Data Warehouse Environment runs usually in millions of dollars and takes years to complete. The cost, as big as it is, is not the primary problem for a given corporation. The risk that all money allocated for planning, design and implementation of the Data Warehouse and Business Intelligence Environment may not bring the result expected, fare out way the cost of entire effort [2,10]. The combination of the two above factors is the main reason that Data Warehouse/Business Intelligence is often single most expensive and most risky IT endeavor for companies [13]. That situation was the main author's inspiration behind founding of Infobright Corp and later on the concept of Data Warehouse Discovery Framework.

Keywords: Data Warehousing, Data Warehouse Discovery (DW Discovery), Business Intelligence, Benchmarking Practices, Product Selection, Evaluation Matrix, Preference Matrix, Priority Matrix, Infobright Enterprise Edition.

1 Introduction

1.1 Background

For many years the author was retained by major corporations and governments agencies to investigate the corporate environment and lay down the foundations for the Data Warehouse and Business Intelligence Strategies.

The biggest challenge facing today's corporations is proper identification and definition of the business requirements due to a large number of data sources and major difficulties in communication between the business community and the IT group [2].

The inability to translate the requirements voiced by the business user community into a defined technological framework for the Data Warehouse and gaining a full understanding of the proposed solution prepared by the IT group to the business community, are the greatest challenges facing company and created a serious risk for the corporation as a whole.

The Business Intelligence is usually a very costly and time consuming effort of building the Data Warehouse and the BI environment with no guarantee that the end-result would meet the needs of the business community [10,13]. This risk was compounded by the fact that the design and implementation process is always a long process, often estimated to last at least between two to three (2-3) years.

Y. Zhang et al. (Eds.): DTA/BSBT 2010, CCIS 118, pp. 142–154, 2010.

During thi period, users were not expected to gain any benefits from this ongoing effort, from either data or BI reports. The success of this project initiative would not be determined until its completion, with no guarantee that it would meet the current/future needs of the organization.

To mitigate these risks and to enable the user to consume some BI information as it was being developed, a new concept of DW Discovery was proposed. It was built on the basis that a "production prototype" would be created during the definition phases that would allow users to sample the data being captured and review the functionality that will be captured in the delivered system.

The greatest challenge facing the concept was to identify and utilize existing technologies that would have the capability to fast load the data in an "as is" fashion and allow for easy querying without the need for physical data base design. These challenges became the catalyst for the methodology being presented in this document, for it enabled to overcome them and create a new approach in Data Warehouse Discovery and development.

It was not until the emergence of ne column-based RDBMS's, notably the Infobright Inc., that the full methodology could be realized.

After considerable time and effort invested into refining and polishing this methodology, it was accepted and implemented, by the author, in one Canadian Government Agency for the very high profile, multi-million dollar project.

The DW Discovery Framework environment creates a database repository that would be used to load a portion of a production data in an "as is" format, with little or no ETL effort being applied to the data [1]. Furthermore, this repository would not require any physical design, tuning or any significant database administration effort. The source data would be loaded to support the initial and ongoing architecture design. There would be no creation or design of objects, such as indexes or table spaces, this all contributing to the low cost associated with this type of environment. The data loaded initially can be loaded into "large and flat" tables if necessary or if such a construct will be closer to the source data. It will use a relatively small amount of space. This environment can be then queried with no performance penalty. Eliminating the need for a physical design combined with "light or no" ETL processes as well as no need for expensive performance tuning and maintenance, will all contribute together producing rapid development, low cost and the mitigation of the risks associated with project of this nature – making the DW Business Discovery Framework an invaluable tool in today's data warehouse and decision support systems market. One of major obstacles faced when implementing DW Discovery was utilizing the right database technology. Only after utilizing columnar database offered by Infobright Inc. and considerable time and effort invested into refining and polishing this methodology, it was accepted one of the Canadian Government Agency and was further implemented by the author on behave of the agency in very high profile, multimillion dollar project.

This paper is organized as follows:

Introduction	The overview of the whole paper.
Methodology	This will lay down the Methodology Framework and indicate the where the DW Discovery enriches the traditional approach.

DW Discovery	The description of the concept.
Transition to Production	Provides guidelines of transition of DW Discovery from a prototype-like environment to the environment tightly integrated with the production environment.

To great extent, this paper is a continuation of the research reported in [2], where the author introduced Benchmark Project for DW Platform selection. The Benchmark Project not only made apparent that the Agency would greatly benefit from adopting The DW Discovery methodology, but also created the right environment for this deployment. The need for DW Discovery became apparent not only to the author, but to the sponsors of the project.

On the other hand, in this paper, we focus mainly on the methodology itself, and the benefits it may brings, which has not been studied previously.

1.2 Methodology

The methodology adopted by this document follows the BI Requirements gathering process and documents production and delivery developed as a part of Data Warehouse Discovery Framework.

DW Discovery steps:

1. Initial investigation of Data Sources
2. Initial Review of User needs
3. Set up DW Discovery database based on the data sources
4. Initiate dialog with the user community
5. Present the "Prototype"
6. Verify needs
7. Proceed with the design

1.3 BI/DW Program Initiation and Development

At the core of solution development is the DW Discovery Methodology supported by the Discovery repository and consisting of a comprehensive set of activities for the full life-cycle development of a Data Warehouse and BI Solution. The best aspect of this approach is being able to see the warehouse before its build. Then an iterative agile development follows to allow for the information consumption while the Data Warehouse is still being constructed.

2 Data Warehouse Discovery

2.1 The Problem

As previously mentioned, the process of laying down the proper strategy and further development of Business Intelligence Environment, is usually a very costly and time consuming effort of building the Data Warehouse [13] and the BI Infrastructure with no guarantee that the end-result would meet the needs of the business community. This risk is always compounded by the fact that the design and implementation process often lasts at least between two to three years with and even sometimes even 5 years before any information can be provided to the user for consumption.

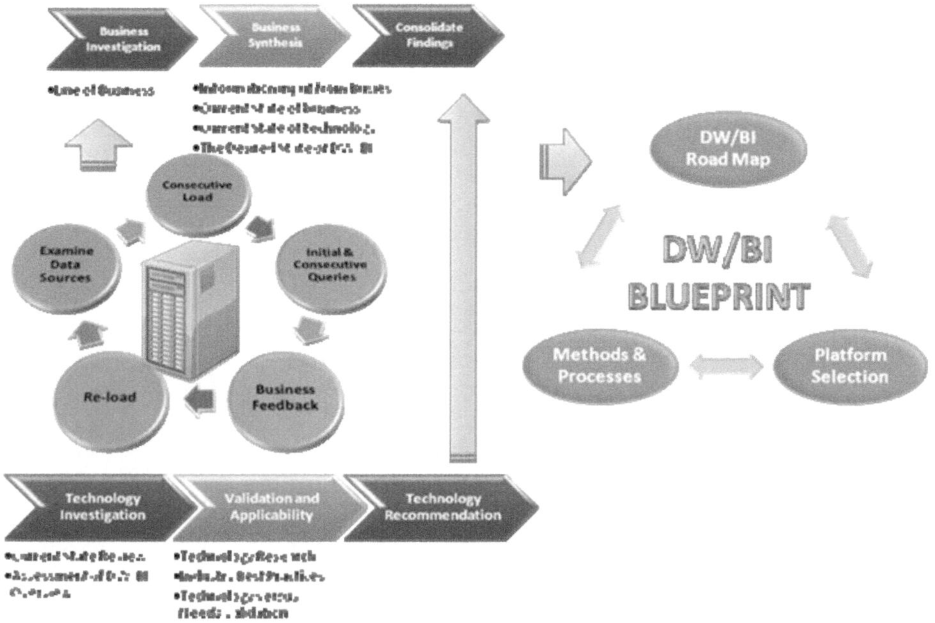

Fig. 1. Organizational Preparedness and DW/BI Blueprint

As much as the cost of this undertaking could be a large portion of the organization's expenditures, probably the biggest challenge facing organization is ability to properly identify and define the current business requirements and information needs as well as to anticipate the future BI evolution due to constant, and often volatile, organizational changes.

The inability to translate the requirements voiced by the business user community into a defined technological framework for the Data Warehouse and gaining a full understanding of the proposed solution prepared by the IT group to the business community, are the greatest challenges facing company and created a serious risk for the corporation as a whole. This risk is further exacerbated by large and disperse number of data sources and major difficulties in communication between the business community and the IT group.

Also, during the design and development period, users are not expected to gain any benefits from this ongoing effort in any form, from data or BI reports. The success of DW/BI initiative would not be determined until its completion, with no guarantee that it would meet the current and future needs of the organization.

2.2 The Concept

To mitigate these risks and to enable user to consume some BI information as it was being developed, a new concept of DW Discovery was proposed. It was built on the basis that a "production prototype" would be created during the definition phases that would allow users to sample the data being captured and review the functionality that will be captured in the delivered system.

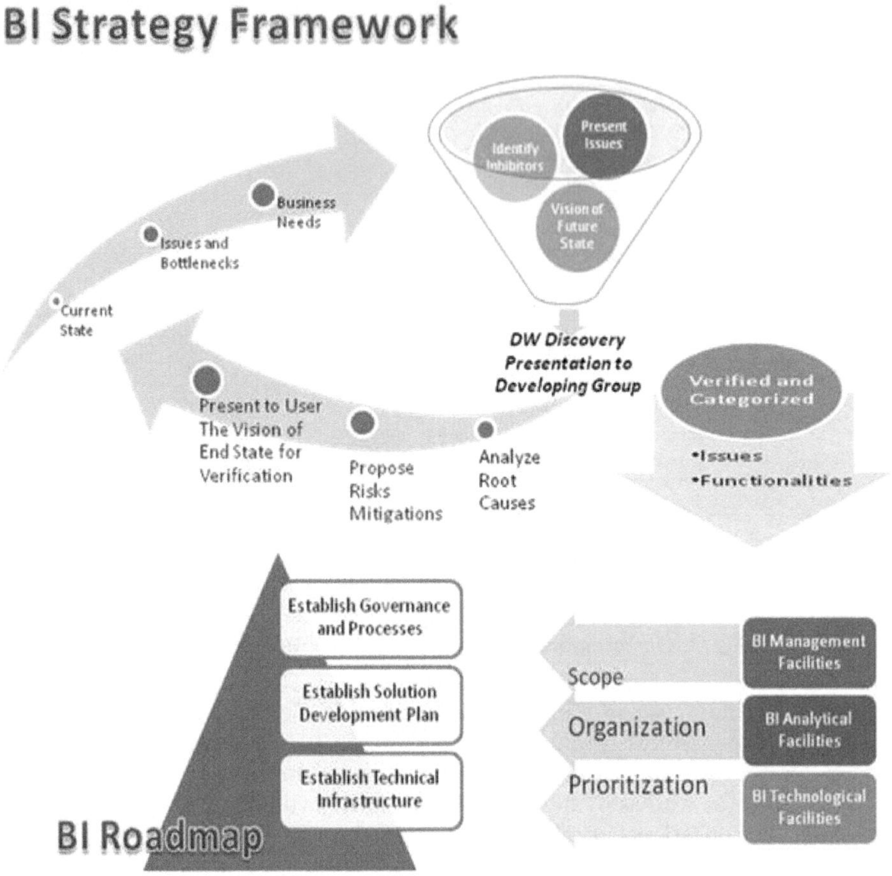

Fig. 2. BI/DW Program Initiation and Development

DW Discovery Framework environment creates a database repository that would be used to load a portion of a production data in an "as is" format, with little or no ETL effort being applied to the data. Furthermore, this repository would not require any physical design, tuning or any significant database administration effort. The source data would be loaded to support the initial and ongoing architecture design. There would be no creation or design of objects, such as indexes or table spaces, this all contributing to the low cost associated with this type of environment. The data loaded initially can be loaded into "large and flat" tables if necessary or if such a construct will be closer to the source data. It will use a relatively small amount of space. This environment can be then queried with no performance penalty. Eliminating the need for a physical design combined with "light or no" ETL processes as well as no need for expensive performance tuning and maintenance, will all contribute together producing rapid development, low cost and the mitigation of the risks associated with project of this nature – make the DW Business Discovery Framework an invaluable tool in today's data warehouse and decision support systems market.

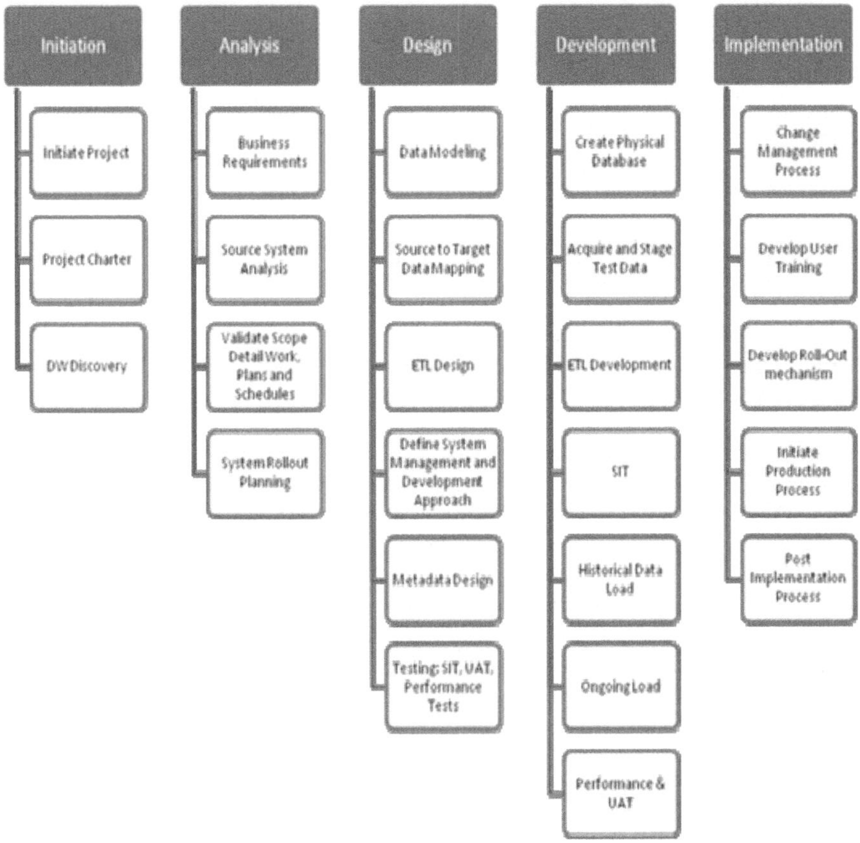

Fig. 3. BI/DW Program Initiation and Development Steps

The primary purpose of deploying a DW "Business" Discovery Environment is typically as a business-driven function and it is created as an adjunct to large-scale BI environments to ensure that there can be a nimble and immediate response to rapidly and frequently evolving user requirements.

The greatest challenge facing the concept was to identify and utilize existing technologies that would have the capability to fast load the data in an "as is" fashion and allow for easy querying without the need for physical data base design. These challenges became the catalyst for the methodology being presented in this document, for it enabled to overcome them and create a new approach in Data Warehouse Discovery and development.

The following features were instrumental in its development and implementation:

- Ability to support wide and de-normalized data tables
- Superior compression to support these tables
- Fast load
- Ad-hoc query capability across renormalized tables
- Low maintenance
- No need for physical design
- Low technology footprint

2.2.1 The Traditional RDBMS

The Relational Data Base Management Systems (RDBMS) are in use in Information Technology for decades now. There are well-established technologies aimed at dealing with large amount of data stored in a tabular format. The Structural Query Language or SQL was developed to manipulate and access data. It is proven and has already matured way to deal with data.

There are numerous commercially available RDBMS vendors and platforms - each of them have common core of functionalities and standards, they all have a specific *flavors* or special characteristics and straights/weaknesses. Behind the common standardized interface and modus operandi they all have elaborate ways and mechanisms of dealing with physical storage, indexing and complicated system of optimization and performance regulations.

While each product has different designs and methods to deal with the above issues, all will require a significant effort of physical design, management and maintenance of the physical components of the above engines. This applies to the initial setup effort as well as ongoing changes and maintenance.

That need for an elaborate system of design and maintenance makes RDBMS not the right candidate for the DW Discovery engine.

2.2.2 The Columnar RDBMS

It was not until the emergence of column-based RDBMS's [3,4,16], the Data Warehouse Discovery Framework could finally be implemented in its entirety and the corporation could reap full benefits that is associated with its deployment.

The columnar databases appear to the user as a regular relational database due to that fact that most of them use a SQL front-end to communicate with the user.

2.3 DW Discovery Environment - Main Characteristics

- Helps discover new insights by addressing rapidly and frequently evolving user requirements combined with a sense of urgency;
- Provides a basis for developing a technology plan that satisfies cross-functional business needs;
- Improves cooperation and understanding between the business users and the Data Management and Data Warehouse Divisions;
- Is an integral part of the architecture serving a technical discovery need;
- Will enable the business and technology to have a 'test bed of emerging solutions' that can be vetted and modified prior to making significant commitments in resources and development.

2.4 The Purpose

The Data Warehouse Discovery Framework is a dual purpose, live concept, production environment that enables the creation of new business solutions and validation of technical concepts. These purposes include:

1. *The primary purpose is for developing and evaluating new concepts (business and technical).*
2. *The secondary purpose is as serving as a nimble, high performance information application centre, enabling an immediate response to rapidly and frequently evolving user requirements and concepts.*

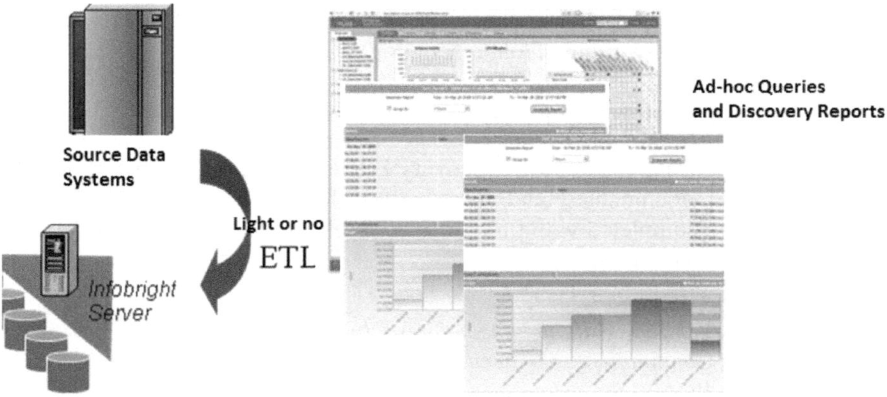

Fig. 4. DW Discovery Environment

A major challenge in developing worthwhile and business relevant solutions is to accurately capture the underlying or hidden 'real' business information requirements. This is a difficult task as initially the majority of business end-users "don't know what they don't know" making it difficult to articulate what they really need.

Data Warehouse Discovery Framework provides a basis for identifying, evolving and validating business end-users requirements as well as developing a technology plan that satisfies cross-functional business needs and a 'test bed of emerging solutions' that can be vetted and modified prior to making significant commitments in resources and development.

On the surface gathering requirements for informational systems may appear to be a somewhat trivial task. However, in many cases this exercise is most effective when an individual experienced in decision support environments leads the business community using "peel-the-onion" techniques to effectively draw out their requirements. This involves much more than merely asking the business community to articulate what they think they need. The best way to conduct such a exercise is to have flexible data repository with ability to query ad-hoc and pre-define in real time dialog with the business user. The individual leading the exercise needs to have a detailed perspective of the nature of the end-state solution.

2.5 The Short Term Benefits

- Enhance the Data Relationship Knowledge / Business Rule Design - allows to identify more detailed BI (query) needs in a low cost environment
- Mitigate the risk of one shot / big build that does not meet user requirements
- Validate schema and evolutionary design
- Validate and vet detailed needs / analyze queries before undertaking a costly and lengthy development process

2.6 The Long Term Benefits

- As business conditions change, it allows quick vetting out of implications
- Rapid designs analysis - (build cycle in DW Discovery Environment could be 1 week versus 12 months)

- Low Cost / Rapid Deployment – easy to maintain and light ETL (little to no integration logic, minimal data modeling)

2.7 DW Discovery Information Flow

When utilizing the Data Warehouse Discovery Framework all the source systems will be identified and documented and then a "fast load" of data to the Discovery Repository will be completed with little or no ETL effort. At that point the repository will be ready to for any ad-hoc queries to:

1. *Articulate User needs*
2. *Vet and verify the needs interactively by user*
3. *Communicate the needs to the IT team*
4. *Present the proposed solution to the user*

The Information Flow, produced by the Framework, represents the high-level conceptual view of that flow. The environment itself will be comprised of data acquisition processes that will feed the data repository by data extracted from the source application and then properly transformed. Subsequently, the Data Delivery layer will facilitate the information consumption by the end user community.

During the DW Discovery process, the Business Analyses and User Requirements gathering will benefit from rapid and immediate verification that will be performed in a prototype fashion, hence, giving immediate vetting mechanism of needs versus proposed solution. Fig. 6 illustrates that process.

The DW Discovery will serve to fill BI Matrices (Fig. 7). They are a comprehensive framework to design and build the planned DW/BI environment [11,12,15]. The research, analysis and discovery deliverables produced by the DW Discovery will be sufficient to complete these matrices, hence enabling the user to complete the estimate, design and deliver the final system with well-defined plans and budgets.

2.8 DW Discovery Data Modeling

The cornerstone of the DW Discovery is that there is no need for any physical design and planning. Therefore, the only modeling that needs to be completed for this component would be based on the source system models with little or no modifications, therefore the modeling effort would be minimal at least for this phase.

2.9 DW Discovery Technology Platform

One of the key aims of the Data Warehouse Discovery Framework was to reduce risk and lowering the cost associated with these types of initiatives. The construction of the Framework should not contribute to any new costs while providing an excellent performance for these phases of the methodology. This is achieved by using an Open Source technology for the Discovery Framework environment. The code within the Framework is contained within its own environment and does not require to be loaded on client processing environments. A broad range of matured Open Source products are now available and enable the Data Warehouse Discovery Framework to be readily deployed at any client environment.

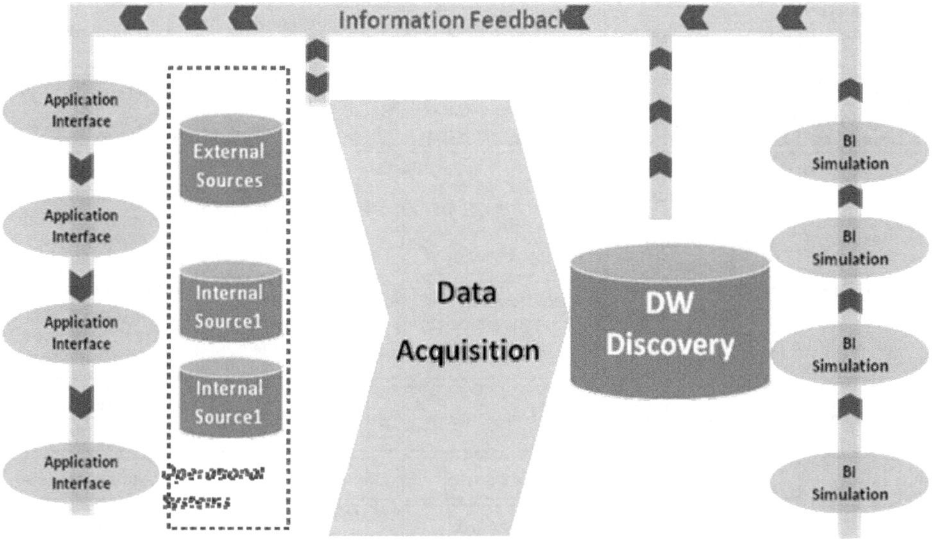

Fig. 5. DW Discovery Information Flow

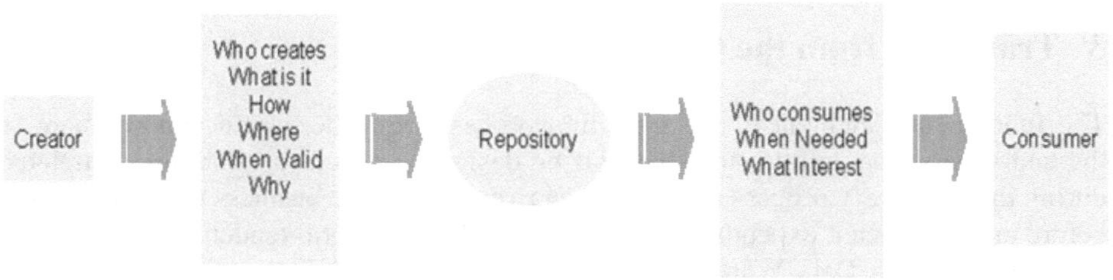

Fig. 6. DW Discovery of Information Usage

2.9.1 DW Discovery Load

The Load to DW Discovery Framework can be achieved by utilizing any technology readily available or being currently used by client or by obtaining an Open Source tool. All these methods are focused on reducing the initial cost of the initiative.

2.9.2 RDBMS

A Columnar database will be used with an Open Source approach for cost-effectiveness or if cost is not an issue an Enterprise Edition will be utilized.

2.9.3 Front-End and Query Tool

A simple query tool can be utilized, such as Toad or similar packages. Freeware or Open Source tools and utilities could be used, depending on the situation, as a means for cost-effectiveness. Either approach will provide sufficient simulation to the BI end-result.

BI Business End-user Domains	Report Recipients	Business Information Users	Information Analysts	Statistical Analysts
Reporting & Analysis Functions	Report Distribution	Performance Measurement	Query & Reporting	Data Mining
Sample Business Information Requirements	*• What reports are available for a given business subject or topic of interest? • Inform me when the latest version of a given report is available.*	*• How are we tracking against Plan? • What are my/our KPI results for the latest period and how are we trending?*	*• What are the drivers behind the results being produced? • What are the key factors that support/explain our results?*	*• What is the probability that a transaction should be referred? • What combination of factors indicate an improvement in the detection of contraband goods?*
Objective	Data Accessibility	Personalization	Analytical Flexibility	Probability & Outcomes
Analysis Characteristics or Class	Standard	Scorecards	Multi-Dimensional	Predictive Models
	Production	Dashboards	Investigative/ Iterative	Descriptive Models

Fig. 7. BI Functional Information Matrix

3 Transition from the Current State

The fundamental element of the transition process from the current environment to the end-state is the architecture that will be design based on the findings completed during the Discovery process and it has been verified by the business user community before any significant expenditure is committed. It is a natural tendency to design the first iteration of a Data Warehouse in a single department-centric manner. This, in effect, would create what's known as an information silo environment where each consecutive Data Mart is a disjoined and isolated system. The DW Discovery will help in preventing this unfortunate process by the following these steps:

- Conformed Dimensions – information Integration
- Hardware selection needs to scale to meet needs of the organization
- Integration of further builds (avoid departmental disjoins)
- Development with the view of an enterprise data warehouse (EDW)
- Establishment of a cross-functional team of business and IT representatives
- New data marts should be mandated to follow corporate strategic standards
- Development of a Data retention and decommission policy
- Modeling the information into the DW as a part of the enterprise rather than departmental
- Limiting the number of subject areas being implemented at one time

In addition, establishment of an Agency-wide Master Data Management (MDM) initiative as a part of Data Management mandate is highly recommended [8].

4 Conclusions

DW/BI efforts are a very costly and time consuming process. Any design and architectural inaccuracy have usually very profound and long lasting adverse effects. An ability to mitigate the risk of delivering incorrect system is extremely crucial and should be part of any BI strategy. DW Discovery has been proven in reducing this risk almost completely by providing the view of the Data Warehouse before any line of code has been written or single design artifacts done. In addition, the method is very successful in significant reduction of overall cost of Data Warehouse and BI Environment by shortening design and development process while streamlining and focusing the effort in the right areas of development.

The success of the presented implementation should be encouraging for organizations to adopt this method as a risk mitigation and cost saving tool and for DW/BI practitioners to include this method to their arsenal of tools and methodologies. On the other hand, careful selection of right technology for the DW Discovery Framework makes the concept a reality.

References

[1] Apanowicz, C.: Data Warehouse Discovery Framework - The Case Study. In: Proc. of DTA 2010, Springer, Heidelberg (2010)

[2] Apanowicz, C.: Data Warehousing and Business Intelligence: Benchmark Project for the Platform Selection. In: Proc. of DTA 2009, Springer, Heidelberg (2009)

[3] Apanowicz, C., Eastwood, V., Ślęzak, D., Synak, P., Wojna, A., Wojnarski, M., Wróblewski, J.: Method and System for Data Compression in a Relational Database. US Patent Application, 2008/0071818 A1 (2008)

[4] Apanowicz, C., Eastwood, V., Ślęzak, D., Synak, P., Wojna, A., Wojnarski, M., Wróblewski, J.: Method and System for Storing, Organizing and Processing Data in a Relational Database. US Patent Application, 2008/0071748 A1 (2008)

[5] Buretta, M.: Data Replication, Tools and Techniques for Managing Distributed Information. John Wiley & Sons, Chichester (1997)

[6] Davenport, T.H., Harris, J.G.: Competing on Analytics: The New Science of Winning. Harvard Business School Press, Boston (2007)

[7] Hughes, R.: Agile Data Warehousing: Delivering World-Class Business Intelligence Systems Using Scrum and XP. Ceregenics Inc. (2008)

[8] Imhoff, C., Galemmo, N., Geiger, J.G.: Mastering Data Warehouse Design: Relational and Dimensional Techniques. Wiley Publishing Inc., Chichester (2009)

[9] Infobright Inc., `http://www.infobright.com`, `http://www.infobright.org`, `http://en.wikipedia.org/wiki/Infobright`

[10] Inmon, W.H., Imhoff, C., Sousa, R.: Corporate Information Factory. Wiley Computer Publishing, Chichester (2001)

[11] Inmon, W.H., O'Neil, B., Fryman, L.: Business Metadata: Capturing Enterprise Knowledge. Morgan Kaufman Publishers, San Francisco (2008)

[12] Inmon, W.H., Strauss, D., Neushloss, G.: DW 2.0: The Architecture for the Next Generation of Data Warehousing. Morgan Kaufman Publishers, San Francisco (2008)

[13] Kimball, R., Ross, M., Thornthwaite, W., Mundy, J., Becker, B.: The Data Warehouse Lifecycle Toolkit
[14] Liautaud, B.: e-Business Intelligence: Turning Information into Knowledge into Profit. Amazon.com (2000)
[15] Moss, L.T., Atre, S.: Business Intelligence Roadmap. Addison-Wesley, Reading (2003)
[16] Ślęzak, D., Wróblewski, J., Eastwood, V., Synak, P.: Rough Sets in Data Warehousing. In: Chan, C.-C., Grzymala-Busse, J.W., Ziarko, W.P. (eds.) RSCTC 2008. LNCS (LNAI), vol. 5306, pp. 505–507. Springer, Heidelberg (2008)
[17] Vercellis, C.: Business Intelligence: Data Mining and Optimization for Decision Making. Wiley Computer Publishing, Chichester (2009)
[18] Williams, S., Williams, N.: The Profit Impact of Business Intelligence. Morgan Kaufman Publishers, San Francisco (2007)

Data Warehouse Discovery Framework: The Case Study

Cas Apanowicz

IT Horizon, Canada
cas.apanowicz@it-horizon.com

Abstract. The cost of building an Enterprise Data Warehouse Environment runs usually in millions of dollars and takes years to complete. Even bigger than cost is the risk that all the design and development of the Data Warehouse and Business Intelligence Environment may not bring the result expected by the user. This was the main incentive behind author's effort of laying down the foundation for new methodology called Data Warehouse Discovery [1]. The foundation met with acceptance by some scientific groups on one hand and industry interest on the other. At that point, the author faced a major challenge. In order to get industry full acceptance as viable tool for the development and maintenance of a robust DW/BI environment, an actual implementation of the methodology in production was necessary. The DW/BI Strategy and Design Project that author was just conducting for the Canadian Federal Government was a perfect opportunity to propose and implement the methodology. This paper is presenting the conduct and results of that business case.

Keywords: Data Warehousing, Data Warehouse Discovery (DW Discovery), Business Intelligence, Benchmarking Practices, Product Selection, Evaluation Matrix, Preference Matrix, Priority Matrix, Infobright Enterprise Edition.

1 Introduction

The biggest challenge that corporations face these days is proper identification and definition of business requirements and business needs for information consumption. Due to a large number of data sources and major difficulties in communication between the business community and the IT group this becomes increasingly difficult.

The inability to properly convey all needs and requirements of the business community to the delivery group, that usually is in-house or outside IT organization - on one hand, and same difficulty of presetting and verification of the end-state solution by the IT group back to the user group on the other - create significant risk for the process of formulating and implementation of the right BI strategy. The lack of sufficient bridge between the two above groups, often leads to design and execution of the content and technological DW/BI framework that does not serve properly the organization needs. This, in turn, leads to the greatest challenge for the enterprise and creates a serious risk for the corporation as a whole.

The above risk is greatly exacerbated by the fact that the design and implementation process, is always a long - in average lasting between 2 and 3 years - period, during which, user is not expected to obtain any access to information. The success of DW/BI initiative would not be determined until its full completion, hence no guarantee that it would meet the current and future needs of the organization.

Y. Zhang et al. (Eds.): DTA/BSBT 2010, CCIS 118, pp. 155–166, 2010.

In [1], we presented the new methodology, which is aimed at mitigation of the above risk and to lower the overall cost of adopting a proper BI strategy and delivery of the DW/BI Environment. This paper outlines the details of the practical implementation and at the same time proves the methodology in production application.

The case study presented in this paper shows how to turn the theoretical foundation into practical and clear blueprints of a successful methodology, ready for any enterprise to employ, and for any DW/BI practitioner to include into the arsenal of tools and methods. The methodology has been met with significant interest in both, scientific and industry circles.

The biggest challenge for successful implementation is to create a technological framework that meets all of the requirements and do not add any substantial cost and effort to the overall BI endeavor. It was imperative, that there was no additional work and cost devoted to design, tweaking and tuning of the platform and processes that would, in effect, defeat the purpose of the method. This is the main reason we put so much consideration to measure and test the performance and amount of work needed to complete the technical framework - hence the reach amount of statistics i the case.

2 The Implementation Example

A large Canadian Government agency went through a substantial reorganization and was split into two separate agencies. One of the newly created agencies was faced the challenge of developing a new BI strategy and to create a new BI platform. The author was initially retained by the agency to assist in devising the strategy of vendor selection for the new platform. That effort was described in the research reported in [2], where the author presented his experience and findings gained during Benchmarking Project.

Subsequently, the author was mandated by the agency with the task of creating *Data Warehouse Information & Technology Design Framework*. Very early in the process, during several preliminary research and consultation with the upper management and IT architectural group the DW Discovery Framework methodology was proposed as a invaluable strategic tool to devise the BI Strategy and associated policies. With the success of the first initiative, the DW Discovery Framework has been adopted by the agency and has become a fundamental component of the DW/BI Strategy and Architecture Framework.

This paper will document the complete process and will include some interesting statistic that resulted from that work.

2.1 The DW Discovery Process

2.1.1 Analyses

The first step after project initialization and orientation was the Analysis phase. This step consisted of User Requirements Gathering and Updating followed by identifying all Source Systems Analysis and the BI framework assessment of the current environment. All agency Libraries and Knowledge Repositories were examined to identify and review all the work completed to this point. The basic components of the current BI framework were indentified and all major business subject areas, as well as the main source data were identified.

2.1.2 DW Discovery Schema Creation

Next the models and business metadata of the source system were analyzed and summarized. Based on the initial analysis a subset of the tables from the source system were selected and minimally modified. Some operation tables were highly renormalized mostly for performance reasons. This resulted in some tables not always reflecting the original source-based entities. In essence, by normalizing the initially flat tables, operation systems assign some meanings to the entities, not on the basis of BI user needs, but on the basis of physical RDBMS performance.

In those cases, we denormalized the tables to more closely resemble the original data sources and to avoid any predetermination in the importance hierarchy of the fields. In particular, since often a data element can be a fact in one business context and a dimension in the other, we wanted to avoid predetermination between facts and dimensions. The most important reason for attempting to flatten the tables into one or few big flat tables is to simplify database schema, very closely, if not completely, resembling the source system schemas. The resulting schema was comprised of a single table encompassing all the entities and all data from the source system.

2.1.3 DW Discovery Database Population

In this phase, we prepared a batch of several bulk load jobs to populate DW Discovery Framework database with source data. Understanding that both the source and target were very close, the process of creating bulk load jobs was relatively quick and with minimal effort.

All load data sets were divided into three, roughly equal-sized, batches in order to measure the load scalability as well. The load process was divided into three phases PHASE1, PHASE2, PHASE3 - each loading the one of three load data sets. This way, we could observe what, if any, effect on the load performance had the amount of data already in table. This effect is known as Load Scalability (some databases slow down significantly as the size of tables grows – the load is not scalable) [5]. The load performance was measured and is presented in the summary statistics in Fig. 2.

```
CREATE TABLE `TCR4BIG_NEW` (
   `SUROGATE_Key6`        int(11) DEFAULT NULL,
   `SUROGATE_Key5`        int(11) DEFAULT NULL,
   `SUROGATE_Key3`        int(11) DEFAULT NULL,
   `DRV_REQ_NBR`          int(11) NOT NULL,
   `AX_CLNT_NBR`          int(11) NOT NULL,
   `SBRN_PGM_ID`          char(2) COLLATE ascii_bin NOT NULL,
   `SBRN_PGM_ACNT_NBR`    int(11) NOT NULL,
   `SO_ID`                int(11) NOT NULL,
   `REQ_VERS_NBR`         int(11) NOT NULL,
   `CLNT_REQ_SCDE`        int(11) NOT NULL,
`AUDT_INS_TSTMP`          timestamp NOT NULL DEFAULT CURRENT_TIMESTAMP ON
                                    UPDATE CURRENT_TIMESTAMP,
   `AUDT_INS_DATE`        char(10) COLLATE ascii_bin NOT NULL,
   `AUDT_INS_YEAR`        int(11) NOT NULL,
   `AUDT_INS_MONTH`       int(11) NOT NULL,
   `AUDT_INS_DAY`         int(11) NOT NULL,
   `AUDT_INS_HOUR`        int(11) NOT NULL,
   `AUDT_INS_MINUTE`      int(11) NOT NULL,
   `STAT_XDT`             date NOT NULL,
   `STAT_EDT`             date NOT NULL,
   `CLNT_REQ_STAT_TCDE`   int(11) NOT NULL,
   `REQ_STAT_CAT_CDE`     int(11) NOT NULL,
   `REQ_STAT_RCDE`        char(2) COLLATE ascii_bin NOT NULL COMMENT 'lookup',
   `OP_CELL_TCDE`         char(2) COLLATE ascii_bin NOT NULL,
```

```
`AUDT_USER_ID`        char(8) COLLATE ascii_bin NOT NULL,
`AUDT_PGM_ID`         char(15) COLLATE ascii_bin NOT NULL,
`AUDT_TSTMP           timestamp NOT NULL DEFAULT '0000-00-00 00:00:00',
`AUDT_DATE`           char(10) COLLATE ascii_bin NOT NULL,
`AUDT_YEAR`           int(11) NOT NULL,
`AUDT_MONTH`          int(11) NOT NULL,
`AUDT_DAY`            int(11) NOT NULL,
`AUDT_HOUR`           int(11) NOT NULL,
`AUDT_MINUTE`         int(11) NOT NULL,
`AUDT_ACTN_CDE`       char(1) COLLATE ascii_bin NOT NULL,
`RCVR_IND`            char(1) COLLATE ascii_bin NOT NULL,
`ORIG_REQ_INS_TSTMP`  timestamp NOT NULL DEFAULT '0000-00-00 00:00:00',
`ORIG_REQ_INS_DATE`   char(10) COLLATE ascii_bin NOT NULL,
`ORIG_REQ_INS_YEAR`   int(11) NOT NULL,
`ORIG_REQ_INS_MONTH`  int(11) NOT NULL,
`ORIG_REQ_INS_DAY`    int(11) NOT NULL,
`ORIG_REQ_INS_HOUR`   int(11) NOT NULL,
`ORIG_REQ_INS_MINUTE` int(11) NOT NULL,
`INS_USER_ID`         char(8) COLLATE ascii_bin NOT NULL,
`INS_WLOC_CDE`        int(11) NOT NULL,
`CLNT_REQ_FUNC_TCDE`  char(2) COLLATE ascii_bin NOT NULL,
`EXT_CUS_CLNT_ID`     varchar(30) COLLATE ascii_bin NOT NULL,
`CLNT_SPLY_REQ_ID`    varchar(50) COLLATE ascii_bin NOT NULL,
`PAUTH_NBR`           char(15) COLLATE ascii_bin NOT NULL COMMENT 'lookup',
`PRSH_GD_IND`         char(1) COLLATE ascii_bin NOT NULL,
`DRIV_WAIT_IND`       char(1) COLLATE ascii_bin NOT NULL,
`RGSTN_SEQ_NBR`       char(4) COLLATE ascii_bin NOT NULL,
`CMNCTN_MD_CDE`       char(2) COLLATE ascii_bin NOT NULL,
`CLNT_REQ_SUBM_TCDE`  char(2) COLLATE ascii_bin NOT NULL,
`EDI_REQ_ID`          int(11) NOT NULL,
`REQ_INIT_ENT_DT`     date DEFAULT NULL,
`DCMT_BTCH_ID`        char(5) COLLATE ascii_bin NOT NULL,
`REQ_DB_LOC_CDE`      char(2) COLLATE ascii_bin NOT NULL,
`REQ_DB_LOC_DT`       date NOT NULL,
`B3_TRANS_NBR`        char(14) COLLATE ascii_bin NOT NULL,
`FOLLOWUP_IND`        char(1) COLLATE ascii_bin NOT NULL,
`CMRC_INVC_NBR`       varchar(50) COLLATE ascii_bin NOT NULL,
`OTR_REF_TXT`         varchar(160) COLLATE ascii_bin NOT NULL,
`TRSHP_CNTRY_CDE`     char(2) COLLATE ascii_bin NOT NULL COMMENT 'lookup',
`SALE_COND_TXT`       varchar(50) COLLATE ascii_bin NOT NULL,
`DSHPMT_PLC_TXT`      varchar(50) COLLATE ascii_bin NOT NULL,
`INVC_TAMT`           decimal(15,2) NOT NULL,
`INCL_TRPRT_AMT`      decimal(13,2) NOT NULL,
`PROD_SPLY_IND`       char(1) COLLATE ascii_bin NOT NULL,
`ROYL_PAYMT_IND`      char(1) COLLATE ascii_bin NOT NULL,
`EXCL_EXPRT_PAC_AMT`  decimal(13,2) NOT NULL,
`EXCL_COMSN_AMT`      decimal(13,2) NOT NULL,
`EXCL_TRPRT_AMT`      decimal(13,2) NOT NULL,
`INCL_EXPRT_PAC_AMT`  decimal(13,2) NOT NULL,
`INCL_ASBLY_AMT`      decimal(13,2) NOT NULL,
`ORIG_CNTRY_CDE`      char(2) COLLATE ascii_bin NOT NULL COMMENT 'lookup',
`EXPRT_CNTRY_CDE`     char(2) COLLATE ascii_bin NOT NULL COMMENT 'lookup',
`DSHPMT_DT`           date DEFAULT NULL,
`CMRC_INVC_PREP_DT`   date DEFAULT NULL,
`PAYMT_TRM_TXT`       varchar(50) COLLATE ascii_bin NOT NULL,
`TAMT_CURCY_CDE`      char(3) COLLATE ascii_bin NOT NULL COMMENT 'lookup',
`EXPRT_STE_CDE`       char(2) COLLATE ascii_bin NOT NULL COMMENT 'lookup',
`ORIG_STE_CDE`        char(2) COLLATE ascii_bin NOT NULL COMMENT 'lookup',
`INVC_PG_NBR`         int(11) NOT NULL,
`INVC_LN_NBR`         int(11) NOT NULL,
`GD_QTY`              decimal(18,4) NOT NULL,
`GD_QTY_UOM_CDE`      char(3) COLLATE ascii_bin NOT NULL COMMENT 'lookup',
`GD_TTL_PRC`          decimal(15,4) DEFAULT NULL,
`GD_UNT_PRC`          decimal(15,4) NOT NULL,
```

```
  `PKG_CNT`              decimal(11,2) NOT NULL,
  `PKG_UOM_CDE`          char(3) COLLATE ascii_bin NOT NULL COMMENT 'lookup',
  `GD_CMRC_DESC`         char(150) COLLATE ascii_bin NOT NULL,
  `HS_CHAP_NBR`          char(2) COLLATE ascii_bin NOT NULL,
  `HS_HD_SFX`            char(2) COLLATE ascii_bin NOT NULL,
  `HS_SBHD_SFX`          char(2) COLLATE ascii_bin NOT NULL,
  `TRF_SFX`              char(2) COLLATE ascii_bin NOT NULL,
  `STAT_SFX`             char(2) COLLATE ascii_bin NOT NULL,
  `GD_VAL`               decimal(13,2) DEFAULT NULL,
  `GD_WT`                decimal(13,4) NOT NULL,
  `GD_WT_UOM_CDE`        char(3) COLLATE ascii_bin NOT NULL COMMENT 'lookup',
  `TTL_PRC_CURCY_CDE`    char(3) COLLATE ascii_bin NOT NULL COMMENT 'lookup',
  `AGR_CMDTY_CLS_CDE`    char(6) COLLATE ascii_bin NOT NULL COMMENT 'lookup',
  `AGR_CNTRY_ORIG_CDE`   char(2) COLLATE ascii_bin NOT NULL,
  `AGR_ORIG_STE_CDE`     char(2) COLLATE ascii_bin NOT NULL COMMENT 'lookup',
  `AGR_IMPRT_RQRMT_ID`   int(11) NOT NULL,
  `AGR_RQRMT_VERS_NBR`   int(11) NOT NULL,
  `AGR_FNL_PROV_CDE`     char(2) COLLATE ascii_bin NOT NULL COMMENT 'lookup',
  `AGR_CMDTY_USE_CDE`    char(3) COLLATE ascii_bin NOT NULL COMMENT 'lookup',
  `AGR_COND_CLS_CDE`     char(3) COLLATE ascii_bin NOT NULL COMMENT 'lookup',
  `XPRT_UOM_CD`          char(2) COLLATE ascii_bin NOT NULL,
  `XPRT_VIN_HIN`         varchar(50) COLLATE ascii_bin NOT NULL
  `EXPRT_CNTRY_EDesc`    varchar(30) COLLATE ascii_bin NOT NULL COMMENT
                                    'lookup',
  `ORIG_CNTRY_EDesc`     varchar(30) COLLATE ascii_bin NOT NULL COMMENT
                                    'lookup',
  `ORIG_STE_Edesc`       varchar(30) COLLATE ascii_bin NOT NULL COMMENT
                                    'lookup',
  `EXPRT_STE_Edesc`      varchar(30) COLLATE ascii_bin NOT NULL COMMENT
                                    'lookup',
) ENGINE=BRIGHTHOUSE DEFAULT        charSET=ascii COLLATE=ascii_bin;
```

Fig. 1. The Single Target Table containing all entities and all data

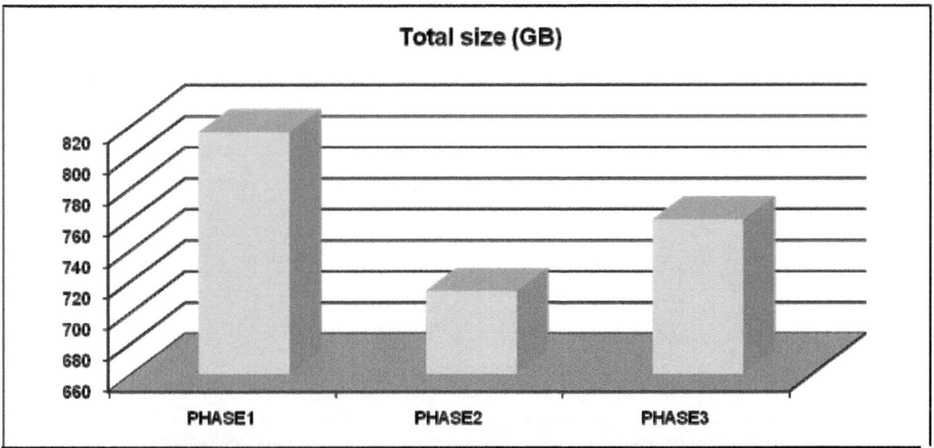

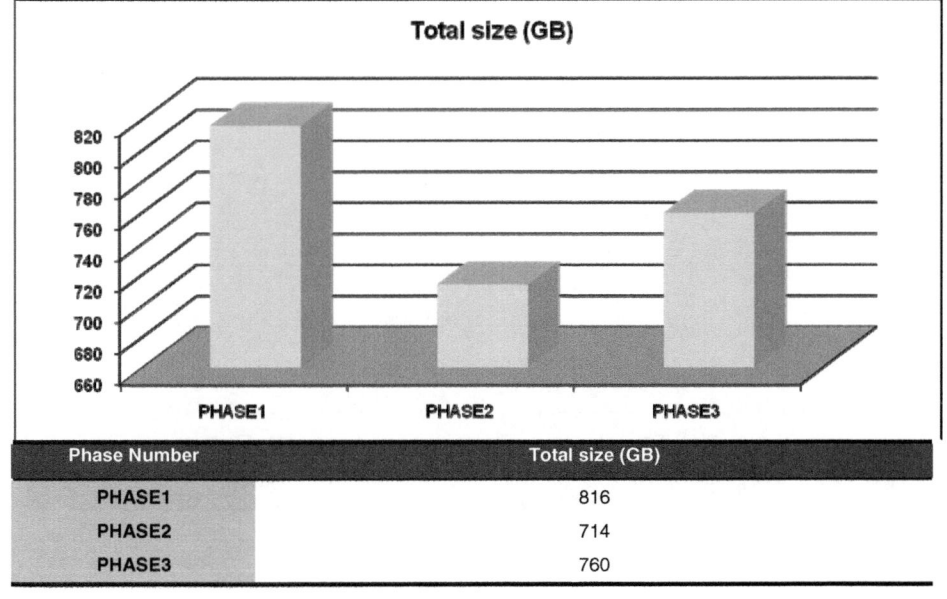

Phase Number	Total size (GB)
PHASE1	816
PHASE2	714
PHASE3	760

Fig. 2a. Load Sets Size by PHASES

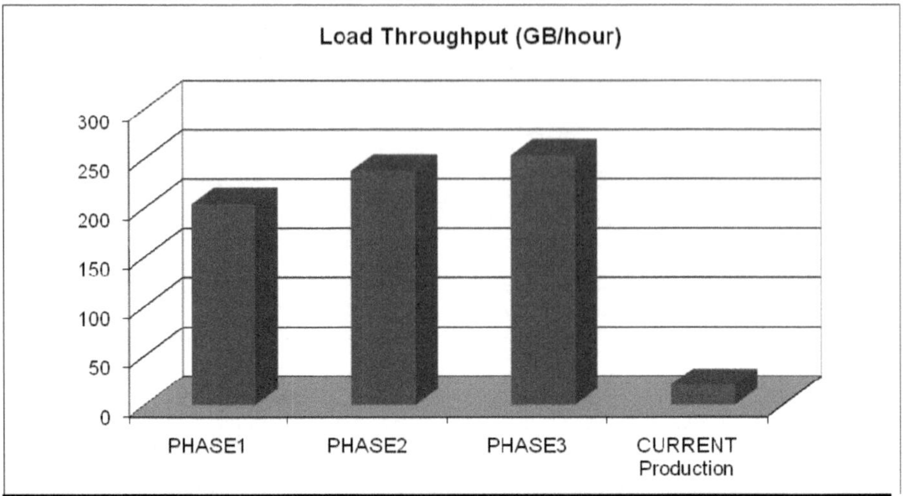

Phase Number	Total size (GB)
PHASE1	816
PHASE2	714
PHASE3	760
Current Production	20

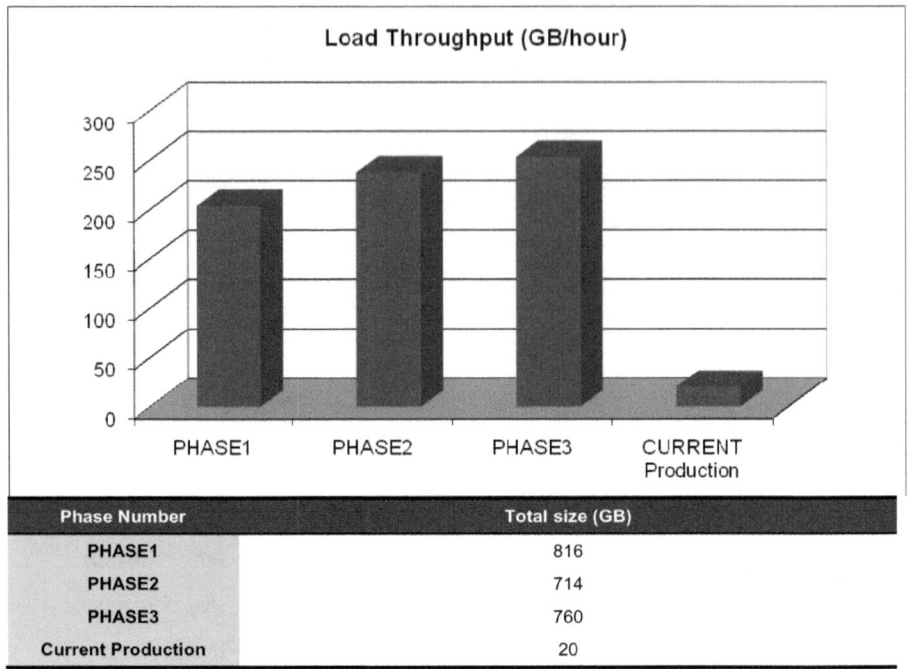

Fig. 2b. Load Throughput (GB/Hour) By PHASES

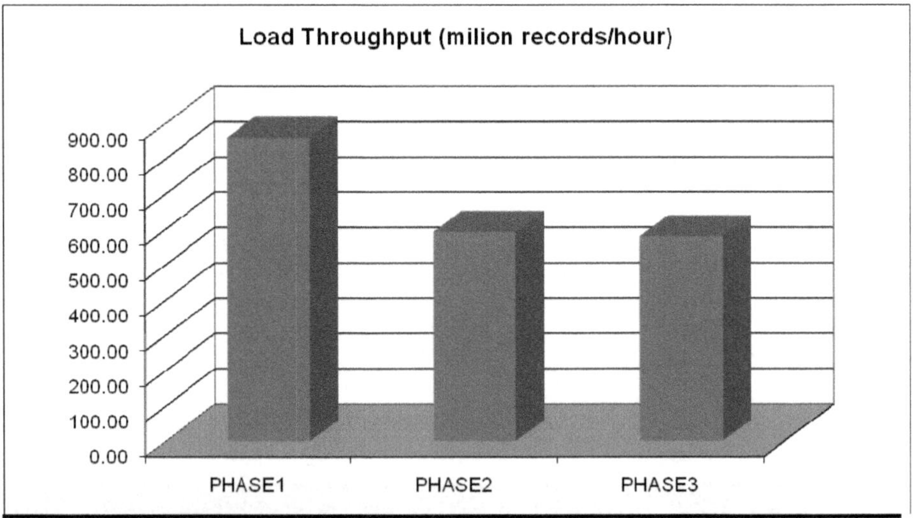

Phase Number	Load Throughput (million records/hour)
PHASE1	859
PHASE2	758
PHASE3	830

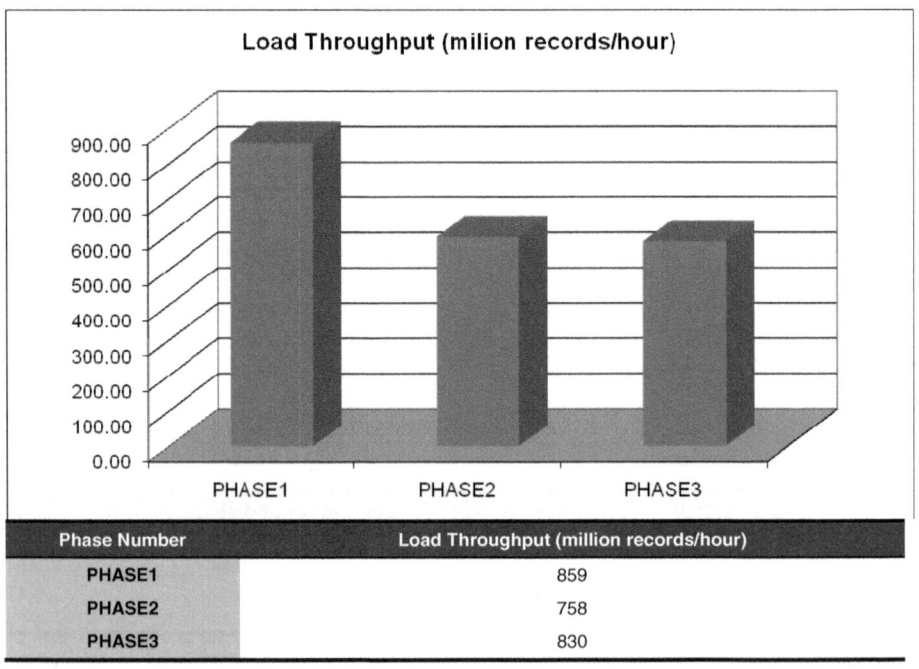

Fig. 2c. Load Throughput (Millions of Records/hour) By PHASES

It is worth of noticing the very significance of the platform used in this project. Some leading databases on the market have very good load performance and scalability. However this project was the first time when author experienced performance better than linear: NOTE in Fig 2b. The load of PHASE3 is faster than PHASE1 and 2. Apart from the scalability, the very throughput of the Infobright [7] engine seems to be the highest the author come across. Please also notice that the load performance of the Discovery environment was significantly higher than current production loads.

2.1.4 DW Discovery Interactive Querying

The next step involved examination and review of all user requirements in order to generate SQL queries that simulate future BI Queries and Reports. The extensive process of collecting current production queries and reports, ad-hoc and pre-defined, was conducted and full set of queries/reports were presented to the business user for verification. During real time dialog with the business users all the queries were presented and modified to reflect current and future needs of the BI users.

This was very crucial step for it allowed a fully interactive and real-time interaction with actual BI users. We were able to collect and create a first fully representative version of queries and reports before single line of code or single design artifacts of the final Data Warehouse was started. Yet, we were able to run all these queries and reports and present the result to the users in an interactive fashion.

BI User Domain	Line of Business	# of Queries/Day
Business Information Users	Branch 1 Reports and Queries	3,867
	Branch 2 Reports and Queries	6,177
Information Analysts	Tactical Analysts Queries	21,000
Statistical Analysts	Strategic Intelligence Queries	450
Total Non-operational		**31,494**

Fig. 3. Query Classification and Volumetrics

Query Type	% per Day
Simple	30%
Moderate	50%
Complex	20%

Fig. 4. Query Classification

In return, the users were able to see the results and made correction or verification or postulate new requirements, before any effort was made to design and create DW. At this point, user could make any necessary modifications or enhancements to see how it would affect the final results. In the end, we were able to present the final design without the design itself, understanding that is completely virtual.

This interactive process was conducted for as long as all the user's needs were meet and verified, and more importantly, presented in the form of actual reports. Once again, all this was completed before any line of code or single piece of design was done. At this point all facts and dimensions were identified and due to the interactive process the actual performance criteria - such as number of concurrent queries, the query response, latency required and the refresh period - were identified, established and user verified. Lastly full classification of query/reports class was determined. The full statistics and summary of requirements are presented in the following section.

2.1.5 DW Discovery Technology and Query Performance

The most important factor when using DW Discovery methodology is to select a technology platform that will not add any significant cost while providing flexibility and the level of performance that would make the methodology both effective and feasible. If the technology is not properly selected, than the time and effort necessary to implement the DW Discovery framework may defeat the purpose of using it at the first place. Thus, time of implementation and performance without any physical design is crucial. We used Infobright engine on off-shelf Intel 4 quad processor and ray of SAN disks. The initial set up was very minimal (approximately one week) because of utilization of the infrastructure from Benchmark Project presented in [2]. Had the environment been set up without prior Benchmark Project, the total time of setting up and building and loading the database would be one month. All the queries run during the process were performing seconds while the same production queries run in single modes for several hours. Fig. 5a/b contains some performance statistics. To run the process mimicking the future production workload, we created script that would run for 24 hours submitting all the collected queries. This script would have parameter indicating the number of parallel queries to be run simultaneously.

There were 26 individual session running each time employing different simultaneous queries varying from 10 to 50 queries. The controlling script would record execution time for all queries so we can later plot the scalability graph. Fig.5a represents such a graph. The axis represents the number of queries running simultaneously and axis Y the total time in seconds needed to complete them. We plotted line X=Y indicating the linear performance: if 1 query runs n sec, 2 queries run for 2*n sec, 3 - 3*n sec and so on we would have the top graph. Linear scalability was the desirable threshold. To our surprise, the scalability was extremely good (see the lower line on the graph) and was rapidly improving as the number of concurrent queries increased (increasing the number of queries by factor of 5 increased the total elapsed time just little over 2 times. Note that some of the queries run on other (production) platform for hours.

To better understand performance and scalability, we plotted another graph. Fig. 5b represents the percentage of time used by a single query as we increased the number of concurrent runs. We see that when we run 10 of them it took 400sec to complete, which translates to 40sec per query or 10% of total elapsed time. But if we ran 50 simultaneous queries, each requires only 5sec or less than 2% of total elapsed time.

After finalizing composition of queries and reports, the second round of reports was executed to present them to the user. Fig. 6 illustrates the modified queries that were confirmed by business users in the second iteration of DW Discovery review and performance was again compared with requirements. Note that this time the scenario included higher scalability running more simultaneous queries in shorter time.

2.1.6 DW Discovery Hardware Consideration

The cornerstone of the DW Discovery Framework is the absence of any physical design or platform tuning. This was the reason that the least expensive off-shelf server 4x4 CPU with 132GB RAM and an effective 10TB SAN attachment were selected. The Red Hat Linux 5 enterprise was installed to further reduce costs. The above installation was only fraction of other alternatives available, yet was much capacity that the DW Discovery required. Another driving force in selecting this configuration was already acquired for other project (Benchmark Project described in [2]).

In order to show that the methodology would equally successful on much moderated platform, several tests were run with a different settings using smaller amount of memory and CPU. Fig. 7 presents the performance with different memory settings.

The axis X represents number of concurrent queries (from 3 to 63). The major advantage of the Infobright server is no need of settings of any sort of physical parameters. The exception are two memory usage parameters: ServerMainHeapSize and ServerCpressedHeapSize. Using these parameters we tested the effect of overall CPU usage on the overall query performance.

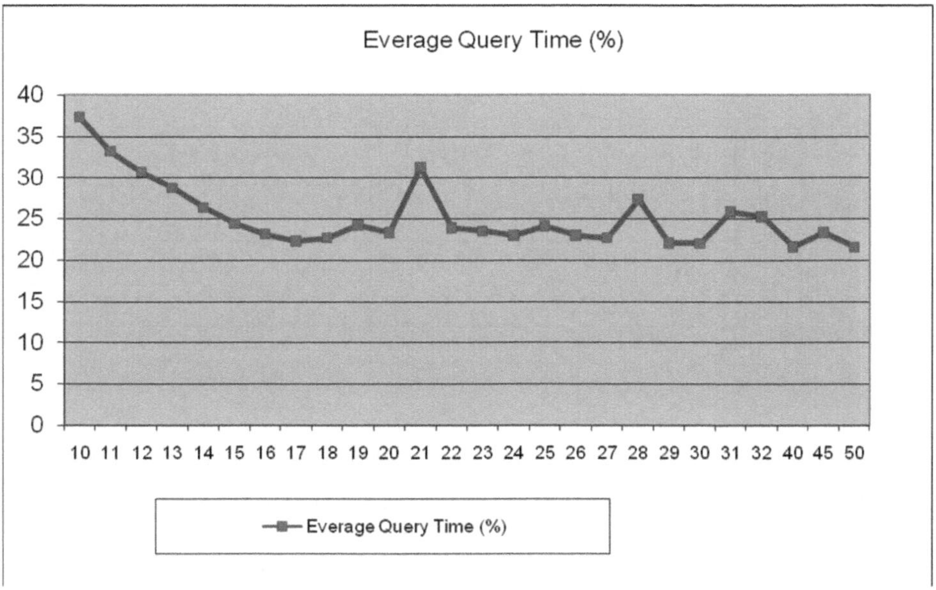

Fig. 5a. DW Discovery Query Performance – Sub-linear performance

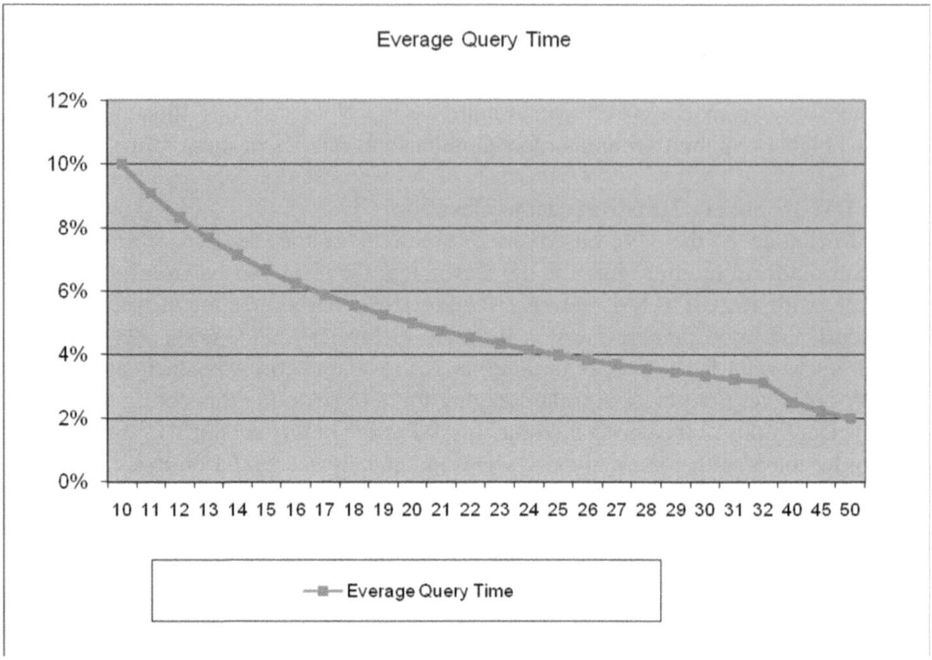

Fig. 5b. DW Discovery Query Performance – Sub-linear Scalability

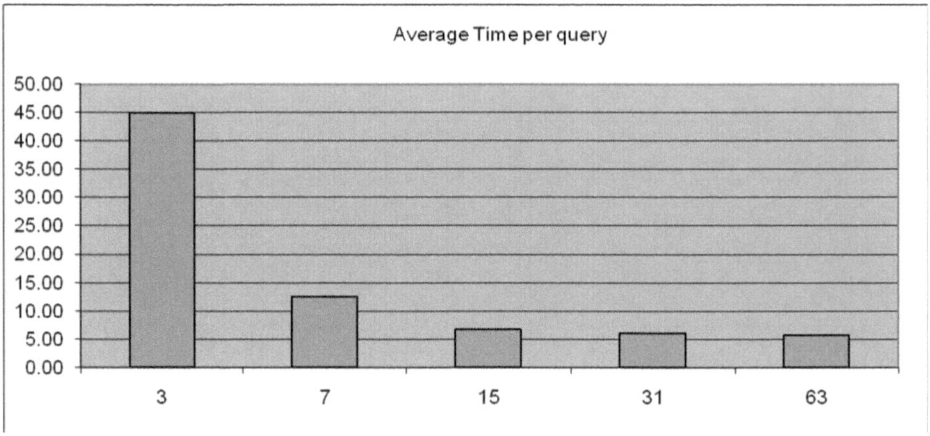

Fig. 6. DW Discovery Query Performance second iteration - Sub-linear Scalability

The conclusion of this test was two-folded: First, it proved that the choice of technology for the DW Discovery was the most optimal and secondly, The DW Discovery method can be implemented on very inexpensive off-shelf server.

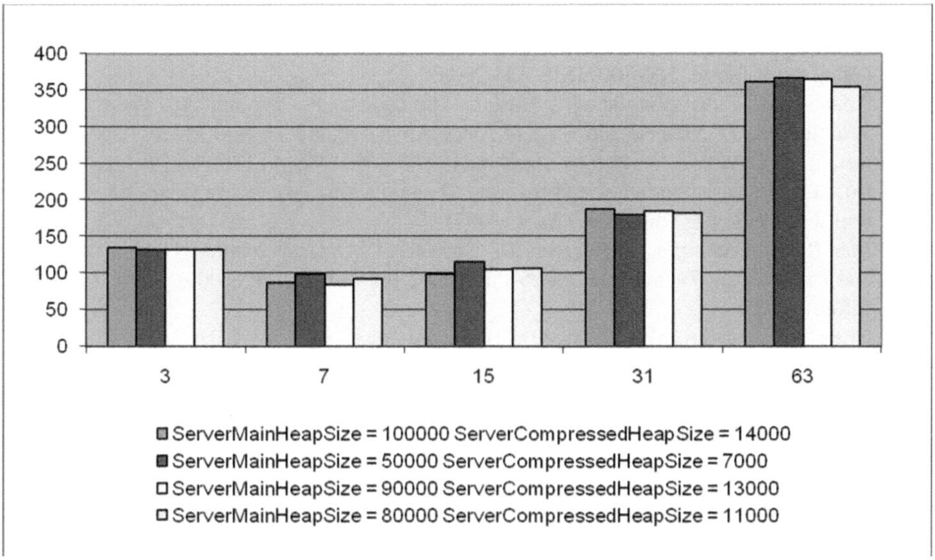

Fig. 7. DW Discovery - Neglectable effect of CPU setting on the performance

3 Conclusions

The above presented DW/BI efforts are usually considered very costly and time consuming processes and any design and architectural inaccuracies have very profound and long lasting adverse effects. The ability to mitigate the risk of delivering the "wrong" system is extremely crucial and should be a key aspect of any good BI strategy. The Case Study presented in this paper clearly proves the great potential that can be derived from the DW Discovery Framework methodology as well as the ease and minimal costs associated with its implementation. The government agency, who adopted DW Discovery Framework into its strategy, has since reduced the budget by millions of dollars. There were many factors playing the role in the savings, but it is clear that proposed methodology and its implementation was a significant component in this success. The DW Discovery Framework methodology has been proven to reduce risk by providing the view and interaction with the Data Warehouse before a single line of code is written or single design artifacts completed.

The success of presented approach can be encouraging for companies to mitigate risks and reduce cost. At the same time it can be a big incentive for the experienced practitioners to include the techniques and guidelines of the DW Discovery methodology into their own practice and be able to offer great value to their clients.

References

[1] Apanowicz, C.: Data Warehouse Discovery Framework - The Foundation. In: Proc. of DTA 2010, Springer, Heidelberg (2010)

[2] Apanowicz, C.: Data Warehousing and Business Intelligence: Benchmark Project for the Platform Selection. In: Proc. of DTA 2009, Springer, Heidelberg (2009)

[3] Apanowicz, C., Eastwood, V., Ślęzak, D., Synak, P., Wojna, A., Wojnarski, M., Wróblewski, J.: Method and System for Data Compression in a Relational Database. US Patent Application, 2008/0071818 A1 (2008)
[4] Apanowicz, C., Eastwood, V., Ślęzak, D., Synak, P., Wojna, A., Wojnarski, M., Wróblewski, J.: Method and System for Storing, Organizing and Processing Data in a Relational Database. US Patent Application, 2008/0071748 A1 (2008)
[5] Buretta, M.: Data Replication, Tools and Techniques for Managing Distributed Information. John Wiley & Sons, Chichester (1997)
[6] Imhoff, C., Galemmo, N., Geiger, J.G.: Mastering Data Warehouse Design: Relational and Dimensional Techniques. Wiley Publishing Inc., Chichester (2009)
[7] Infobright Inc.,
`http://www.infobright.com,http://www.infobright.org`
`http://en.wikipedia.org/wiki/Infobright`
[8] Inmon, W.H., Imhoff, C., Sousa, R.: Corporate Information Factory. Wiley Computer Publishing, Chichester (2001)
[9] Inmon, W.H., O'Neil, B., Fryman, L.: Business Metadata: Capturing Enterprise Knowledge. Morgan Kaufman Publishers, San Francisco (2008)
[10] Inmon, W.H., Strauss, D., Neushloss, G.: DW 2.0: The Architecture for the Next Generation of Data Warehousing. Morgan Kaufman Publishers, San Francisco (2008)
[11] Kimball, R., Ross, M., Thornthwaite, W., Mundy, J., Becker, B.: The Data Warehouse Lifecycle Toolkit
[12] Moss, L.T., Atre, S.: Business Intelligence Roadmap. Addison-Wesley, Reading (2003)

An Elitist-Ant System for Solving the Post-Enrolment Course Timetabling Problem

Ghaith M. Jaradat and Masri Ayob

Data Mining and Optimization Research Group, Centre of Artificial Intelligence Technology, The National University of Malaysia, Faculty of Information Science and Technology, 43600 UKM Bangi, Selangor, Malaysia
gm45973@ftsm.ukm.my, masri@ftsm.ukm.my

Abstract. Ant System algorithms are nature-inspired population-based meta-heuristics derived from the field of swarm intelligence. Seemingly, the ant system has a lack of search diversity control since it has only a global pheromone update that intensifies the search. Hence, one or more assistant mechanisms are required to strengthen the search of the ant system. Therefore, we propose, in this study, an elitist-ant system to strike a balance between search diversity and intensification while maintaining the quality of solutions. This process is achieved by employing two diversification and intensification mechanisms to assist both pheromone evaporation and elite pheromone updating, in order to gain a good control over the search exploration and exploitation. The diversification mechanism is employed to avoid early convergence, whilst the intensification mechanism is employed to exploore the neighbors of a solution more effectively. In this paper, we test our algorithm on post-enrolment course timetabling problem. Experimental results show that our algorithm produces good quality solutions and outperforms some results reported in the literature (with regards to Socha's instances) including other ant system algorithms. Therefore, we can conclude that our elitist-ant system has performed an efficient problem's specific knowledge exploitation, and an effective guided search exploration to obtain better quality solutions.

Keywords: Elitist-ant system, intensification and diversification, post-enrolment course timetabling problem.

1 Introduction

The course timetabling problems mainly comprise of assigning a set of courses, students and lecturers to a fixed number of timeslots and rooms in a week, while satisfying some constraints (Petrovic and Burke; 2004). The university course timetabling problems are considered as NP-hard problems (Even *et al.*; 1976), which are difficult to solve for optimality.

The two types of constraints are hard and soft ones. In order to produce a feasible timetable, all of the hard constraints must be satisfied, whereas the violation of the soft constraints must be minimized in order to produce a good quality timetable. Each violation of soft constraints will incur a penalty cost, where lower penalty values indicate good quality solutions. The ultimate goal of this work is to construct a feasible timetable and to satisfy soft constraints as much as possible.

Y. Zhang et al. (Eds.): DTA/BSBT 2010, CCIS 118, pp. 167–176, 2010.

In this paper, we tested our algorithm on benchmark post-enrolment course timetabling instances introduced by Socha *et al.* (2002) without consideration to lecturer assignments. The hard constraints for Socha's instances are listed as follows:

H1: No student attends more than one course at the same time;
H2: The room is big enough for all the attending students and satisfies all the features required by the course;
H3: Only one course can be scheduled in each room at any timeslot;

A quality of timetable is measured by penalising equally each violation of the following soft constraints (i.e. penalty cost=1 for each violation):

S1: A student should have no classes in the last slot of the day;
S2: A student should have no more than two classes consecutively;
S3: A student should have no single class on a day.

More information about the problem's instances and the problem formulation can be found in (Socha *et al.*; 2002).

1.1 Related Work

The Ant Colony Optimization (ACO) is a family of population-based foraging behaviour metaheuristics, first proposed by Dorigo *et al.* (1991), namely as the Ant Systems (AS). The AS is inspired by the foraging behaviour of real ants that communicate indirectly via the distribution and dynamic change of information known as the pheromone trails. The weight of these trails reflects the collective search experience, which will be exploited by future ants in their attempts to solve a given problem instance more effectively. The AS has been successfully applied to various combinatorial optimization problems as it evolves gradually to a higher level of development and extensions. Examples on successful works that inspired our own effort include Gambardella *et al.* (1999) who developed a max-min ant system, a powerful extension of the AS and Elitist-AS to solve the quadratic assignment problem. An accumulated experience ant colony was developed and applied to the travelling salesman problem by Montgomery and Randall (2003). Finally, an ant colouring system was adopted and applied to the examination timetabling problem by Eley (2007).

According to Blum and Roli (2008), the power of population-based methods is depending on the capability of recombining solutions to obtain new ones. They also stated that, the evolutionary computational algorithms (e.g. scatter search and genetic algorithms) perform explicit re-combinations to generate new solutions by one or more re-combination operators. That is a crossover, for instance, that generates a new solution from two other good quality (and/or diverse) solutions (a structured combination) which provides useful information about the structure or location of the global optima solution. However, ACO re-combinations are implicit because new solutions are generated by using a distribution over the search space, which is a function of earlier populations (i.e. pheromone updating). They claimed that the implicit recombination enables the search process to perform a guided sampling of the search space (i.e. pheromone updating, and/or roulette wheel selection). Nonetheless, this usually results in a rugged exploration (a landscape with large average fitness differences which will be usually characterized by many local optima) that can effectively find promising areas of the search space.

Recently, various ACO approaches have been applied to solve course timetabling problem. Some of the works using basic ACO approach applied on Socha's instances are as follows: the max-min ant system and ant colony system applied in Socha *et al.* (2002; 2003), an ant colony system in Rossi-Doria *et al.* (2003), an ant system-based behaviour in Ejaz and Javed (2007), and an ant system in Mayer *et al.* (2008).

Socha *et al.* (2002; 2003) and Rossi-Doria *et al.* (2003) applied a Max-Min Ant System and an Ant Colony System with a construction graph method for solving the course timetabling problem. They compared these approaches against other metaheuristics, genetic algorithm, tabu search, simulated annealing, and iterated local search. The ant system was ranked third in terms of feasibility and quality after the tabu search and iterated local search. Their ACO algorithm is able to produce good quality results. Socha (2003) discussed about how to fine-tune and determine the best suitable parameters in a certain stage during the run time in solving the problem. That is, variable parameters are modified depending on the possible solution and a predefined sequence of values or adaptively changed due to the occurrence of a certain state of the algorithm.

A Die Hard Co-Operative Ant Behaviour Approach was introduced by Ejaz and Javed (2007). They produced higher quality solutions than the previous Ant Colony methods (Socha *et al.*; 2002, 2003, Rossi-Doria *et al.*; 2003) for most of the instances, by enforcing a group of ants to walk through the desired path toward a feasible and good quality solution. This is done by dispatching more ants to the same visited path to help others if an assignment is unable to be placed into the solution after trying to place it for a certain number of attempts.

Mayer *et al.* (2008) developed an ant system that utilizes two pheromone matrices representing possible timeslots and rooms for a course to be assigned. This pheromone model eases the selection of assignments (heuristics). It is designed in contrast to the one used in Socha *et al.* (2003), where, a course is assigned to a timeslot first then a matching algorithm handles the assignment of a course-timeslot pair into a feasible room. Socha's model (Socha *et al.*; 2003) works well when considering similarities in solution components to maintain good search diversity, since it is easy to measure the similarity of solutions by counting the same timeslots occupied by the same courses. Mayer's model work is mainly based on the roulette wheel selection procedure to update the pheromone (search experience) which increases the diversity of the search by providing equal probabilities of exploring different areas of the search space. Basically, Mayer's model manipulates both course-timeslot and course-room assignments in two matrices at one time in the solution construction phase. This model proved to be quite flexible and better than Socha's model, where both assignments (matrices) are manipulated individually in the construction phase, course-timeslot assignment first followed by the course-room assignment using a matching algorithm to select the suitable room for the pair course-timeslot.

In Gambardella *et al.* (1999), two mechanisms (diversification and intensification) were employed in a hybrid ant colony system to maintain a balance between diversity and quality of the search, which proved to be effective in producing good quality results. These diversification and intensification mechanisms are adopted and used in our work in order to assist the pheromone evaporation in controlling the search diversity and maintaining the quality of the search. Subsequently, Bernardino *et al.* (2010) employed the same mechanisms in their hybrid ant colony system, but they claimed that a diversification mechanism has no significant impact on the search efficiency due to the local search used in their work.

Hence, in this study, we present a variant of ACOs known as the Elitist-Ant System (similar to Mayer *et al.*; 2008) and we incorporated it with an Iterated Local Search routine and two intensification and diversification controlling mechanisms (from Gambardella *et al.*; 1999). We utilize the pheromone model in Mayer *et al.* (2008) for exploiting the problem-specific knowledge effectively and rapidly in the construction phase, and to enhance the diversity of constructing new solutions and the exploration of solution's neighbourhoods in the improvement phase.

2 The Algorithm

Generally, Gambardella *et al.* (1999) transposed the real ants' behaviour into a hybrid ant colony system algorithm by making an analogy between the real ants search and the set of feasible solutions to the problem; the amount of food in a source and the fitness function; and the pheromone trail and an adaptive memory. They used pheromone trails for exploration and exploitation. In the former, the pheromone chooses the component used to construct a solution probabilistically. In the latter, the pheromone chooses the component that maximizes a mixture of pheromone trail values and partial objective function evaluations.

In this work, we extended an Elitist-ant system algorithm proposed by Gambardella *et al.* (1999), to solve the course timetabling problem, which utilizes the intensification and diversification mechanisms as in Gambardella *et al.* (1999). We also utilized an external memory with a variable size as in (Montgomery and Randall; 2003) to store the elite solutions found in each iteration in order to determine whether to update the pheromone trail or restart the search with new population generated from elite solutions. By employing both mechanisms, we are able to manage the rugged exploration caused by implicit re-combinations. The pseudo code of our Elitist-Ant System algorithm is shown in Figure 1.

In this work, we used the same formulas presented in Mayer *et al.*'s (2008) work regarding the pheromone model, the pheromone proportional sampling, pheromone trails matrix update, pheromone evaporation, and the pheromone update rules. For further information and description of the other pheromone model used in solving the course timetabling problem, kindly refer to Socha *et al.* (2002, 2003), Socha (2003), Rossi-Doria *et al.* (2003), and Mayer *et al.* (2008).

The algorithm starts by initializing all parameters as in Table 2 (*Step 1*). Then (*Step 2*), each ant constructs solution from scratch (empty timetables) using roulette wheel selection (as in Mayer *et al.*; 2008) guided by two pheromone trail matrices ($\boldsymbol{\tau}_j^{\,t}$, $\boldsymbol{\tau}_j^{\,r}$) which are represented as the lists of available timeslots and rooms for a course (refer to Mayer *et al.*; 2008). We use a combination of least saturation degree and largest weighted degree heuristics to represent an ant for constructing a solution (recommended by Qu *et al.*; 2009). In other words, the unscheduled courses are ordered based on the largest number of enrolments and the least number of available timeslots. During the construction phase, we represent pheromone matrices $\boldsymbol{\tau}_j^{\,t}$ and $\boldsymbol{\tau}_j^{\,r}$ as the weight of the preferable (least penalty cost) assignment of course j into timeslot p, and course j into room k based on history of previous assignments (i.e. experience of previous ants). As for the improvement phase, we represent pheromone as the weight of the preferable (significant permutation improvement) neighbourhood structures sequence of a solution (intensification mechanism).

```
Step 1: Initialization phase
While StoppingCriterion is not met do
    Step 2: Construction phase
        for each ant //solution construction
            Assign all courses into feasible timeslots and rooms using roulette wheel selection
                mechanism guided by pheromone information
        end for
    Step 3: Improvement phase
        while non-improvement criterion is not met, do //max number of stagnation iterations n_stagiter
            Locally improve each constructed solution // iterated local search routine
            Update size and content of external memory // update size & contents
        end while
    Step 4: Intensification phase
        Apply intensification mechanism
    Step 5: Global pheromone update phase
        Update pheromone trails for assignments appearing in solution
    Step 6: Diversification phase
        Evaporate pheromone matrix // diversity control
        Apply diversification mechanism
End while
Step 7: Return best ant // best solution
```

Fig. 1. The pseudo code of our Elitist-AS algorithm

Once all ant solutions are feasible, we will improve them using an iterated local search (ILS) (as in Rossi-Doria *et al.*; 2003) guided by the pheromone trail matrix (*Step 3*). The ILS is simply implemented five types of perturbation moves (see Table 1), where each move is applied for a predetermined number of iterations (we use 100 in this work) to determine the strength of the perturbation (by increasing or decreasing pheromone values). We use simple descent heuristic (accept first improvement). A new solution is accepted if it is better than the current solution. After applying the local search, we update the external memory's contents and size containing elite solutions found (*Step 3*) so far as in Mayer *et al.* (2008). An elite solution is a solution that has the best fitness value and the least number of unscheduled courses (if any). The external memory stores the elite solutions found. It is used to maintain good search experience information in order to guide the search efficiently. It is also used in constructing new solutions in successive iterations from elite solutions (by performing some perturbations to elite solutions) rather than constructing them from scratch. After each iteration, ant with the lowest number of unscheduled courses (if any) and a better than average fitness values-score will add certain amount of pheromone to guide the subsequent ant searching for better solution.

In (*Step 4*), the intensification mechanism is employed to explore the neighbour of good solutions more effectively after the best solution found so far has been improved. This is done using a random descent heuristic, where some neighbours from all neighbourhoods (see Table 1) of a solution are generated and the best one is chosen until no more improvement is possible. Then all ants start their next iteration with the best permutation performed to construct new solutions from previous ones (as in *Step 2* but by performing perturbations to elite solutions rather than constructing new solutions from scratch).

Only the best ant (elite solution) updates the pheromone trails matrix (global update) (*Step 5*). In addition, the best performed neighbour (that results an elite solution) of a best solution in the current iteration, will be used as the starting permutation of solutions in the next iteration in the improvement phase. This is to provide further intensification around promising regions when constructing new solutions. In this work, we use five neighbourhood structures as illustrated in Table 1. Neighbourhood structures are simply explored using a simple descent heuristic (*Step 3*). At each iteration, one neighbourhood (selected randomly) is employed. If there is an improvement, the neighbourhood will be employed again in successive iterations until no improvement, then a different randomly selected neighbourhood (rather than the previous one) is employed. This step proceeds for a predetermined number of iterations.

Table 1. Problem specific neighborhoods

N_s	Description
N_1	Move a randomly selected course to a random feasible room and timeslot (Socha; 2003).
N_2	Swap timeslots and rooms of two randomly selected courses (with maintaining feasibility) (Socha; 2003).
N_3	Randomly select two timeslots and rooms and swap all courses between these timeslots and rooms so that they are still in feasible timeslot and room pair where the course will not violate any hard constraints (Socha; 2003).
N_4	Select randomly two timeslots, say t_i and t_j (where $j>i$) where the timeslots are ordered t_1, t_2, .., t_{45}. Take all courses in t_i and assign them in t_j. Move all courses in t_j and assign them in t_{j-1}. Then assign all courses that were in t_{j-1} to t_{j-2} and so on until all courses that were in t_{i+1} are assigned to t_i and terminate the process (Alvarez-Valdes *et al.* 2002).
N_5	Move the highest penalty cost course from a random 10% selection of the courses to a random feasible timeslot (Abdullah *et al.*; 2007a).

The diversification mechanism (*Step 6*), also called pheromone evaporation, is employed after performing a predefined number of non-improvement iterations of the best solution found so far in the local search routine. This helps avoid early algorithm convergence. The mechanism periodically (the diversification mechanism is employed each time) erases all the pheromone trails. We re-initialize pheromone values once the intensification mechanism failed to improve all ant solutions. In other words, if all the neighbourhoods in Table 1 are explored and no solution improvement is done, then the pheromone trails must be erased to track a new path toward better quality solutions. Each ant will generate a new solution from the elite solution in the external memory for all ants (except the one with best solution found so far) by performing some perturbations to the elite solutions. The whole process of our algorithm is repeated until the stopping criterion is met; either the best solution has been found or the number of iterations reached its limit.

3 Experimental Results

As recommended by Mayer *et al.* (2008) and our preliminary experiments, we have set our parameters as shown in Table 2 below.

In this work, we tested our algorithm on well known benchmark post-enrolment course timetabling instances introduced by Socha et al. (2002).

We ran our algorithm 25 times on each instance for runtime ranges between 30 seconds and 12 hours for each run depending on the instance size and complexity to obtain the possible global optimum solution (a much longer time than the one allowed for the competition). The experiments were performed on Intel Pentium Core2 Duo 2.16 GHz processor, 2GB RAM, and implemented in Java NetBeans IDE v 6.8.

Table 2. Parameters settings of our algorithm

Parameter	Description and Value
n_{ant}	number of dispatched ants (solutions) per iteration =20
n_{iter}	number of Iterations= 100,000
$n_{stagiter}$	number of non-improvement iterations =100
τ^r and τ^t	pheromone matrices initial values of rooms and timeslots =0.5
ρ	pheromone evaporation rate =0.25 , $\rho \in [0,1]$
α	importance of pheromone controlling (exploration) ratio =1.0, $\alpha \in [1.0,1.1]$
ψ	importance of soft constraints (penalty) =0.3
N_s	number of employed neighborhood structures per solution =5
k_m	initial external memory size =5 (elite solutions)
Elitism	use the best ant (for each iteration) to update global pheromone.

Table 3 shows the results obtained from our AS algorithm based on our parameters presented in Table 2, compared to other ACO approaches, population-based approaches and the best known results obtained by other methodologies applied over Socha's test instances. Best results are presented in bold type. The statistical readings are based on the following performance indicators under a relaxed stopping condition, which is the number of iterations: the best obtained result (*best*), the median (*m*), the standard deviation (σ), the lower (*Q1*) and upper (*Q2*) quartiles of results' distributions and the worst cost value (*w*) obtained out of the total number of runs for each instance. In addition, the best results (*best**) obtained by our algorithm under the competition's stopping condition, which is the limited time equal to 474 seconds. Some of the methodologies presented in Table 3 share common features with each other and somewhat with our algorithm, such as population-based features, large number of iterations (ranges from 50,000 to 200,000), variable neighbourhood structures, controlling mechanisms, and relaxed stopping condition, in which we tried to make a reasonable, fair and meaningful comparisons.

Results from Table 3 showed that our Elitist-AS algorithm is capable of producing feasible timetables for all instances on every run. Although the running time of our algorithm was somewhat short when considering the competition's running time limit allowed, our results are somewhat competitive to ACO approaches. Of course if we run the algorithm for a longer time we will definitely be able to obtain better cost value of a solution or even obtaining the global optimum solution. Therefore, we have extended the running time to obtain better results. Also it can be shown that our algorithm is relatively consistent in producing converged optimal results for all small sized instances. Our algorithm has produced feasible and good quality results for all

Table 3. Computational statistics of our algorithm applied on the Socha's test instances

Instance	*Our Elitist-Ant System algorithm*							*ACO based approaches*		*Population-based approaches*		
	*best**	*best*	*Q1*	*σ*	*m*	*Q2*	*w*	*MM AS*	*DH CA*	*EGS GA*	*EM GD*	*HEA*
small1	**0**	**0**	0	.79	1	2	2	1	5	**0**	**0**	**0**
small2	**0**	**0**	0	.913	1	2	3	3	5	**0**	**0**	**0**
small3	**0**	**0**	1	1.262	1	2	4	1	3	**0**	**0**	**0**
small4	**0**	**0**	0	.586	0	1	2	1	3	**0**	**0**	**0**
small5	**0**	**0**	0	.5	0	1	1	**0**	**0**	**0**	**0**	**0**
medium1	190	**84**	87	6.837	92	97	110	195	176	139	96	221
medium2	223	**82**	85	8.542	90	98	109	184	154	92	96	147
medium3	259	123	128	11.876	134	146	160	248	191	**122**	135	246
medium4	127	**62**	68	10.649	76	83	99	164.5	148	98	79	165
medium5	132	**75**	82	7.659	87	92	102	219.5	166	116	87	130
large	869	690	719	50.868	750	783	877	851.5	798	615	683	**529**

Approaches' results:

- MMAS: the Max-Min Ant System in (Socha *et al.*; 2002).
- DHCA: the Die Hard Cooperative Ant behavior algorithm in (Ejaz and Javed; 2007).
- EGSGA: the Extended Guided Search Genetic Algorithm in (Jat and Yang; 2010).
- EMGD: the hybrid of Electromagnetic-Like mechanism with force decay Rate Great Deluge in (Abdullah *et al.*; 2009).
- HEA: the Hybrid Evolutionary Algorithm – GA operators and Randomized Iterative Improvement in (Abdullah *et al.*; 2007b).

instances, and their quality outperformed all the population-based approaches shown in the table for all instances except *medium3* and *large* instances. Our algorithm obtained the best results dedicated to all medium sized instances compared only to the approaches in the table.

In addition, the results of our Elitist-AS algorithm obtained under the competition's stopping condition reached optimality for all small instances within time less than 300 seconds as well as the case of results obtained under the relaxed stopping condition, whilst all medium and large instances results are relatively better than some of the approaches in obtaining a good quality results within shorter computational time in most cases. Our algorithm obtained better quality results than all illustrated ant system-based and population-based approaches in the table above, with respect to the best known results reported in the literature. Moreover, the statistical readings from Table 3 demonstrate that our algorithm (Elitist-AS) is effective and reliable.

4 Conclusions

The overall goal of this study was to investigate the performance of the Elitist-Ant System when being incorporated with two mechanisms to gain a balance between diversification and intensification of the search. Both mechanisms recover the drawbacks of the ant system.

The diversification mechanism recovers restarting the search efficiently and avoids premature convergence, while these may not sufficiently be achieved by the pheromone trail evaporation in order to enforce ants embark different paths. The intensification mechanism recovers exploring neighbors of good solutions effectively, whilst this may not be achieved easily when employing only a global pheromone trail update rule in the ant systems.

In addition to those mechanisms, elitism is required to organize the ant system's pheromone trail updates, where only best ant in each iteration is able to update the pheromone trail.

Both mechanisms proved to be greatly effective in guiding the search in the ant system. Unlike the case as in Bernardino *et al.* (2010), we consider employing both mechanisms as a significantly crucial guidance of the ant system's search process. The respected authors have intentionally ignored the diversification mechanism in their ant colony system, since they claimed that this mechanism has no use when using the problem's specific knowledge (heuristic information) presented by local and global pheromone trail updates.

The results obtained by our Elitist-Ant System add more evidence of effective and robust performance, especially when compared to recent ACO and population-based approaches.

References

1. Abdullah, S., Burke, E.K., McCollum, B.: Using a randomised iterative improvement algorithm with composite neighbourhood structures for university course timetabling. In: Metaheuristics: Progress in complex systems optimization. Operations Research/Computer Science Interfaces Series, vol. ch. 8, Springer, Heidelberg (2007a)
2. Abdullah, S., Burke, E.K., McCollum, B.: A hybrid evolutionary approach to the university course timetabling problem. In: IEEE Congres on Evolutionary Computation, Singapore, September 25-28, pp. 1764–1768 (2007b) ISBN: 1-4244-1340-0
3. Abdullah, S., Turabeih, H., McCollum, B.: Electromagnetism-like mechanism with force decay rate great deluge for CTP. In: Wen, P., Li, Y., Polkowski, L., Yao, Y., Tsumoto, S., Wang, G. (eds.) RSKT 2009. LNCS, vol. 5589, pp. 497–504. Springer, Heidelberg (2009)
4. Alvarez-Valdes, R., Crespo, E., Tamarit, J.M.: Design and implementation of a course scheduling system using Tabu Search. The Proceedings of the Production, Manufactoring and Logistics: European Journal of Operational Research 137(2002), 512–523 (2002)
5. Bernardino, A.M., Bernardino, E.M., Sanchez-Perez, J.M., Gomez-Pulido, J.A., Vega-Rodriguez, M.A.: A Hybrid Ant Colony Optimization Algorithm for Solving the Ring Arc-Loading Problem. In: Konstantopoulos, S., Perantonis, S., Karkaletsis, V., Spyropoulos, C.D., Vouros, G. (eds.) SETN 2010. LNCS (LNAI), vol. 6040, pp. 49–59. Springer, Heidelberg (2010)
6. Blum, C., Roli, A.: Hybrid Metaheuristics: An Introduction, Studies in Computational Intelligence. In: C. Blum, M.J.B. Aguilera, A. Roli, M. Samples (Eds.), Hybrid Metaheuristics: An Emerging Approach to Optimization, SCI, vol. 114, Springer-Verlag Berlin, Heidelberg, pp. 1-30, (2008)
7. Dorigo, M., Maniezzo, V., Colorni, A.: The Ant System: An autocatalytic optimization process, Technical Report 91-016 revised, Dipartimento di Elettronica e Informazione, Policecnico di Milano, Italy (1991)

8. Ejaz, N., Javed, M.: An Approach for Course Scheduling Inspired by Die-Hard Co-Operative Ant Behavior. In: Proceedings of the IEEE International Conference on Automation and Logistics, Jinan, China, August 18 - 21 (2007)
9. Eley, M.: Ant Algorithms for the Exam Timetabling Problem. In: Burke, E.D., Rudova, H. (eds.) PATAT 2007. LNCS, vol. 3867, pp. 364–382. Springer, Heidelberg (2007)
10. Even, S., Itai, A., Shamir, A.: On the Complexity of Timetable and Multi commodity Flow Problem. SIAM J. Comput. 5, 691–703 (1976)
11. Gambardella, L.M., Taillard, E.D., Dorigo, M.: Ant Colonies for the quadratic assignment problem. Journal of the Operational Research Society 50, 167–176 (1999)
12. Mayer, A., Nothegger, C., Chawatal, A., Raidl, G.: Solving the Post Enrolment Course Timetabling Problem by Ant Colony Optimization. In: The Proceeding of the 7thInternational Conference on the Practice and Theory of Automated Timetabling (PATAT 2008), Montreal, Canada (2008)
13. Montgomery, J., Randall, M.: The accumulated experience ant colony for the travelling salesman problem. International Journal of Computational Intelligence and Applications 3(2), 189–198 (2003)
14. Petrovic, S., Burke, E.K.: University timetabling. In: Leung, J. (ed.) Handbook of Scheduling: Algorithms, Models and Performance Analysis, ch. 45, CRC Press, Boca Raton (2004)
15. Qu, R., Burke, E.D., McCollum, B.: Adaptive automated construction of hybrid heuristics for exam timetabling and graph colouring. Discrete Optimization, European Journal of Operational Research 198, 392–404 (2009)
16. Rossi-Doria, O., Samples, M., Birattari, M., Chiarandini, M., Dorigo, M., Gambardella, L.M., Knowels, J., Manfrin, M., Mastrolilli, M., Paechter, B., Paquete, L., Stultzle, T.: A Comparison of the Performance of Different Metaheuristics on the Timetabling Problem. In: Burke, E.K., De Causmaecker, P. (eds.) PATAT 2002. LNCS, vol. 2740, pp. 329–354. Springer, Heidelberg (2003)
17. Socha, K., Knowles, J., Samples, M.: A max-min ant system for the university course timetabling problem. In: Dorigo, M., Di Caro, G.A., Sampels, M. (eds.) Ant Algorithms 2002. LNCS, vol. 2463, pp. 1–13. Springer, Heidelberg (2002)
18. Socha, K., Samples, M., Manfrin, M.: Ant algorithms for the university course timetabling problem with regard to the state-of-the-art. In: Raidl, G.R., Cagnoni, S., Cardalda, J.J.R., Corne, D.W., Gottlieb, J., Guillot, A., Hart, E., Johnson, C.G., Marchiori, E., Meyer, J.-A., Middendorf, M. (eds.) EvoIASP 2003, EvoWorkshops 2003, EvoSTIM 2003, EvoROB/EvoRobot 2003, EvoCOP 2003, EvoBIO 2003, and EvoMUSART 2003. LNCS, vol. 2611, pp. 334–345. Springer, Heidelberg (2003)
19. Socha, K.: The Influence of Run-Time Limits on Choosing Ant System Parameters. In: Cantu-Paz, E., et al. (eds.) GECCO 2003. LNCS, vol. 2723, pp. 49–60. Springer, Heidelberg (2003)
20. Yang, S., Jat, S.N.: Genetic Algorithms with Guided and Local Search Strategies for University Course Timetabling. IEEE Transactions on Systems, Man, and Cybernetics—Part C: Applications and Reviews, 1–14 (2010), doi:10.1109/TSMCC.2010.2049200

An Improved Back Propagation Neural Network Algorithm on Classification Problems

Nazri Mohd Nawi[1], R.S. Ransing[2], Mohd Najib Mohd Salleh[1], Rozaida Ghazali[1], and Norhamreeza Abdul Hamid[1]

[1] Faculty of Information Technology and Multimedia
Universiti Tun Hussein Onn Malaysia
86400, Parit Raja, Batu Pahat, Johor, Malaysia
[2] Civil and Computational Engineering Centre, School of Engineering
University of Wales Swansea
Singleton Park, Swansea, SA2 8PP, United Kingdom
{nazri,najib,rozaida}@uthm.edu.my,
r.s.ransing@swansea.ac.uk, gi090007@siswa.uthm.edu.my

Abstract. The back propagation algorithm is one the most popular algorithms to train feed forward neural networks. However, the convergence of this algorithm is slow, it is mainly because of gradient descent algorithm. Previous research demonstrated that in 'feed forward' algorithm, the slope of the activation function is directly influenced by a parameter referred to as 'gain'. This research proposed an algorithm for improving the performance of the back propagation algorithm by introducing the adaptive gain of the activation function. The gain values change adaptively for each node. The influence of the adaptive gain on the learning ability of a neural network is analysed. Multi layer feed forward neural networks have been assessed. Physical interpretation of the relationship between the gain value and the learning rate and weight values is given. The efficiency of the proposed algorithm is compared with conventional Gradient Descent Method and verified by means of simulation on four classification problems. In learning the patterns, the simulations result demonstrate that the proposed method converged faster on Wisconsin breast cancer with an improvement ratio of nearly 2.8, 1.76 on diabetes problem,, 65% better on thyroid data sets and 97% faster on IRIS classification problem. The results clearly show that the proposed algorithm significantly improves the learning speed of the conventional back-propagation algorithm.

Keywords: Back-propagation Neural Networks; Gain; Activation Function; Learning Rate; Training Efficiency.

1 Introduction

A neural network is a computing system made up of a number of simple, interconnected processing neurons or elements, which process information by its dynamic state response to external inputs [1]. The development and application of neural networks are unlimited as it spans a wide variety of fields. This could be attributed to the fact that these networks are attempts to model the capabilities of human. It had successfully

Y. Zhang et al. (Eds.): DTA/BSBT 2010, CCIS 118, pp. 177–188, 2010.

implemented in the real world application which are accounting and finance [2,3], health and medicine [4,5], engineering and manufacturing [6,7], marketing [8,9] and general applications [10,11,12]. Most papers concerning the use of neural networks have applied a multilayered, feed-forward, fully connected network of perceptions [13,14]. Reasons for the use of simple neural networks are done by the simplicity of the theory, ease of programming, good results and because this type of NN represents an universal function in the sense that if the topology of the network is allowed to vary freely it can take the shape of any broken curve [15]. A standard multi-layer feed forward neural network has an input layer of neurons, a hidden layer of neurons and an output layer of neurons. Every node in a layer is connected to other node in the adjacent forward layer. Several types of learning algorithm have been used in the literature. However, back-propagation algorithm is the most popular, effective, and easy to earn model for complex, multilayered networks. This algorithm is used more than all other combined and used in many different types of applications [16]. A back-propagation is a supervised learning technique that uses a gradient descent rule which attempts to minimize the error of the network by moving down the gradient of the error curve [1]. When using the back-propagation to train a standard multi-layer feed forward neural network, the designer is required to arbitrarily select parameters such as the network topology, initial weights and biases, a learning rate value, the activation function, and a value for the gain in the activation function. Improper selection of any of these parameters can result in slow convergence or even network paralysis where the training process comes to a virtual standstill. Another problem is the tendency of the steepest descent technique, which is used in the training process, can easily get stuck at local minima. Hence, improving the application of back-propagation remains an important research issue.

In recent years, a number of research studies have attempted to overcome these problems. These involved the development of heuristic techniques, based on studies of properties of the conventional back-propagation algorithm. These techniques include such idea as varying the learning rate, using momentum and gain tuning of activation function. Perantonis et al. [17] proposed an algorithm for efficient learning in feed forward neural networks using momentum acceleration. Nevertheless, it could be found that more training time is spent on the computation of constrained conditions in the algorithm. Kamarthi and Pittner [18] presented a universal acceleration technique for the back-propagation algorithm based on extrapolation of each individual interconnection weight. This requires the error surface to have a smooth variation along the respective axes, therefore extrapolation is possible. For performing extrapolation, at the end of each epoch, the converge behaviour of each network weight in back-propagation algorithm is individually examined. They also focused on the use of standard numerical optimization techniques. Though, this technique often must be tuned to fit a particular application. Møller [19] explained how conjugate gradient algorithm could be used to train multi-layer feed forward neural networks. In this algorithm a search is performed along conjugate directions, which generally leads to faster convergence than steepest gradient descent directions. The error function is guaranteed not to increase consequently of the weights update. However, if it reaches a local minimum, it remains forever, as there is no mechanism for this algorithm to escape. Lera et al. [20] described the use of Levenberg-Marquardt algorithm for training multi-layer feed forward neural networks. Though, the training times required strongly depend on neighbourhood size.

Nazri et al. [21] demonstrated that changing the 'gain' value adaptively for each node can significantly reduce the training time. Based on [21], this paper proposed an improved algorithm that will change the gain value adaptively which significantly improve the performance of the back-propagation algorithm. In order to verify the efficiency of the proposed algorithm, and to compare it with the Gradient Descent Method (GDM) proposed by Nazri et al. [21], some simulation experiments was performed on four benchmark classification problems including Wisconsin breast cancer [22], diabetes [23], thyroid [24] and IRIS [25].

The remaining of the paper is organised as follows. In Section 2, using activation function with adaptive gain is reviewed. While in section 3 presents the proposed algorithm. The performance of the proposed algorithm is tested on benchmark problems are conducted in Section 4. This paper is concluded in the final section.

2 Using Activation Function with Adaptive Gain

An activation function is used for limiting the amplitude of the output of neuron. It generates an output value for a node in a predefined range as the closed unit interval $[0,1]$ or alternatively $[-1,1]$. This value is a function of the weighted inputs of the corresponding node. The most commonly used activation function is the logistic sigmoid activation function. Alternative choices are the hyperbolic tangent, linear, step activation functions. For the jth node, a logistic sigmoid activation function which has a range of $[0,1]$ is a function of the following variables, viz

$$o_j = \frac{1}{1+e^{-c_j a_{net,j}}} \tag{1}$$

where,

$$a_{net,j} = \left(\sum_{i=1}^{l} w_{ij} o_i \right) + \theta_j \tag{2}$$

where,

o_j Output of the jth unit.

o_i Output of the ith unit.

w_{ij} weight of the link from unit i to unit j.

$a_{net,j}$ net input activation function for the jth unit.

θ_j bias for the jth unit.

c_j gain of the activation function.

The value of the gain parameter, c_j, directly influences the slope of the activation function. For large gain values $(c \leq 1)$, the activation function approaches a 'step function' whereas for small gain values $(0 < c \leq 1)$ the output values change from zero to unity over a large range of the weighted sum of the input values and the sigmoid function approximates a linear function.

Most of the application oriented papers on neural networks tend to advocate that neural networks operate like a 'magic black box', which can simulate the "learning from example" ability of our brain with the help of network parameters such as weights, biases, gain, hidden nodes, etc. Also, a unit value for gain has generally been used for most of the research reported in the literature but a few authors have researched the relationship of the gain parameter with other parameters which used in back-propagation algorithms. The recent results [27] show that learning rate, momentum constant and gain of the activation function have a significant impact on training speed. Unfortunately, higher values of learning rate and/or gain cause instability [28]. Thimm et al. [29] also proved that a relationship between the gain value, a set of initial weight values, and a learning rate value exists. Looney [30] suggested to adjust the gain value in small increments during the early iterations and to keep it fixed somewhere around halfway through the learning. Eom et al. [31] proposed a method for automatic gain tuning using a fuzzy logic system. Nazri et al. [21] proposed a method to change adaptively gain value on other optimisation method such as conjugate gradient.

3 The Proposed Algorithm

In this section, an improved algorithm for improving the training efficiency of back-propagation is proposed. The proposed algorithm modifies the initial search direction by changing the gain value adaptively for each node. The following subsection describes the proposed algorithm. The advantages of using an adaptive gain value have been explored. Gain update expressions as well as weight and bias update expressions for output and hidden nodes have also been proposed. These expressions have been derived using same principles as used in deriving weight updating expressions.

There are two different ways in which this gradient descent can be implemented which are incremental mode and batch mode. In this paper, the batch mode was used for training process. As in batch mode training, the weights, biases and gains are updated after one complete presentation of the entire training set. An epoch is defined as one complete presentation of the training set. A sum squared error value is calculated after the presentation of the training set and compared with the target error. Training is done on an epoch-by-epoch basis until the sum squared error falls below the desired target value.

3.1 Algorithm

The following iterative algorithm is proposed by the authors for the batch mode of training. Weights, biases and gains are calculated and updated for the entire training set, which is being presented to the network.

The gain update expression for a gradient descent method is calculated by differentiating the following error term E with respect to the corresponding gain parameter.

```
For a given epoch,
    For each input vector,
         Step 1.
         Calculate the weight and bias values using the
         previously converged gain value.
         Step 2.
         Use the weight and bias value calculated in Step
         (1) to calculate the new gain value.
Repeat Steps (1) and (2) for each example on an epoch-
by-epoch basis until the error on the entire training
data set reduces to a predefined value.
```

The network error E is defined as follows

$$E = \frac{1}{2}\sum \left(t_k - o_k\left(o_j, c_k\right)\right)^2 \tag{3}$$

For output unit, $\frac{\partial E}{\partial c_k}$ needs to be calculated whereas for hidden units. $\frac{\partial E}{\partial c_j}$ is also required. The respective gain values would then be updated with the following equations.

$$\Delta c_k = \eta \, (-\frac{\partial E}{\partial c_k}) \tag{4}$$

$$\Delta c_j = \eta \, (-\frac{\partial E}{\partial c_j}) \tag{5}$$

$$\frac{\partial E}{\partial c_k} = -\left(t_k - o_k\right) o_k \left(1 - o_k\right)\left(\sum w_{jk} o_j + \theta_k\right) \tag{6}$$

Therefore, the gain update expression for links connecting to output nodes is:

$$\Delta c_k(n+1) = \eta \left(t_k - o_k\right) o_k \left(1 - o_k\right)\left(\sum w_{jk} o_j + \theta_k\right) \tag{7}$$

$$\frac{\partial E}{\partial c_j} = \left[-\sum_k c_k w_{jk} o_k \left(1 - o_k\right)\left(t_k - o_k\right)\right] o_j \left(1 - o_j\right)\left(\left(\sum_j w_{ij} o_i\right) + \theta_j\right) \tag{8}$$

Therefore, the gain update expression for the links connecting hidden nodes is;

$$\Delta c_j(n+1) = \eta\left[-\sum_k c_k w_{jk} o_k \left(1 - o_k\right)\left(t_k - o_k\right)\right] o_j \left(1 - o_j\right)\left(\left(\sum_j w_{ij} o_i\right) + \theta_j\right) \tag{9}$$

Similarly, the weight and bias expressions are calculated as follows:

The weight update expression for the links connecting to output nodes with a bias is:

$$\Delta w_{jk} = \eta \left(t_k - o_k\right) o_k \left(1 - o_k\right) c_k o_j \tag{10}$$

Similarly, the bias update expressions for the output nodes would be:

$$\Delta \theta_k = \eta \left(t_k - o_k\right) o_k \left(1 - o_k\right) c_k \tag{11}$$

The weight update expression for the links connecting to hidden nodes is:

$$\Delta w_{ij} = \eta \left[\sum_k c_k w_{jk} o_k (1 - o_k)(t_k - o_k) \right] c_j o_j (1 - o_j) o_i \tag{12}$$

Similarly, the bias update expressions for the hidden nodes would be:

$$\Delta \theta_j = \eta \left[\sum_k c_k w_{jk} o_k (1 - o_k)(t_k - o_k) \right] c_j o_j (1 - o_j) \tag{13}$$

3.2 Advantages of Using Adaptive Gain

An algorithm has been proposed in this paper for the efficient calculation of the adaptive gain value in batch modes of learning. We proposed that the total learning rate value can be split into two parts – a local (nodal) learning rate value and a global (same for all nodes in a network) learning rate value. The value of parameter gain is interpreted as the local learning rate of a node in the network. The network is trained using a fixed value of learning rate equal to 0.3 which is interpreted as the global learning rate of the network. However, as the gain value was modified, the weights and biases were updated using the new value of gain. This resulted in higher values of gain which caused instability [29]. To avoid oscillations during training and to achieve convergence, an upper limit of 2.0 is set for the gain value. This will be explained in detail in our next publication.

The method has been illustrated for Gradient Descent training algorithm using the sequential and batch modes of training. An advantage of using the adaptive gain procedure is that it is easy to introduce into a back-propagation algorithm and it also accelerates the learning process without a need to invoke solution procedures other than the Gradient Descent method. The adaptive gain procedure has a positive effect in the learning process by modifying the magnitude, and not the direction, of the weight change vector. This greatly increases the learning speed by amplifying the directions in the weight space that are successfully chosen by the Gradient-Descent method. However, the method will also be advantageous when using other faster optimization algorithms such as Conjugate-Gradient method and Quasi-Newton method. These methods can only optimize an equivalent of the global learning rate (the step length). By introducing an additional local learning rate parameter, further increase in the learning speed can be achieved. Work is currently under progress to implement this algorithm using other optimization methods.

4 Results and Discussion

The performance criterion used in this research focuses on the speed of convergence, measured in number of iterations and CPU time. The benchmark problems used to verify our algorithm are taken from the open literature. Four classification problems have been tested including Wisconsin breast cancer [22], diabetes [23], thyroid [24] and IRIS [25]. The simulations have been carried out on a Pentium IV with 3 GHz PC Dell, 1 GB RAM and using MATLAB version 6.5.0 (R13).

On each problem, the following three algorithms were analyzed and simulated.

1) The standard Gradient descent with momentum (GDM) [21]
2) The proposed Gradient descent with momentum and Adaptive Gain (GDM/AG)

To compare the performance of the proposed algorithm with gradient descent method [21], network parameters such as network size and architecture (number of nodes, hidden layers etc), values for the initial weights and gain parameters were kept the same. For all problems the neural network had one hidden layer with five hidden nodes and sigmoid activation function was used for all nodes. All algorithms were tested using the same initial weights, initialized randomly from range $[0,1]$ and received the input patterns for training in the same sequence.

For all training algorithms, the learning rate value is 0.3 and the momentum term value is 0.7. The initial value used for the gain parameter is one. For each run, the numerical data is stored in two files - the results file, and the summary file. The result file lists data about each network. The number of iterations until convergence is accumulated for each algorithm from which the mean, the standard deviation and the number of failures are calculated. The networks that fail to converge are obviously excluded from the calculations of the mean and standard deviation but are reported as failures. For each problem, 100 different trials were run, each with different initial random set of weights. For each run, the number of iterations required for convergence is reported. For an experiment of 100 runs, the mean of the number of iterations (mean), the standard deviation (SD), and the number of failures are collected. A failure occurs when the network exceeds the maximum iteration limit; each experiment is run to ten thousand iterations; otherwise, it is halted and the run is reported as a failure. Convergence is achieved when the outputs of the network conform to the error criterion as compared to the desired outputs.

4.1 Breast Cancer Classification Problem

This dataset was created based on the 'Breast Cancer Wisconsin' problem dataset from UCI repository of machine learning databases from Dr. William H. Wolberg [22]. This problem tries to diagnosis of breast cancer by trying to classify a tumor as

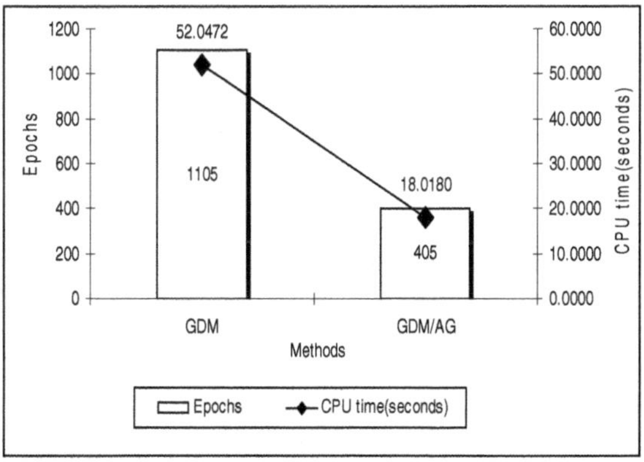

Fig. 1. Performance comparison of GDM/AG with GDM for Breast Cancer Classification Problem

either benign or malignant based on cell descriptions gathered by microscopic examination. The selected architecture of the Feed- forward Neural Network is 9-5-2. The target error is set as to 0.02.

Figure 1 shows that the proposed algorithm (GDM/AG) exhibit very good average performance in order to reach target error. Furthermore the number of failures for the proposed algorithm is smaller as compared to GDM. The proposed algorithm (GDM/AG) needs only 405 epochs to converge as opposed to the standard GDM at about 1105 epochs. Apart from speed of convergence, the time required for training the classification problem is another important factor when analyzing the performance. For numerous models, training process may suppose a very important time consuming process. The results in Figure 1 clearly show that the proposed algorithm (GDM/AG) outperform GDM with an improvement ratio, nearly 2.8, for the total time of converge.

4.2 IRIS Classification Problem

This is a classical classification dataset made famous by Fisher [25], who used it to illustrate principles of discriminant analysis. This is perhaps the best-known database to be found in the pattern recognition literature. Fisher's paper is a classic in the field and is referenced frequently to this day. The selected architecture of the Feed- forward Neural Network is 4-5-3 with target error was set as 0.05

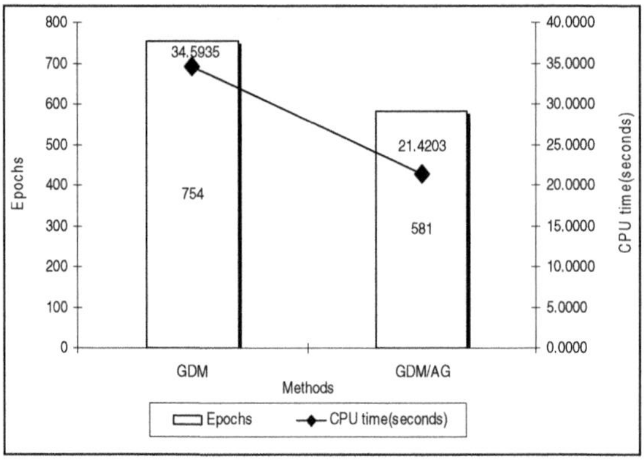

Fig. 2. Performance comparison of GDM/AG with GDM for IRIS Classification Problem

Figure 2 shows that the proposed algorithm (GDM/AG) still outperforms other algorithms in term of CPU time and number of epochs. The proposed algorithm (GDM/AG) only required 581 epochs in 21.4203 seconds CPU times to achieve the target error, whereas GDM required 754 epochs in 34.5935 seconds CPU times. As we can see that the number of success rate for the proposed algorithm (GDM/AG) was 97% as compared to GDM in learning the patterns. Furthermore, the average number of learning iterations for the proposed algorithm was reduced up to 1.3 times faster as compared to GDM.

4.3 Diabetes Classification Problem

This dataset was created based on the 'Pima Indians Diabetes' problem dataset [23] from the UCI repository of machine learning database. From the dataset doctors try to diagnose diabetes of Pima Indians based on personal data (age, number of times pregnant) and the results of medical examinations (e.g. blood pressure, body mass index, result of glucose tolerance test, etc.) before decide whether a Pima Indian individual is diabetes positive or not. The selected architecture of the Feed-forward Neural Network is 8-5-2. The target error is set to 0.01.

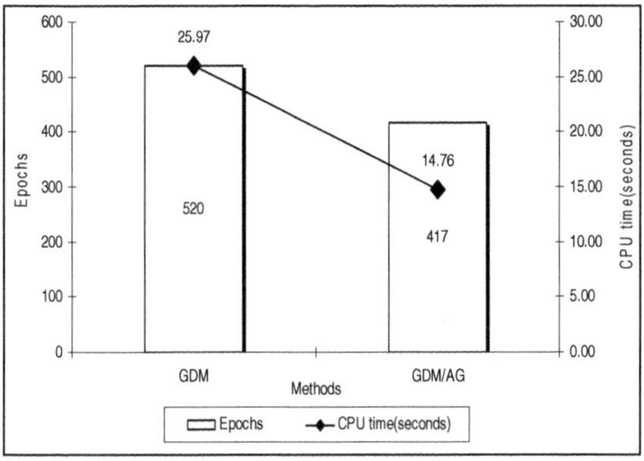

Fig. 3. Performance comparison of GDM/AG with GDM for Diabetes Classification Problem

From Figure 3, it is worth noticing that the performance of the proposed GDM/AG is almost 1.8 faster as compared to GDM. GDM/AG took only 417 epochs to reach the target error as compared to GDM which took about 520 epochs to converge. Still the proposed algorithm outperform the traditional GDM in term of the total time of converge.

4.4 Thyroid Classification Problem

This dataset was created based on the 'Thyroid Disease' problem dataset [24] from the UCI repository of machine learning database. The dataset was obtained from the Garvan Institute. This dataset deals with diagnosing a patient thyroid function. Each pattern has 21 attributes and can be assigned to any of three classes which were hyper-, hypo- and normal function of thyroid gland. The selected architecture of the Feed-forward Neural Network is 21-5-3. The target error is set to 0.05.

Figure 4 reveal that GDM need 316.4905 seconds with 3441 epochs to converge. However, the proposed algorithm (GDM/AG) performed significantly better with only 111.631 seconds and 1114 epochs to converge. The result shown that the GDM/AG perform better as compared to GDM.

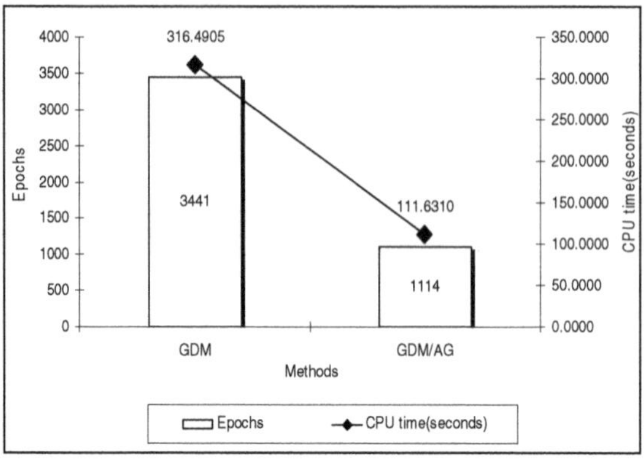

Fig. 4. Performance comparison of GDM/AG with GDM for Thyroid Classification Problem

Consequently, from the previous results, we can claim that in general, the proposed algorithm (GDM/AG) presents better performance than GDM. The experimental results from four classification problems allow to compares the proposed algorithm (GDM/AG) with the traditional GDM, in terms of speed of convergence, measured in number of iterations and CPU time. Moreover, when comparing the proposed algorithm with GDM, it has been empirically demonstrated that the proposed algorithm (GDM/AG) performed highest accuracy than GDM. This conclusion enforces the usage of the proposed algorithm as alternative training algorithm of back propagation neural networks.

5 Conclusion

While the back-propagation algorithm is used in the majority of practical neural networks application and has been shown to perform relatively well, there still exist areas where improvement can be made. We proposed an algorithm to adaptively change the gain parameter of the activation function to improve the learning speed. It was observed that the influence of variation in the gain value is similar to the influence of variation in the learning rate value. It changes the gain value adaptively for each node. The effectiveness of the proposed algorithm has been compared with the Gradient Descent Method (GDM) [21], verified by means of simulation on four classification problems including Wisconsin breast cancer with an improvement ratio nearly 2.8 for the total time of converge, diabetes almost 1.76 faster respectively, thyroid took almost 65% less time to converge and IRIS the proposed algorithm outperformed the traditional GDM with 97% faster in learning the patterns.

Acknowledgment

The authors would like to thank the Ministry of Science, Technology and Innovation, Malaysia for supporting this research under the Science Fund Research Grant.

References

1. Majdi, A., Beiki, M.: Evolving Neural Network Using Genetic Algorithm for Predicting the Deformation Modulus of Rock Masses. International Journal of Rock Mechanics and Mining Science, vol 47(2), 246–253 (2010)
2. Lee, K., Booth, D., Alam, P.: A Comparison of Supervised and Unsupervised Neural Networks in Predicting Bankruptcy of Korean Firms. Expert Systems with Applications 29(1), 1–16 (2005)
3. Landajo, M., Andres, J.D., Lorca, P.: Robust Neural Modeling for the Cross-Sectional Analysis of Accounting Information. European Journal of Operational Research 177(2), 1232–1252 (2007)
4. Razi, M.A., Athappily, K.: A Comparative Predictive Analysis of Neural Networks (NNs), Nonlinear Regression and Classification and Regression Tree (CART) Models. Expert Systems with Applications 2(1), 65–74 (2005)
5. Behrman, M., Linder, R., Assadi, A.H., Stacey, B.R., Backonja, M.M.: Classification of Patients with Pain Based on Neuropathic Pain Symptoms: Comparison of an Artificial Neural Network against an Established Scoring System. European Journal of Pain 11(4), 370–376 (2007)
6. Yesilnacar, E., Topal, T.: Landslide Susceptibility Mapping: A Comparison of Logistic Regression and Neural Networks Methods in a Medium Scale Study, Hendek region (Turkey). Engineering Geology 79(3–4), 251–266 (2005)
7. Dvir, D., Ben-Davidb, A., Sadehb, A., Shenhar, A.J.: Critical Managerial Factors Affecting Defense Projects Success: A Comparison between Neural Network and Regression Analysis. Engineering Applications of Artificial Intelligence 19, 535–543 (2006)
8. Gan, C., Limsombunchai, V., Clemes, M., Weng, A.: Consumer Choice Prediction: Artificial Neural Networks Versus Logistic Models. Journal of Social Sciences 1(4), 211–219 (2005)
9. Chiang, W.K., Zhang, D., Zhou, L.: Predicting and Explaining Patronage Behavior toward Web and Traditional Stores Using Neural Networks: A Comparative Analysis with Logistic Regression. Decision Support Systems 41, 514–531 (2006)
10. Chang, L.Y.: Analysis of Freeway Accident Frequencies: Negative Binomial Regression versus Artificial Neural Network. Safety Science 43, 541–557 (2005)
11. Sharda, R., Delen, D.: Predicting box-office success of motion pictures with neural networks. Expert Systems with Applications 30, 243–254 (2006)
12. Nikolopoulos, K., Goodwin, P., Patelis, A., Assimakopoulos, V.: Forecasting with cue information: A comparison of multiple regression with alternative forecasting approaches. European Journal of Operational Research 180(1), 354–368 (2007)
13. Curteanu, S., Cazacu, M.: Neural Networks and Genetic Algorithms Used For Modeling and Optimization of the Siloxane-Siloxane Copolymers Synthesis. Journal of Macromolecular Science, Part A 45, 123–136 (2007)
14. Lisa, C., Curteanu, S.: Neural Network Based Predictions for the Liquid Crystal Properties of Organic Compounds. Computer-Aided Chemical Engineering 24, 39–45 (2007)
15. Fernandes, Lona, L.M.F.: Neural Network Applications in Polymerization Processes. Brazilian Journal Chemical Engineering 22, 323–330 (2005)
16. Mutasem, K.S.A., Khairuddin, O., Shahrul, A.N.: Back Propagation Algorithm: The Best Algorithm Among the Multi-layer Perceptron Algorithm. International Journal of Computer Science and Network Security 9(4), 378–383 (2009)

17. Perantonis, S.J., Karras, D.A.: An Efficient Constrained Learning Algorithm with Momentum Acceleration. Neural Networks 8(2), 237–249 (1995)
18. Kamarthi, S.V., Pittner, S.: Accelerating Neural Network Training using Weight Extrapolations. Neural Networks 12, 1285–1299 (1999)
19. Møller, M.F.: A Scaled Conjugate Gradient Algorithm for Fast Supervised Learning. Neural Networks 6(4), 525–533 (1993)
20. Lera, G., Pinzolas, M.: Neighborhood based Levenberg-Marquardt Algorithm for Neural Network Training. IEEE Transaction on Neural Networks 13(5), 1200–1203 (2002)
21. Nawi, N.M., Ransing, M.R., Ransing, R.S.: An Improved Conjugate Gradient Based Learning Algorithm for Back Propagation Neural Networks. International Journal of Computational Intelligence 4(1), 46–55 (2007)
22. Mangasarian, O.L., Wolberg, W.H.: Cancer Diagnosis via Linear Programming. SIAM News 23(5), 1–18 (1990)
23. Smith, J.W., Everhart, J.E., Dickson, W.C., Knowler, W.C., Johannes, R.S.: Using the ADAP Learning Algorithm to Forecast the Onset of Diabetes Mellitus. In: Proceedings of the Symposium on Computer Applications and Medical Care, pp. 261–265. IEEE Computer Society Press, Los Alamitos (1988)
24. Coomans, D., Broeckaert, I., Jonckheer, M., Massart, D.L.: Comparison of Multivariate Discrimination Techniques for Clinical Data—Application to The Thyroid Functional State. Methods of Information Medicine 22, 93–101 (1983)
25. Fisher, R.A.: The Use of Multiple Measurements in Taxonomic Problems. Annals of Eugenics 7, 179–188 (1936)
26. Holger, R.M., Graeme, C.D.: The Effect of Internal Parameters and Geometry on the Performance of Back-Propagation Neural Networks. Environmental Modeling and Software 13(1), 193–209 (1998)
27. Hollis, P.W., Harper, J.S., Paulos, J.J.: The Effects of Precision Constraints in a Backpropagation Learning Network. Neural Computation 2(3), 363–373 (1990)
28. Thimm, G., Moerland, F., Fiesler, E.: The Interchangeability of Learning Rate and Gain in Backpropagation Neural Networks. Neural Computation 8(2), 451–460 (1996)
29. Looney, C.G.: Stabilization and Speedup of Convergence in Training Feed Forward Neural Networks. Neurocomputing 10(1), 7–31 (1996)
30. Eom, K., Jung, K., Sirisena, H.: Performance Improvement of Backpropagation Algorithm by Automatic Activation Function Gain Tuning Using Fuzzy Logic. Neurocomputing 50, 439–460 (2003)

Great Deluge Algorithm for Rough Set Attribute Reduction

Salwani Abdullah and Najmeh Sadat Jaddi

Data Mining and Optimisation Research Group (DMO)
Center for Artificial Intelligence Technology,
Universiti Kebangsaan Malaysia, 43600 Bangi, Selangor, Malaysia
salwani@ftsm.ukm.my, najmehjaddi@yahoo.com

Abstract. Attribute reduction is the process of selecting a subset of features from the original set of features that forms patterns in a given dataset. It can be defined as a process to eliminate redundant attributes and at the same time is able to avoid any information loss, so that the selected subset is sufficient to describe the original features. In this paper, we present a great deluge algorithm for attribute reduction in rough set theory (GD-RSAR). Great deluge is a meta-heuristic approach that is less parameter dependent. There are only two parameters needed; the time to "spend" and the expected final solution. The algorithm always accepts improved solutions. The worse solution will be accepted if it is better than the upper boundary value or "level". GD-RSAR has been tested on the public domain datasets available in UCI. Experimental results on benchmark datasets demonstrate that this approach is effective and able to obtain competitive results compared to previous available methods. Possible extensions upon this simple approach are also discussed.

Keywords: Great Deluge, Rough Set, Attribute Reduction.

1 Introduction

Attribute reduction in rough set theory can be defined as an attribute subset selection process. This process tries to find minimal reducts while avoiding information loss [11,12], and is known as a NP-hard problem. Attribute reduction has been applied in many areas such as pattern recognition, machine learning and signal processing. Attribute reduction process will remove redundant and misleading attributes with an aim to obtain optimal attributes that can be determined by relevancy and redundancy factors. An attribute is said to be relevant if a decision is depending on it, otherwise it is irrelevant. Whilst, an attribute is said to be redundant if it is highly correlated with other attributes. Hence, the aim of an attribute reduction process is to search for attributes that are strongly relevant and correlate with the decision attribute.

Lately, attribute reduction in rough set theory has gained a great attention. A number of approaches are available in the literature that can be classified as single-based and population-based approaches. Examples of single-based approaches are simulated annealing by Jensen and Shen [6]; and tabu search by Hedar et al. [5]. Population-based approaches can be found in genetic algorithm by Jensen and Shen

Y. Zhang et al. (Eds.): DTA/BSBT 2010, CCIS 118, pp. 189–197, 2010.

[6]; scatter search by Wang et al. [15]; and ant colony by Jensen and Shen [6] and Ke et al. [8]. Interested readers can find other available approaches and surveys on rough set attribute reduction in [4,7,16]. In this work, we propose a great deluge algorithm which is an alternative to Simulated Annealing but less parameter dependent to cater attribute reduction problem. The algorithm is tested on 13 standard benchmark datasets taken from UCI (available at http://www.ics.uci.edu/~mlearn). The minimal reduct is obtained using rough set theory.

The paper is organised as follows; Section 2 provides a brief introduction on rough set theory. Section 3 describes the detailed implementation of the great deluge algorithm. The simulation results are presented in Section 4. Finally, the paper is concluded by discussing the effectiveness of the technique and the potential future research approaches.

2 Rough Set Theory

Rough Set Theory (RST) is an extension of classical set theory that represents incomplete knowledge [11,12]. The following description of RST is based on the example given in Table. 1. The table contains three conditional attributes (*a,b,c*) and one decision attribute (*d*) and nine objects labeled as (0,…,8).

Table 1. Example of dataset

U/A	*a*	*b*	*c*	*d*
0	0	2	2	0
1	1	1	1	2
2	0	0	1	1
3	1	0	2	2
4	0	2	0	1
5	2	0	1	1
6	1	1	1	2
7	1	1	0	1
8	2	1	0	2

Let (U, A)be an information system where consider U is a non empty set of finite set of objects and A be a non empty finite set of attributes that $\alpha: U \rightarrow V_\alpha$ for each $\alpha \in A$. With any $P \subset A$ exists equivalence relation $IND\ (P)$:

$$IND\ (P) = \{(x, y) \in U^2 \mid \forall\, \alpha \in P, \alpha(x) = \alpha(y)\}$$

The part of U, produced by $IND\ (P)$ is as: $U/IND\ (P) = \otimes\ \{\alpha \in P: U/IND(\{\alpha\})\}$ where: $A \otimes B = \{X \cap Y: \forall\, X \in A, \forall\, Y \in B, X \cap Y \neq \emptyset\}$.If $(x, y) \in IND\ (P)$, then x and y be indiscernible by attributes from P. The equivalence classes of the P-indiscernible relation are labeled $[x]_P$. For example if $P = \{a, b\}$, then objects 1, 6 and 7 are indiscernible; as are objects 0 and 4. $IND\ (P)$ generates the partition of U as:

$$U/IND\,(P) = U/IND\,(a) \otimes U/IND\,(b)$$
$$= \{\{0, 2, 4\},\{1, 3, 6, 7\},\{5, 8\}\} \otimes \{\{2, 3, 5\},\{1, 6, 7, 8\},\{0, 4\}\}$$
$$= \{\{2\}, \{0, 4\}, \{3\}, \{1, 6, 7\}, \{5\}, \{8\}\}$$

Let $X \subseteq U$, the P-lower approximation $\underline{P}X$ of set X is defined as: $\underline{P}X = \{x | [x]_p \subseteq X\}$ and P-upper approximation $\bar{P}X$ is specified as $\bar{P}X = \{ x | [x]_p \cap X \neq \emptyset \}$. Let P and Q be equivalence relations over U, then the positive regions can be specified as:

$$POS_P(Q) = \bigcup_{X \in U/Q} \underline{P}X$$

The positive region consists of all objects of U that can be classified to classes of U/Q using attributes P. For example let $P = \{a, b\}$ and $Q = \{d\}$, then

$$POS_{IND(P)}(Q) = \cup\{ \emptyset,\{2, 5\},\{3, 8\}\}=\{2, 3, 5, 8\}$$

This means that objects 2, 3, 5 and 8 are certainly belong to a class in attribute d, when attributes a and b are considered. The rest of the objects cannot be classified as the information that can make them discernible. An important issue in data analysis is finding dependencies between attributes. For $P, \subset A$, it means Q depends on P in a degree k $(0 \leq k \leq 1)$.

$$k = \gamma_{P}(Q) = \frac{|POS_P(Q)|}{|U|}$$

In the example above, degree of dependency of attribute $\{d\}$ from the attributes $\{a, b\}$ is calculated as:

$$k = \gamma_{\{a,b\}}(\{d\}) = \frac{|\{2,3,5,8\}|}{|\{0,1,2,3,4,5,6,7,8\}|} = \frac{4}{9}$$

The reduction of attribute can be calculated by comparing equivalence relations that are presented by a set of attributes. For this case, by using dependency degree, attributes are removed from the considered set of conditional attributes. The reduced set thus provides the same classification quality as the original dataset. The basic and costly solution to find such a subset is to find all possible subsets and choose those that have dependency degree equals to full attribute set. Only one reduction is needed most of the time. Therefore, an efficient algorithm can be used to find the optimum solution, which considers the maximum dependency degree and minimum cardinality of reducts found.

3 Great Deluge Algorithm for Attribute Reduction (GD-RSAR)

3.1 Solution Representation and Construction

In this work, a solution that is initially constructed by a constrictive heuristic is represented in one dimensional vector, where the length of the vector is based on the best number of attribute found so far in the literature for each dataset. Each cell in the vector contains the index of the attributes. For example, in Fig. 1, the solution contain attributes $\{4,17,12,6,9,19\}$, where the number of attributes is 6.

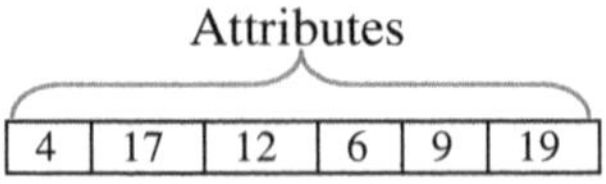

Fig. 1. Solution representation

3.2 Great Deluge List (GDList)

Great Deluge List (GDList) is used to keep the sequential of the best solutions (attributes) found at each iteration. The purpose of introducing GDList is to assist the great deluge algorithm to employ different neighborhood structures when creating a trial solution.

3.3 Neighborhood Structures

The neighborhood of a current solution is generated by adding or removing attribute from the current solution to generate a trial solution:

N_1: if the current solution is the best solution found by the previous iteration, then one attribute (with the lowest occurrence in the GDList) will be removed from the current solution.
N_2: if the current solution is a worse solution updated by the previous iteration, then one attribute (with the highest occurrence in the GDList) will be added to the current solution.
N_3: otherwise, an alteration of an attribute will be considered by replacing it with another attribute randomly.

3.4 Solution Quality Measure

The qualities of the solutions are measured based on the dependency degree, denoted as γ. Given two solutions: current solution, *Sol*, and trial solution, *Sol**. The solution *Sol** is accepted if there is an enhancement in the dependency degree (i.e. if $\gamma(Sol^*) > \gamma(Sol)$). If the dependency degree for both solutions are the same (i.e. $\gamma(Sol^*) = \gamma(Sol)$), then the solution with less cardinality (number of attributes) will be accepted. Cardinality of the solution, *Sol* is denoted as $|Sol|$.

- $\gamma_{Sol^*}(D) > \gamma_{Sol}(D)$
- $|Sol^*| < |Sol|$ if $\gamma_{Sol^*}(D) = \gamma_{Sol}(D)$

3.5 The Great Deluge Algorithm

The great deluge algorithm was introduced by Dueck in 1993 [3]. It is a local search procedure which has certain similarities with simulated annealing but has been introduced as an alternative. This approach is far less dependent upon parameters than simulated annealing. It needs just two parameters: the amount of computational time that the user wishes to "spend" and an estimate of the quality of solution that a user requires. Apart from accepting a move that improves the solution quality, the great deluge algorithm also accepts a worse solution if the quality of the solution is less than (for the case of minimization) or equal to some given upper boundary value (in the paper by Dueck [3], it was called *level*). Normally, the *level* is initially set to be the objective function value of the initial solution. During its run, the *level* is iteratively lowered by a constant β where β is a decreasing rate. This approach has been successfully applied on various optimization problems such as examination timetabling problem [1,2] and course timetabling problem [9,10]. A hybridization of great deluge and other approaches have shown some outstanding improvements when tested on the same instances. For example, see [13,14].

In this work, we present a standard great deluge for a maximization problem. The objective function is calculated based on the dependency degree. The higher the dependency degree, the better the solution is. The sample of main structure of GD-RSAR applied in this work is presented in Fig. 2.

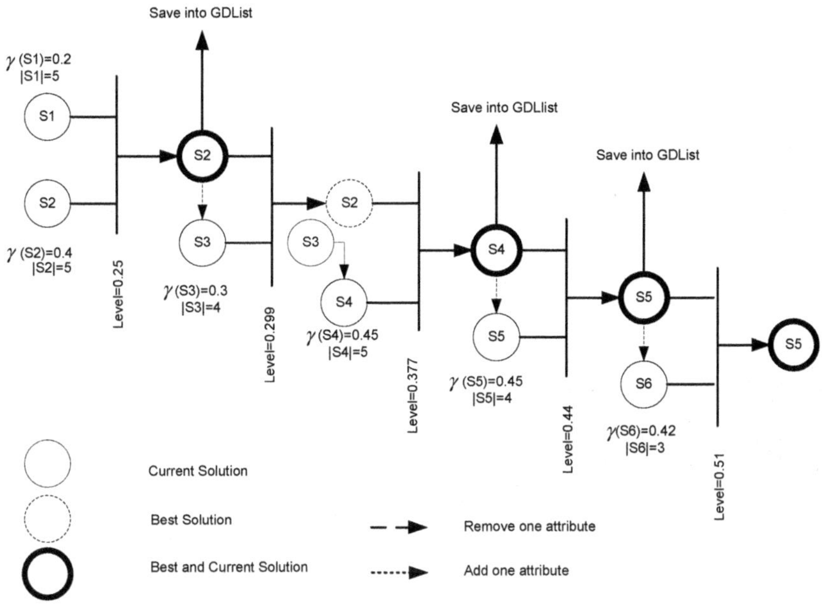

Fig. 2. The example of main structure of GD-RSAR

The number of iterations was defined as *NumOfIte,* the estimated quality of the final solution as *EstimatedQuality*, and an increasing rate as β that is calculated as below:

$$\beta = (EstimatedQuality - \gamma\ (Sol))\ /\ NumOfIte$$

The *level* is equal to the quality of the initial solution, $\gamma(Sol)$ at the start, and will be increased by the value β. In the *do-while* loop, a neighbor is defined by inserting and removing the attribute from the current solution using GDList to obtain a trial solution, *Sol**. After that, the cost function (dependency degree) value is calculated ($\gamma(Sol^*)$). Then, the dependency degree of the trial solution $\gamma(Sol^*)$ is compared with the cost of best solution $\gamma(Sol_{best})$. Note that the dependency degree is measured using rough set theory [12,13]. If there is an improvement, that is $\gamma(Sol^*) > \gamma(Sol_{best})$, then the trial solution, *Sol*,* is accepted and the current and best solutions are updated

($Sol \leftarrow Sol^*$; $Sol_{best} \leftarrow Sol^*$; $\gamma(Sol) \leftarrow \gamma(Sol^*)$; $\gamma(Sol_{best}) \leftarrow \gamma(Sol^*)$). Otherwise, the quality of the trial solution will be compared with the best solution. If there is no change in the dependency degree between these solutions, that is $\gamma(Sol^*) = \gamma(Sol_{best})$), then the number of attribute will be compared. If $|Sol^*| < |Sol|$, then the current solution and the best solution will be updated ($Sol \leftarrow Sol^*$; $Sol_{best} \leftarrow Sol^*$; $\gamma(Sol) \leftarrow \gamma(Sol^*)$; $\gamma(Sol_{best}) \leftarrow \gamma(Sol^*)$). A worse solution will be accepted as the current solution if the quality of the trial solution, $\gamma(Sol^*)$ is greater than the *level*. By inserting the best solution into GDList, the GDList will be updated when the best solution is updated. The process continues until the end of the iteration. The pseudo code for GD-RSAR and calculating dependency degree are presented in Fig. 3.

```
Set initial solution as Sol, obtained from constructive heuristic;
Set best solution, Sol_best ← Sol;
Calculate the initial and best cost function, γ (Sol) and γ
(Sol_best);
Set estimated quality of final solution (i.e. maximum
dependency degree value), EstimatedQuality ← 1;
Set number of iteration, NumOfIte;
Set initial level: level ← γ (Sol);
Set increasing rate β;
Set iteration ← 0;
do while (iteration < NumOfIte)
      Add or remove one attribute to/from the current solution
      (Sol) to obtain a trial solution, Sol*;
      Evaluate trial solution, γ (Sol*);
      if (γ (Sol*) > γ (Sol_best))
            Sol ← Sol*; Sol_best ← Sol*;
            γ (Sol) ← γ (Sol*); γ (Sol_best) ← γ (Sol*);
      else
            if ((γ (Sol*) == γ (Sol_best))
               Calculate cardinality of trial solution, |Sol*|;
               Calculate cardinality of best solution, |Sol_best|;
               If (|Sol*| < |Sol_best|)
                  Sol ← Sol*; Sol_best ← Sol*;
                  γ (Sol) ← γ (Sol*); γ (Sol_best) ← γ (Sol*);
            else
                 if (γ (Sol*) ≥ level)
                        Sol ← Sol*; γ (Sol) ← γ (Sol*);
            endif
      endif
      level = level + β;
      iteration++;
end do;
Calculate cardinality of best solution, |Sol_best|;
return |Sol_best|, γ (Sol_best), Sol_best;
```

Fig. 3. Pseudo code for GD-RSAR

4 Simulation Results

The proposed algorithm was programmed using Java and simulations were performed on the Intel Pentium 4 2.33 GHz. In this paper, we considered 13 well-known UCI datasets as shown in Table. 2. The preprocessing techniques such as filling missing values, removing noise and discretization were applied on the raw data of these datasets. This was to improve the quality of the data and to make them suitable for the model. For each dataset, the algorithm was run 20 times. The stopping conditions were the number of iterations when exceeds *NumOfIte* and the *level* when exceeds 1.

Table. 3 shows the parameters for the great deluge algorithm as employed in other available approaches. The parameters that GD-RSAR needs can be understood easily by users. Having less number of parameters and higher understandability makes the GD-RSAR method more comfortable to use. The correctness of the proposed method is proved by discovering the reducts with dependency degree equal to one (quality function of full conditional attribute set in these tested datasets).

Table 2. UCI datasets

Datasets	No of Attributes	No. of Objects
M-of-N	13	1000
Exactly	13	1000
Exactly2	13	1000
Heart	13	294
Vote	16	300
Credit	20	1000
Mushroom	22	8124
LED	24	2000
Letters	25	26
Drem	34	366
Drem2	34	358
WQ	38	521
Lung	56	32

Table 3. Parameter setting

Parameters	**Value**
Number of iteration	250
Estimated final quality	1

Table. 4 provides the comparison of our results with other results in the literature for these benchmark datasets. We compare our approach with other attribute reduction methods that are available in the literature on the thirteen instances. The algorithms compared in the table are described as follows: tabu search (TSAR) by Hedar et al. [5]; ant colony optimization (AntRSAR) by Jensen and Shen [6]; genetic algorithm (GenRSAR) by Jensen and Shen [6]; simulated annealing (SimRSAR) by Jensen and Shen [6]; ant colony optimization (ACOAR) by Ke et al. [8]; and scatter search (SSAR) by Wang et al. [15].

The entries in Table 4 represent the number of attributes in the minimal reducts obtained by each method. The superscripts in parentheses represent the number of

Table 4. Comparison of results

Datasets	GD-RSAR	TSAR	SimRSAR	AntRSAR	GenRSAR	ACOAR	SSAR
M-of-N	$6^{(10)}7^{(10)}$	6	6	6	$6^{(6)}7^{(12)}$	6	6
Exactly	$6^{(7)}7^{(10)}8^{(3)}$	6	6	6	$6^{(10)}7^{(10)}$	6	6
Exactly2	$10^{(14)}11^{(6)}$	10	10	10	$10^{(9)}11^{(11)}$	10	10
Heart	$9^{(4)}10^{(16)}$	6	$6^{(29)}\ 7^{(1)}$	$6^{(18)}\ 7^{(2)}$	$6^{(18)}7^{(2)}$	6	6
Vote	$9^{(17)}10^{(3)}$	8	$8^{(15)}\ 9^{(15)}$	8	$8^{(2)}9^{(18)}$	8	8
Credit	$11^{(11)}12^{(9)}$	$8^{(13)}\ 9^{(5)}\ 10^{(2)}$	$8^{(18)}\ 9^{(1)}\ 11^{(1)}$	$8^{(12)}\ 9^{(4)}\ 10^{(4)}$	$10^{(6)}11^{(14)}$	$8^{(16)}9^{(4)}$	$8^{(9)}\ 9^{(8)}\ 10^{(3)}$
Mushroom	$4^{(8)}5^{(9)}6^{(3)}$	$4^{(17)}\ 5^{(3)}$	4	4	$5^{(1)}6^{(5)}7^{(14)}$	4	$4^{(12)}\ 5^{(8)}$
LED	$8^{(14)}9^{(6)}$	5	5	$5^{(12)}\ 6^{(4)}\ 7^{(3)}$	$6^{(1)}7^{(3)}8^{(16)}$	5	5
Letters	$8^{(7)}9^{(13)}$	$8^{(17)}\ 9^{(3)}$	8	8	$8^{(8)}9^{(12)}$	8	$8^{(5)}\ 9^{(15)}$
Drem	$12^{(14)}13^{(6)}$	$6^{(14)}\ 7^{(6)}$	$6^{(12)}\ 7^{(8)}$	$6^{(17)}\ 7^{(3)}$	$10^{(6)}11^{(14)}$	6	6
Drem2	$11^{(14)}12^{(6)}$	$8^{(2)}\ 9^{(14)}\ 10^{(4)}$	$8^{(3)}\ 9^{(7)}$	$8^{(3)}\ 9^{(17)}$	$10^{(4)}11^{(16)}$	$8^{(4)}9^{(16)}$	$8^{(2)}\ 9^{(18)}$
WQ	$15^{(14)}16^{(6)}$	$12^{(1)}\ 13^{(13)}\ 14^{(6)}$	$13^{(16)}\ 14^{(4)}$	$12^{(2)}\ 13^{(7)}\ 14^{(11)}$	16	$12^{(4)}13^{(12)}14^{(4)}$	$13^{(4)}\ 14^{(16)}$
Lung	$4^{(5)}5^{(2)}6^{(13)}$	$4^{(6)}\ 5^{(13)}\ 6^{(1)}$	$4^{(7)}\ 5^{(12)}\ 6^{(1)}$	4	$6^{(8)}7^{(12)}$	4	4

runs that achieved the minimal reducts. The number of attribute without superscripts means that the method could obtain this number of attribute for all runs. Note that we used the same number of runs as other methods except SimRSAR, which used 30, 30 and 10 runs for Heart, Vote and Drem2 datasets respectively. Generally, the performance of GD-RSAR is comparable to other approaches except for some of datasets (Heart and Derm datasets) where the reducts obtained have slightly higher cardinality compared to other approaches. We believe that this is due to the generated initial solution where we started with a minimal number of reducts as reported in the literature. This might limit the search space for the great deluge algorithm to better explore the solution space in finding a better solution during the optimization process. That may have caused our search algorithm hard to improve the solution. However, our algorithm is less dependent on the parameters (as presented in Table. 5).

Table 5. Comparison of parameter setting

GD-RSAR	TSAR	GenRSAR	SimRSAR	SSAR	AntRSAR
2	7	3	3	5	2

Table 5 shows that unlike GD-RSAR, other methods namely TSAR has 7, GenRSAR and SimRSAR have 3 each, SSAR has 5, and finally AntRSAR has 2 parameters to be tuned in advance. We also believe that better solutions can be obtained if the linear great deluge algorithm applied in this paper is changed to a non-linear great deluge algorithm where the *level* in the non-linear great deluge is increased based on the quality of the current solution. This hypothesis will be considered in our future work.

5 Conclusion and Future Work

The attribute reduction problem in rough set theory has been studied in this paper by using a great deluge algorithm. To our knowledge, this is the first algorithm that addresses such a problem domain. In order to test the performance of our approach, experiments were carried out on UCI datasets and the results were compared with a set of state-of-the-art methods from the literature. It has been found that the algorithm always accepts the best solution and the worse solution will be accepted based on the *level.* This approach is simple yet effective and able to produce comparable results as compared to other approaches studied in the literature. Unlike other available methods, our algorithm however is less dependent on the parameters. In our future work, efforts will be made to establish, compare and report on timings in relation to previously reported literature. With the increasing complexity (in terms of the size) of the problems, we believe that the proposed approach can be easily adapted.

References

1. Abdullah, S., Burke, E.K.: A Multi-start large neighbourhood search approach with local search methods for examination timetabling. In: International Conference on Automated Planning and Scheduling (ICAPS 2006), Cumbria, UK, pp. 334–337 (2006)
2. Burke, E.K., Bykov, Y., Newall, J.P., Petrovic, S.: A time-predefined local search approach to exam timetabling problem. IIE Transactions 36(6), 509–528 (2004)
3. Dueck, G.: New Optimization Heuristics. The Great Deluge Algorithm and the Record-to-Record Travel. Journal of Computational Physics 104(1), 86–92 (1993)
4. Duntsch, I., Gediga, G.: Uncertainty measures of rough set prediction. Artificial Intelligence 106(1), 109–137 (1998)
5. Hedar, A., Wang, J., Fukushima, M.: Tabu search for attribute reduction in rough set theory. Technical Report 2006- 08, Department of Applied Mathematics and Physics, Kyoto University. Soft Computing 12(9), 909–918 (2006)
6. Jensen, R., Shen, Q.: Finding Rough set reducts with ant colony optimization. In: Proc. of 2003 UK Workshop Computational Intelligence, pp. 15–22 (2003)
7. Jensen, R., Shen, Q.: Semantics-Preserving Dimensionality Reduction: Rough and Fuzzy-Rough Based Approaches. IEEE Transactions on Knowledge and Data Engineering 16(12), 1457–1471 (2004)
8. Ke, L., Feng, Z., Ren, Z.: An efficient ant colony optimization approach to attribute reduction in rough set theory. Pattern Recognition Letters 29(9), 1351–1357 (2008)
9. Landa-Silva, D., Obit, J.H.: Evolutionary Non-linear Great Deluge for University Course Timetabling. In: Corchado, E., Wu, X., Oja, E., Herrero, Á., Baruque, B. (eds.) HAIS 2009. LNCS, vol. 5572, pp. 269–276. Springer, Heidelberg (2009)
10. McMullan, P.: An extended implementation of the great deluge algorithm for course timetabling. In: Shi, Y., van Albada, G.D., Dongarra, J., Sloot, P.M.A. (eds.) ICCS Part I 2007. LNCS, vol. 4487, pp. 538–545. Springer, Heidelberg (2007)
11. Pawlak, Z.: Rough sets. Internat. J. Comput. Inform. Sci. 11(5), 341–356 (1982)
12. Pawlak, Z.: Rough Sets: Theoretical Aspects of Reasoning about Data. Kluwer, Boston (1991)
13. Turabieh, H., Abdullah, S., McCollum, B.: Electromagnetism-like Mechanism with Force Decay Rate Great Deluge for the Course Timetabling Problem. In: Wen, P., Li, Y., Polkowski, L., Yao, Y., Tsumoto, S., Wang, G. (eds.) RSKT 2009. LNCS, vol. 5589, pp. 497–504. Springer, Heidelberg (2009)
14. Turabieh, H., Abdullah, S., McCollum, B.: A hybridization of Electromagnetism-like Mechanism and Great Deluge for Examination Timetabling Problem. In: Blesa, M.J., Blum, C., Di Gaspero, L., Roli, A., Sampels, M., Schaerf, A. (eds.) HM. LNCS, vol. 5818, pp. 60–72. Springer, Heidelberg (2009)
15. Wang, J., Hedar, A., Zheng, G., Wang, S.: Scatter Search for Rough Set Attribute Reduction. International Joint Conference on Computational Sciences and Optimization 1, 531–535 (2009)
16. WenXiu, Z., GuoFang, Q., WeiZhi, W.: A general approach to attribute reduction in rough set theory. Science in China Series F: Information Sciences 50(2), 188–197 (2007)

Syntactic Reordering for Arabic- English Phrase-Based Machine Translation

Arwa Hatem and Nazlia Omar

School of Computer Science, Faculty of Information Science and Technology
Universiti Kebangsaan Malaysia, 43600 Bangi, Selangor, Malaysia
{arwa,no}@ftsm.ukm.my

Abstract. Machine Translation (MT) refers to the use of a machine for performing translation task which converts text or speech in one Natural Language (Source Language (SL)) into another Natural Language (Target Language (TL)). The translation from Arabic to English is difficult task due to the Arabic languages are highly inflectional, rich morphology and relatively free word order. Word ordering plays an important part in the translation process. The paper proposes a transfer-based approach in Arabic to English MT to handle the word ordering problem. Preliminary tested indicate that our system, AE-TBMT is competitive when compared against other approaches from the literature.

Keywords: Rule based, Arabic Syntactic, word reordering.

1 Introduction

Translation is the transformation of natural language into other. Machine Translation (MT) refers to the use of a machine for performing translation task which converts text or speech in one Natural Language (Source Language (SL)) into another Natural Language (Target Language (TL)). The first successful attempt to create machine translation started in the 1949s after the Second World War. In the early 1960s some scientists were starting to lose hope in development MT due to the slow progress. In end of 1970 the Commission of the European Communities (CEC) supported work on the Eurotra system [2]. This was a project which aimed at the development of a multilingual interlingua system for most commonly used languages. Since 1970s to 1980s this projects started and proved to be practically successful. Recently much software became available for translation. Amongst many challenges that Natural Language Processing (NLP) presents the biggest is the inherent ambiguity of natural language. In addition, the linguistic diversity between the source and target language makes MT a bigger challenge. This is particularly true of languages widely divergent in their sentence structure such as Arabic and English language. The major structural difference between Arabic and English languages is that Arabic language are highly inflectional, with a rich morphology, relatively free word order, and default sentence structure as Subject-Object-Verb or Subject-Verb-Object or Verb-Subject-Object or

Y. Zhang et al. (Eds.): DTA/BSBT 2010, CCIS 118, pp. 198–206, 2010.

Verb-Object-Subject whereas, English follows structure as Subject-Verb-Object. As is recognized the world over, with the current state of art in MT, it is not possible to have fully automatic, high quality, and general-purpose machine translation. The major need is to handle ambiguity and other complexities of NLP in practical systems.

2 Related Work

Machine translation (MT) including the translation from and into Arabic has been attracting attention from the researchers and many approaches are applied to enhance the quality of machine translations. Abu Shquier and Sembok [1] asserts that Arabic language differs extremely in terms of its characters, and morphology from other languages. the authors develop MT system from English to Arabic using rule-based approach emphasis given in handling of word agreement and ordering . In addition Attia[2] describes the agreement as one of the features that greatly affect the output of MT. The author used the transfer approach to analysis of English as a source language, problems related to the transfer of English into Arabic, and the generation of Arabic as a target language focusing on implications of the agreement features in machine translation.

On the other hand Omar et al.[3] developed a machine translation system called Npae-Rbmt that translates Arabic noun phrases into English by using transfer-based approach. Yngve [4] reported that Arabic was one of the language besides English, German and French which were subjects of the COMIT (operational grammar encoding project) in the late 1950s. Chafia Mankai and Ali Mili [5] presented an attempt to carry out MT from Arabic language to English language and from Arabic language to French language. They have proved that analyzing Arabic and reordering must be doing to get good results according to Arabic rules. Salem et al[6] introduced an Arabic to English MT system called UniArab, which is based on rule based Role and Reference Grammar model to support rule-based lexical framework. We differ from the other efforts; we deal on utilizing syntactic analysis to overcome the reordering challenges.

This work presents a new system to translate the news titles of Aljazeera[1] website by using Transfer-based method. The motivation for this study is to develop an automated translator sufficient in translating from Arabic phrases into English.

We evaluate our system (AE-TBMT) by applied 100 titles from Aljazeera news website on it and compare it is translation results with translation results from Google translator and Microsoft translator. We chosen the corpus from Aljazeera website due to that Aljazeera has become the most popular in the Arab world.

[1] http://aljazeera.net/portal

3 Arabic Syntactic Matters

Arabic is a morphologically and syntactically complex language with many variations from English .Arabic morphology has been well studied in the context of MT. Preceding results all refer that the tokenization is helpful when translating from Arabic language [8, 9]. When translating from a morphologically rich language, tokenization means that the translation process is passed into multiple steps [10]. Arabic is segmented by simple punctuation tokenization. This rank of tokenization not enough to Syntactic analysis [12]. Chanod and Tapanainen [11] defined the tokenization as a significant issue in natural language processing as it is closely related to the morphological analysis. We used tokenization to get more segmented language. We expect we would achieve better performance by setting up more Arabic segmentation. In this work we focus on syntactic reordering for Arabic- English phrases. Below we describe eight major problems that regarding the translation from Arabic to English which motivates a number of our decisions in this paper.

1. Word order is one of the main differences between Arabic and English is sentence word order .In Arabic there are four structures for the sentence VSO, OVS, SVO or VOS, while English has only SVO order. A reordering rule should move the verb of an Arabic sentence to the right of the subject, for example اعلن الأمين العام للأمم المتحدة بان كي مون [Announced The Secretary General UN Ban Ki-moon] which translate to " The UN Secretary General Ban Ki-moon announced".
2. Arabic adjective usually follows their nouns except in superlative adjectives. Whereas, English adjectives precede their nouns. For example رجل غني [man rich] translate to" a rich man". Reordering rule should move the noun of an Arabic sentence to the right of the adjective, but the superlative case, translates without reordering for example اغنى رجل "richest man".
3. Idafa is one of the syntactic construction in Arabic, which indicates possession and compounding. English has three syntactic constructions. Idafa typically consists of one or more indefinite nouns followed by a definite noun [12]. For example مفاتيح البيت 'keys the house' translates as "the house keys", "the house's keys" and "the keys of the house". Two of the three English constructions require content word reordering.
4. Multiple meanings : this problem occurs because the different structures have different meanings of the same meaning. For example"المانيا تطلب " can translate to "Germany requests" or "Germany calls".
5. Grammatical relations. In some cases, when translating phrases from Arabic to English the sentences become weak distinction to determine which is the subject or object. For example " يزور الرئيس السوداني الرئيس الجزائري" , "Sudanese President visit Algerian President".

6. Addition and deletion: This problem appears because of the limitation of lexicon as the original translation contains extra words that have no equivalent words in the input of the source language.
7. Determiner Agreement: This problem appears in the target language because the noun phrase that are preceded by "a(n)" is translated as if it were preceded by "the".

4 The Approach

In a rule-based machine translation system the source sentence is first analysed morphologically in order to get a syntactic representation. This representation will be useful to find a suitable rule for the source sentence and the equivalent pattern for the target sentence in order to generate an acceptable sentence. The following steps summarize the translation process:

STEP 1: Input the source text (Arabic sentence) and start tokenization process. The task of the tokenizer is to divide the text into tokens.
STEP 2: Start morphological analysis. In this step the morphological analyzer provides morpho-syntactic information.
STEP 3: The syntactic parser builds a syntactic relevant tree, which represents relationships between the words of the phrase.
STEP 4: Lexical transfer will map Arabic lexical elements to their English equivalent. It will also map Arabic morphological features to the corresponding set of English features.
STEP 5: Structure transfer will map the Arabic dependency tree to the equivalent English syntactic structure.
STEP 6: Arabic synthesiser will synthesis the inflected English word-form based on the morphological features and traverse the syntactic tree to produce the surface English phrase.

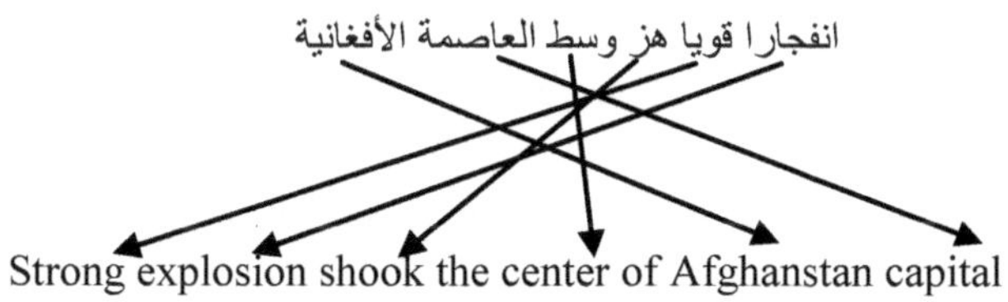
انفجارا قويا هز وسط العاصمة الأفغانية
Strong explosion shook the center of Afghanstan capital

S: [NP VP] == S: [NP:[N Adj] VP:[V N1 Det N2 Adj]] Arabic

S: [NP VP] == S: [NP: [Adj N] VP:[V Det N1 Adj N2]] English

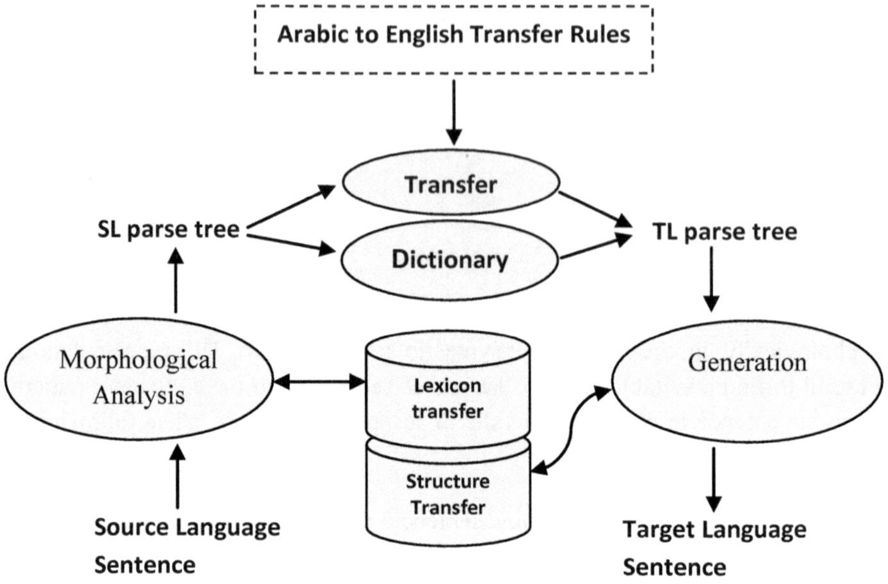

Fig. 1. Implementation process

Fig. 1 illustrates the whole process which includes the following stages:

- Tokenization: This an important step for a syntactic parser to construct a phrase structure tree from syntactic units. After inserting the source sentence in the system the tokenizer divides the text into tokens. The token can be a word, a part of a word, or a punctuation mark. A tokenizer requests to know the white spaces and punctuation marks.
- Morphological analysis: After the tokenization process the morphological analyser will provide the morphological information about words. It provides the grammatical class of the words (parts of speech) and create the Arabic word in its right form depending on the morphological features.
- Lexicon: In this system the lexicon is accountable for inferring morphological and classifying verbs, nouns, adverb and adjectives when needed. It is the main lexicon translation; the source language searches in a dictionary and then chooses the translation.

  A lexicon provides the specific details about every individual lexical entry (i.e. word or phrase) in the vocabulary of the language concerned.

  Lexicon contains grammatical information which are usually have abbreviated form: 'n' for noun, 'v' for verb, 'pron' for pronoun, 'det' for determiner, 'prep' for preposition, 'adj' for adjective, 'adv' for adverb, and 'conj' for conjunction. The lexicon must contain information about all the different words that can be used. If the word is ambiguous, it will be described by multiple entries in the lexicon, one for each different use.

- Parsing: The parser divides the sentence into smaller sets depending on their syntactic functions in the sentence. There are four types of phrases i.e. Verb Phrase (VP), Noun Phrase (NP), Adjective/Adverbial Phrase (AP), and Prepositional Phrase (PP). After the parsing process the sentence is represented in a phrase structure tree. Figure 1 show the phrase structure tree for the sentence الطالب الذكي قرأ الكتاب (the cleaver student reads the book).

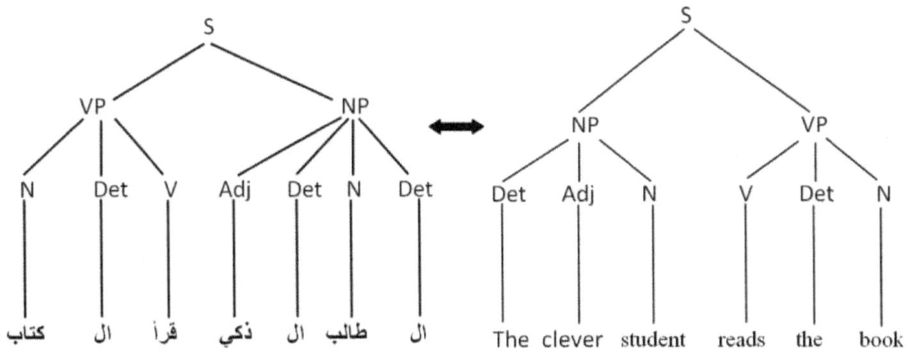

The above tree captures various grammatical relationships like dominance and precedence. Dominance means that some nodes dominate other nodes. In the above example S dominates NP and VP. Precedence means that some nodes precede other nodes. In the above example NP precedes VP. The root node is the node that dominates other nodes. The first NP is the root for Det, Adj and N, which in their turn are said to be children of the first NP and siblings of one another.

- Syntactic rules: A set of Arabic and English rules are fed into the system. In this step the reordering process will be found which will be based on the order of words in a sentence, and how the words are grouped.
- Agreement rules: After syntactic rules the agreement rules applied which are responsible about the additions of prefix and suffix in the sentences.

In order to test the performance of our approach, experiments are carried out to translate the news titles of Aljazeera website. We evaluate our results on the 100 sentences (in different fields i.e. political, sports and economic news) taken from Aljazeera news website and which are available at http://www.aljazeera.net/portal.

5 Results and Discussion

The aim of this experiment is to investigate whether the following machine translation systems, namely, Bing Translator, Google, Systran and AE-TBMT, are suitable for handling the word reordering in the translation from Arabic to English. We evaluate our results on the 100 sentences taken from Aljazeera news website and which are available at http://www.aljazeera.net/portal. The evaluation involves counting the numbers of problems which appear in each of the system.

The weight for every problem is 1 in all sentences, and it can be interpreted as the basic unit of penalty as we count every problem that appears in the target sentence as 1 weakness.

The percentage of the total score for each system is calculated by dividing the total score by 700. We have 100 test examples and each is evaluated out of 7 depending on the numbers of problems. The score is given by human expert in translation and it tests the differences between the human translation and the machine translation systems. Table 2 illustrates part of the result achieved by this experiment. For the first example i.e. "خطة عراقية لإنشاء محطات تكرير", the weakness of Google system is in problems (4,6) and for that , the system score 5 out of 7. The weakness of Bing system was in problem (6) and the system scored 5 out of 7. The weakness of Systran system was in problem (4,6) and the system scored 5 out of 7. AE-TBMT got score 6 out of 7 because there is only one weakness translation in problem (4).

Table 1, 2 and 3 show the comparison results on some different sentences when applied to above translation systems. In table 1 a comparison between the manual translation results and other machine translation systems results are shown.

Table 1. Experiment results and comparison with other systems

Sentence	Manual Translation	Google	Bing	Systran	AE-TBMT
خطة عراقية لإنشاء محطات تكرير	Iraqi plan to build refinaries	Plan for the establishment of an Iraqi refineries	Iraqi plan for refineries	Iraqi plan for establishment of stations repeating	Iraqi plan to establishment refineries
جرحى في انفجارات غزة	Injuries in Gaza blasts	Wounded in the blasts Gaza	Wounded in Gaza bombing	Wound in explosions of Gaza	Wounded in Gaza explosions
مصر ترفض تدخلا تركيا بالمصالحة	Egypt refuses interference of Turkeg in reconciliation	Egypt rejects interference Turkey reconciliation	Egypt rejects Turkish intervention in reconciliation	Egypt refuses Turkish interventions in the reconciliation	Egypt refuses interference of Turkey in reconciliation
انفجارا قويا هز وسط العاصمة الأفغانية	Strong explosion shook the center of Afghanstan capital	A powerful explosion rocked the centre of Afghan capital	Powerful explosion Central	Strong explosions shaking in the middle of the capital Afghan	a strong explosion shooke centre of Afghan capital
تعتقل امريكا شبكة تجسس روسية	America arrested Russian spy net	America spy network arrested Russia	Arrest of Russian spy network America	America arrests Russian net of spying	America arrests Russia spy net
تقترح ألمانيا موعد لمغادرة أفغانستان	German proposes time to leave Afganistan	Germany is proposing a date to leave Afghanistan	Germany proposes to leave Afghanistan	Afghanistan proposes Germany appointment for departure	Germany proposes a date to leave Afghanistan

Table 2 shows the number of problems which appeared after the translation and shows how many penalties each system has been scored. For example in Google when we translate the sentence "خطة عراقية لإنشاء محطات تكرير" problem 4 and 6 occurred in the output and as we give 1 penalty for each problem, Google has two penalties. Bing has only one penalty because only problem 6 occurred and Systran has two penalties similar to Google. AE-TBMT has one penalty when problem 4 occurred in the output.

Table 2. Number of Problems and Penalties

Sentence	Google		Bing		Systran		AE-TBMT	
	Problem	Penalty no.	problem	Penalty no.	Problem	Penalty no.	problem	Penalty no.
خطة عراقية لإنشاء محطات تكرير	4,6	2	4,6	2	4,6	2	4	1
جرحى في انفجارات غزة	3,4,6	3	3,4,6	3	3,4	2	4	1
مصر ترفض تدخلا تركيا بالمصالحة	2,4,6	3	2,4	2	2,4,6,7	4	0	0
انفجارا قويا هز وسط العاصمة الأفغانية	4	1	1,3	2	4,6	2	6	1
تعتقل امريكا شبكة تجسس روسية	1,2,5,6	4	1,4,6	3	1,6	2	1	1
تقترح ألمانيا موعد لمغادرة أفغانستان	1,4,6	3	1,6	2	1,3,4,5,6	5	1,4	2

Table 3. Total Number of Penalties

Translators	Penalties
Google	122
Bing	161
Systran	149
AE-TBMT	113

It can be seen that in general, the result for AE-TBMT is better than other approaches as reported in Table 3 the system has only 113 problems after translating 100 sentences. This shows that AE-TBMT is able to generate best translation from Arabic to English.

We believe that a good translation can be achieved with a systematic lexicon which provides more correct translations. It can be seen that with more complicated sentence structures, the features of the different words are essential in accurate translation. In addition, good reordering rules play an important role in the quality of translation.

6 Conclusion and Future Work

In this paper, we described a set of syntactic reordering rules that exploit systematic differences between Arabic and English word order to transform Arabic sentences to be equivalent to English in terms of their word order. The approach was tested on 100 titles from Aljazeera news website. Preliminary comparisons indicate that our approach is competitive with other approaches in the literature. Our manual evaluation of the reordering accuracy indicated that our approach is helpful at improving the translation quality despite relatively frequent reordering errors. Future research will be aimed at testing our approach on different domain.

References

1. Abu Shquier, M., Sembok, T.: Handling agreement in machine translation from English to Arabic. In: 1st International Conference on Digital Communications and Computer Applications (DCCA 2007), pp. 385–379 (2007)
2. Attia, M.: Mplications of the Agreement Features in Machine Translation, Master thesis. Al-Azhar University (2002)
3. Shirko, O., Omar, N., Arshad, H., Albared, M.: Machine Translation of Noun Phrases from Arabic to English Using Transfer-Based Approach. Journal of Computer Science 6(3), 350–356 (2010)
4. Yngve, V.H.: Early Research at M.I.T. In: John Hutchins, W. (ed.) Research of Adequate Theory Early years in machine translation: memoirs and biographies of pioneers, pp. 39–72 (2000)
5. Mankai, C., Mili, A.: Machine Translation from Arabic to English and French information sciences, vol. 3, pp. 91–109. Elsevier Science Inc., New York (1995)
6. Salem, Y., Hensman, A., Nolan, B.: Implementing Arabic to- English Machine Translation using the Role and Reference Grammar Linguistic Model. In: Proceedings of the Eighth Annual International Conference on Information Technology and Telecommunication (ITT 2008), Galway, Ireland (Runner-up for Best Paper Award) (2008)
7. Lavie, A., Probst, K., Peterson, E., Vogel, S., Levin, L., Font-Llitjos, A., Carbonell, J.: A Trainable Transfer-based Machine Translation Approach for Languages with Limited Resources. In: Proceedings of Workshop of the European Association for Machine Translation, Valletta, Malta, EAMT 2004 (2004)
8. Habash, N., Sadat, F.: Arabic preprocessing schemes for statistical machine translation. In: Proceedings of HLT-NAACL 2006, New York, NY, USA (2006)
9. Lee, Y.: Morphological Analysis for Statistical Machine Translation. In: Proceedings of HLTNAACL 2004, Boston, MA, USA (2004)
10. Badr, I., Zbib, R., Glass, J.: Segmentation for English-to-Arabic statistical machine translation. In: Proceedings of ACL 2008, HLT: Short Papers, Columbus,OH, USA (2008)
11. Jean-Pierre, C., Pasi, T.: A Non-Deterministic Tokenizer for Finite-State Parsing. In: The European Conference on Artificial Intelligence, Workshop on Extended Finite State Models of Language (ECAI 1996), Budapest, Hungary, pp. 10–12 (1996)
12. Elming, J., Habash, N.: Syntactic Reordering for English-Arabic Phrase-Based Machine Translation. In: Proceedings of the EACL, Workshop on Computational Approaches to Semitic Languages, Athens, Greece, pp. 69–77 (2009)

The Concept of Indiscernibility Level of Rough Set to Reduce the Dendrogram Instability

R.B. Fajriya Hakim[1], Subanar[2], and Edi Winarko[2]

[1] Statistics Department, Faculty of Mathematics and Natural Sciences
Universitas Islam Indonesia,
Jalan Kaliurang KM 14.5 Sleman, Jogjakarta, Indonesia, 55584
[2] Mathematics Department, Faculty of Mathematics and Natural Sciences,
Universitas Gadjah Mada
Sekip Utara, Jogjakarta, Indonesia 55528
hakimf@fmipa.uii.ac.id

Abstract. The main concept of rough sets theory is clustering similarities of objects based on the notions of indiscernibility relation. In this paper, we develop the concept of indiscernibility level of rough set theory as an additional measurement for hierarchical clustering. The combination between indiscernibility (quantitative indiscernibility relation) and indiscernibility level are used as a new method for hierarchical clustering. The indiscernibility level quantifies the indiscernibility of pairs of objects among other objects in information system. For comparison, the following four clustering methods were selected and evaluated on a simulation data set : average-, complete- and single-linkage agglomerative hierarchical clustering and Ward's method. The simulation shows that the hierarchical clustering yields dendrogram instability that gives different solutions under permutations of input order of data objects. The result of this paper shows that the new method plays an important role in clustering information system and compared to other method, clustering based on indiscernibility and its indiscernibility level reduces the dendrogram instability.

Keywords: Rough Set, Hierarchical clustering, Indiscernibility, Indiscernibility Level, Instability.

1 Introduction

The goal of cluster analysis is to develop a classification scheme that will partition the set of objects into distinct groups (clusters). To disclose the clustering in the set of objects, a measure of nearness (proximity) needs to be defined. Two proximity measures are the degree of distance and the degree of association. The degree of distance represents the dissimilarity between objects and the degree of associations represents the similarity between objects. Since similarity is used as a proximity measure, the measurement value will increase when two objects become more similar. It is interesting to explore the feasibility of indiscernibility relations of canonical rough sets for similarity measure since each indiscernible relation is also a sort of

Y. Zhang et al. (Eds.): DTA/BSBT 2010, CCIS 118, pp. 207–225, 2010.

cluster, hence indiscernibility relation is used as a measure of similarity instead of distance function to clustering the objects. Quantitative indiscernibility relation as an indiscernibility relation with binary attribute values on information system, which is reflexive and symmetric [1], creates a similarity measure of indiscernible classes among pairs of objects in information system. Quantitative indiscernibility relation and indiscernibility level have been used as a new method for hierarchical clustering introduced by Hakim [2]. Indiscernibility level is the ratio of quantitative indiscernibility relation among pairs of objects to the total quantitative indiscernibility relation of object pairs on the information system. The indiscernibility level shows the level of global agreement of object pairs on the information system. The purpose of this study is to compare the performance of this new method i.e. clustering based on indiscernibility and indiscernibility level with traditional hierarchical cluster analysis including single-, complete-, average linkage and Ward's method. This study shows that the traditional hierarchical cluster analysis has dendrogram instability due to data input order while clustering based on indiscernibility and indiscernibility level can reduce instability of cluster analysis due to data input order.

2 Literature Review about the Comparison of Clustering Techniques

There are various studies within the literature that used different clustering methods and compared their results. Hirano et al.[3] showed how clustering methods work on a practical medical data set and Ward's method give better clusters than other three method including single- , complete-linkage agglomerative hierarchical and rough clustering when clinically reasonable attributes were selected. Mingoti and Lima [4] elaborated studies about comparison of hierarchical and nonhierarchical clustering techniques in his introduction. In their research, a total of 2530 data sets were simulated and they also considered correlated and uncorrelated variables, overlapping and non-overlapping cluster with and without outliers. Their results show that Fuzzy C-means has a very good performance than Self Organizing Map (SOM), K-means and traditional cluster analysis. Budayan et al.[5] in their literature survey had elaborated a lot of study about comparison among some nonhierarchical and hierarchical clustering techniques and their discussion concluded that there is no consensus about the superiority of a given clustering method with respect to the others and insisted that the suitability of clustering methods to a given problem changes with the structure of the data set and purpose of the study. For their strategic grouping, Self Organizing Maps (SOM) and Fuzzy C-means (FCM) gives better performance than traditional cluster analysis. Finch [6] focused on binary data and conducted cluster analysis using four distance measures, Russel/Rao, Jaccard, Matching coefficient and Dice. He analyzed for real data, the Jaccard, Russel/Rao and Dice measures are largerly in agreement in terms of grouping individuals, and appear to do a better job of finding distinct clusters in the data than either the Matching or raw data approaches and suggested that the results may only be applicable to his dataset research. When the binary data contain noise, Hitchcock and Chen [7] suggested a smoothing operation, either on the data themselves or on the dissimilarities as a preliminary step to cluster analysis but there is no guarantee to produces a better partition of the test items and give a few differences than a common hierarchical clustering.

All those papers presented interesting results. However none of them consider the problem of instability of hierarchical cluster analysis due to data input order. Van Der Kloot et al. [8] showed that hierarchical agglomerative cluster analysis yields different solutions (dendrogram) under permutations of the input order of the data. In every hierarchical clustering analysis, a user may choose among any distance functions and similarity coefficients to build a proximity matrix, then the cluster procedure finds the closest objects and merges them into a cluster. Next, the distances between the new cluster and the remaining objects are computed, and again the two entities (cluster or objects) with the smallest distance are merged. Problem arises when two pairs or more have the smallest distance value (i.e. the smallest distance is tied). They disclose the fact that hierarchical clustering may yield solutions that depend on the order of processing ties. This fact has been known for more than 50 years and the ties processing caused the instability. They recommended to repeat the analysis on a large number of random permutations of the rows and column of the proximity matrix and select a solution with the highest goodness-of-fit. At the end of their papers, they agreed that this best-fitting solution is not necessarily the best or most actionable solution and wether it might be possible to develop other criteria to cluster analysis. In this paper we presented a simulation study of new method in hierarchical agglomerative clustering analysis introduced in Hakim et al. [2] and show how this method can reduce the instability of cluster analysis and compared to other method of traditional hierarchical cluster analysis.

3 Clustering Methods: A Brief Explanation

3.1 The Agglomerative Hierarchical Clustering

Hierarchical clustering has been widely applied to cluster analysis to determine the number of cluster of data sets. Hierarchical clustering methods group data objects into a treelike structure of cluster (dendrogram). There are two main approaches of this methods: divisive and agglomerative hierarchical techniques. Hierarchical divisive methods start from single cluster that contains all objects and finally divide it into each object representing its own cluster. Hierarchical agglomerative methods begin with a cluster that contains a single object and finally group objects until all objects form a single cluster. There are at least three main categories that indicate how pairs of the clusters are merged in hierarchical agglomerative method: linkage, centroid and error variance method. A linkage method is used to compare the clusters and to decide which of them should be merged in each step of the clustering processes. Single, complete and average are common procedures in linkage method. In single linkage, the smallest value of the objects distance is defined as the distance between two clusters. This method is also called the nearest neighbor technique. In contrast to the single linkage, the complete linkage considers the largest (objects) distances as the distance between two clusters. It is also called the farthest neighbor technique. The average linkage considers all members of the clusters and evaluates the average distances as the distance between two clusters. Centroid method is quite similar to the average linkage method and uses the cluster centroid as the dissimilarity of two clusters. The Ward's method is based on the sum of squares within a cluster then a new cluster will be formed by determining the smallest increase in overal sum of the squared within-cluster distances among all clusters. Budayan et al. [5] reminded that each above-mentioned method can illustrate different shape of dendrogram with respect to the effects of outliers and the order of the objects in the data set.

In many applications of hierarchical cluster analysis in information system, one begins with a data matrix and then continues to a data matrix; that is, the starting point of a cluster analysis in information system is a data matrix $S = |U| \times |At|$ with u measurements (objects) of a attributes, where U and At is defined at definition 1 later. Given the proximity matrix of order ($|U| \times |U|$) say, the entries may represent dissimilarities or similarities between pair of objects. S is widely known as an information tables, also known as information systems, data tables, attribute-value systems, and investigated by many researchers of rough set theory. It is assumed that data are represented in a table form, where a set of objects (rows) are described by a finite set of attributes (columns).

Definition 1. An information table S is the tuple

$$S = (U, At, \{V_a \mid a \in At\}, \{I_a \mid a \in At\}), \tag{1}$$

where U is a finite non-empty set of objects called universe, At is a finite non-empty set of attributes, V_a is a non-empty set of values for an attribute $a \in At$, and $I_a : U \rightarrow V_a$ is an information function, such that for an object $x_i \in U$, $i \in \{1, 2, .. |U|\}$, and attribute $a \in At$, and a value $v \in V_a$, $I_a(x_i) = v$ means that the object x_i has the value v on the attribute a.

The nature of the observation values plays an important role in the choice of proximity measure for clustering. Nominal values (like binary variables) lead in general to proximity values, whereas metric values lead (in general) to distance matrices. We present here possibilities for S in the binary case, so $I_a(x_i) = v$, $v \in \{0, 1\}$. In order to measure the similarity between objects with binary structure, we always compare pairs of observations (x_i, x_j) where x_i, $x_j \in \{0, 1\}$, and $x_i = 1$ indicate that the object possesses the attribute a, $a \in At$. Obviously there are four cases :

$$x_i = x_j = 1,$$
$$x_i = 0,\ x_j = 1,$$
$$x_i = 1,\ x_j = 0,$$
$$x_i = x_j = 0.$$

Define,

$$a_1 = \sum_{i,j=1}^{|U|} I(x_i = x_j = 1)$$

$$a_2 = \sum_{i,j=1}^{|U|} I(x_i = 0, x_j = 1)$$

$$a_3 = \sum_{i,j=1}^{|U|} I(x_i = 1, x_j = 0)$$

$$a_4 = \sum_{i,j=1}^{|U|} I(x_i = x_j = 0) \tag{2}$$

Then the following proximity measures are used in practice :

$$d_{ij} = \frac{a_1 + \delta\, a_4}{a_1 + \delta\, a_4 + \lambda\,(a_2 + a_3)} \tag{3}$$

where δ and λ are weighting factors. Table 1 below shows some similarity measures for given weighting factors [9].

Table 1. Some similarity measures for clustering

Name	δ	λ	Definition
Jaccard	0	1	$\frac{a_1}{a_1 + a_2 + a_3}$
Tanimoto	1	2	$\frac{a_1 + a_4}{a_1 + 2(a_2 + a_3) + a_4}$
Simple Matching (M)	1	1	$\frac{a_1 + a_4}{\vert A_t \vert}$
Russel and Rao (RR)	-	-	$\frac{a_1}{\vert A_t \vert}$
Dice	0	0.5	$\frac{2a_1}{2a_1 + (a_2 + a_3)}$
Kulezynski	-	-	$\frac{a_1}{a_2 + a_3}$

These measures provide alternative ways of weighting mismatchings and positive (presence of a common attribute) or negative (absence of a common attribute) matchings. In principle, we could also consider the Euclidian distance. However, the disadvantage of this distance is that there will be a lot of element value in the matrix similarity that have the same value since the binary case is used and this produces the instability of dendrogram as a result of hierarchical clustering.

3.2 Clustering Based on Indiscernibility and Indiscernibility Level

3.2.1 Indiscernibility Relation of Rough Set Theory

The notions of indiscernibility of Rough Set Theory [10] have attracted some researcher in clustering. Indiscernibility shows a sort of cluster that the objects share the same properties and it is used as a proximity measure without any distance function for clustering the objects. This notion of indiscernibity is related to the

Leibniz's Law of Indicernibility that objects are indiscernible if it is characterized by the same information in view of the available information about them and in the rough set approach, the indiscernibility of objects information is defined relative to a given set of attributes [11,12]. Parmar et al. [13] proposed an algorithm for clustering categorical based on Rough Set theory, termed Min-Min-Roughness (MMR). MMR algorithm has the ability to handle uncertainty in the clustering process but it can not handle the cluster stability due to data input order. Kumar et al. [14] presented the Indiscernibility-based rough agglomerative hierarchical clustering algorithm for sequential data. Kumar et al. [14] used indiscernibility relations rather than similarity relations with the reason that similarity relations do not give the same kind of partitions of the universe as the indiscernibility relations. Similarity classes of each object x present in the universe U provide the similarity information for the object. An object from one similarity class may be similar to objects from other similarity classes. Therefore the basic granule of knowledge is intermixed.

Upadhyaya et al. [15] used indiscernibility relation as a single measurement for clustering and use it as an approach of similarity measure in cluster analysis which creates indiscernible classes and represent this classes with graph. Indiscernibility and discernibility relations and the two pairs of dual relations: the strong-indiscernibility weak-discernibility relations and the weak-indiscernibility strong-discernibility relations has been used for data analysis by Zhao [1]. Their main methods can simplify the table and rules and still can not show the granules (cluster) of information system. Hirano and Tsumoto have done a lot of research about clustering based on indiscernibility relations and they called their method as Rough Clustering [16,17]. Hirano et al. [18] compared the traditional hierarchical method, Ward's method and Rough Clustering for clininal databases but their experimental results showed that the best clusters were obtained using Ward's method when the clinically reasonable attributes were selected. They also used combination of indiscernibility level as a measure for hierarchical clustering and single linkage algorithm to construct dendrogram [19, 20]. Their indiscernibility level $\gamma(x_i, x_j)$ is defined for pair of objects x_i and x_j that quantifies the ratio of binary classifications that agree to classify x_i and x_j as indiscernible, as follows

$$\gamma(x,y) = \frac{\sum_{k=1}^{|U|} \delta_k^{indis}(x_i, x_j)}{\sum_{k=1}^{|U|} \delta_k^{indis}(x_i, x_j) + \sum_{k=1}^{|U|} \delta_k^{dis}(x_i, x_j)}$$

Where

$$\delta_k^{indis}(x_i, x_j) = \begin{cases} 1, & if\ (x_i \in [x_k]_{R_k} \wedge x_j \in [x_k]_{R_k}) \\ 0, & otherwise \end{cases}$$

and

$$\delta_k^{dis}(x_i, x_j) = \begin{cases} 1, & if\ (x_i \in [x_k]_{R_k} \wedge x_j \notin [x_k]_{R_k})\ or \\ & if\ (x_i \notin [x_k]_{R_k} \wedge x_j \in [x_k]_{R_k} \\ 0, & otherwise \end{cases} \tag{4}$$

This quantity will be used to merge the two objects and it is a stepwise abstraction hierarchical clustering process that goes from bottom to the top. This indiscernibility level needs a method for determining the binary classification which is arbitrary and this level quantifies the ratio of indiscernibility of a pair of objecst to the sum of indiscernibility and discernibility of all objects. The denominator of that level is indiscernibility and discernibility a pair of objects, unlike the above indiscernibility level, Hakim et al. [2] proposed a new quantity of indiscernibility level as the ratio of the indiscernibility of a pair of objects to the total indiscernibility object pairs in information system. This level shows how strong an indiscernibility of a pair of objects compared to all objects in information system.

3.2.2 A New Method : Clustering Based on Indiscernibility and Indiscernibility Level

Hakim et al. [2] introduced a new method of clustering based on Indiscernibility and Indiscernibility Level. The method here can be illustrated by the picture of set of twins. Figure 1 shows what do we mean by the concept of indiscernibility level. Child number 1 and number 2 have much in common, while others also have much in common to others (say, child number 3 and number 12 and so on).

Fig. 1. Set of twins

In Figure 2, we re-arrange the order of the children and we give a name to each child as A, B, C, ..., L. A simple question is asking whether a pair of (A, B) having more similarities among children than a pair of (G, H) or not. (A, B) having more similarities with other children than (G, H). We could say that (A, B) have higher global agreement to children than (G, H) or (A, B) have higher indiscernibility level than (G, H).

Fig. 2. Re-arrange the order of children

Usually many methods in hierarchical clustering only consider with the similarities or indiscernibility between objects and do not pay attention to how many similarity agreement of a pair of objects among other objects in one information table. The similarity agreement of a pair of objects with other objects represents the indiscernibility level of a pair of objects.

The indiscernibility here is also motivated by quantitative indiscernibility relations presented in Zhao et al. [1]. Quantitative indiscernibility relation, which is reflexive and symmetric, will be used to count the number of attributes on which two objects have similar or different values. Quantitative indiscernibility relation is the weaker forms of indiscernibility relation which was assumed to be an equivalence relation. With respect to a non-empty set of attributes $A \subseteq At$, a graded, or quantitative, indiscernibility relation is defined as a mapping from $U \times U$ to the unit interval [0, 1].

Definition 2. Given a subset of attributes $A \subseteq At$ and a pair of objects $(x_i, x_j) \in U \times U$, $i, j \in \{1, 2, ..., |U|\}$, the quantitative indiscernibility relation $ind(A)(x_i, x_j)$ is defined by

$$ind(A)(x_i, x_j) = \frac{|\{a \in A \mid I_a(x_i) = I_a(x_j)\}|}{|A|} \tag{5}$$

where $|.|$ denotes the cardinality of a set and two cases for $ind(A)(x_i, x_j)$ for $i \neq j$,

$x_i = x_j = 1$ and
$x_i = x_j = 0$

A quantitative indiscernibility relation satisfies the following properties:

(i1) $ind(A)(x_i, x_i) = 1$
(i2) $ind(A)(x_i, x_j) = ind(A)(x_j, x_i)$

The properties (i1) and (i2) reflect that a quantitative indiscernibility relation is reflexive and symmetric. We here introduce indiscernibility level, a novel measure that solves the above problems and makes it possible to represent the granularity of objects while keeping the use of independently defined binary classifications. The indiscernibility level, $\gamma(x_i, x_j)$, defined for a pair of objects x_i and x_j, is the ratio of the number that x_i and x_j have same values at the same attributes to the sum of the number that x_i and x_j have same values at the same attributes and the number of other objects (x_k, $k \in \{1, 2, ..., |U|\}$) that have the same attribute values with attributes of x_i and x_j. The higher level of indiscernibility implies that they are likely to be treated as indiscernible:

$$\gamma(x_i, x_j) = \frac{ind(A)(x_i, x_j)}{ind(A)(x_i, x_j) + ind(A)((x_i, x_j), x_k))} \tag{6}$$

or,

$$\gamma(x_i, x_j) = \frac{|\{a \in A \mid I_a(x_i) = I_a(x_j)\}|}{|\{a \in A \mid I_a(x_i) = I_a(x_j)\}| + |\{a \in A \mid (I_a(x_i) = I_a(x_j)) = I_a(x_k)\}|}$$

For any $i, j, k \in \{1, 2, .., |U|\}$, $k \neq i, j$, there are two cases for $ind(A)((x_i, x_j), x_k)$

$$x_i = x_j = x_k = 1 \text{ and } x_i = x_j = x_k = 0$$

The quantitative indiscernibility relations are connected to the level of indiscernibility relations as show in table 2:

Table 2. Criteria for Indiscernibility and Indiscernibility Level

$ind(A)(x_i, x_j)$	$\gamma(x_i, x_j)$	Note
1	1	x_i and x_j strict indiscernibility
1	$0<\gamma(x_i, x_j)<1$	x_i and x_j strong indiscernibility
$0< ind(A)(x_i, x_j) <1$	1	x_i and x_j weak indiscernibility, $\|U\| = 2$
$0< ind(A)(x_i, x_j) <1$	$0<\gamma(x_i, x_j)<1$	x_i and x_j weak indiscernibility
0	0/0	x_i and x_j strict discernibility

$\gamma(x_i, x_j) = 0/0$ is regarded as 0 since the pairs of object (x_i, x_j) do not meet any attributes at all in information system.

Definition 3. Given a subset of attributes $A \subseteq At$ and a pair of objects $(x_i, x_j) \in U \times U$, $i, j \in \{1, 2, ..., |U|\}$, the quantitative discernibility relation $dis(A)(x_i, x_j)$ is defined as the complement of a quantitative indiscernibility relation :

$$dis(A)(x_i, x_j) = 1 - ind(A)(x_i, x_j) = \frac{|\{a \in A \mid I_a(x_i) \neq I_a(x_j)\}|}{|A|} \tag{7}$$

A quantitative discernibility relation satisfies the following properties:

($d1$) $dis(A)(x_i, x_i) = 0$

($d2$) $dis(A)(x_i, x_j) = dis(A)(x_j, x_i)$

The properties ($d1$) and ($d2$) reflect that a quantitative discernibility relation is reflexive and symmetric. The discernibility levels measure the two objects are discernible among all objects given attribute set.

$$\xi(x_i, x_j) = \frac{dis(A)(x_i, x_j)}{dis(A)(x_i, x_j) + dis(A)((x_i, x_j), x_k))} \qquad (8)$$

or

$$\xi(x_i, x_j) = \frac{|\{a \in A \mid I_a(x_i) \neq I_a(x_j)\}|}{|\{a \in A \mid I_a(x_i) \neq I_a(x_j)\}| + |\{a \in A \mid I_a(x_i, x_j) \neq I_a(x_k)\}|}$$

For any $i, j, k \in \{1, 2, .., |U|\}$, $k \neq i, j$, there are six cases for $dis(A)((x_i, x_j), x_k)$

$x_i = x_j = 1, x_k = 0$
$x_i = x_j = 0, x_k = 1$
$x_i = 0, x_j = 1, x_k = 1$
$x_i = 0, x_j = 1, x_k = 0$
$x_i = 1, x_j = 0, x_k = 1$
$x_i = 1, x_j = 0, x_k = 0$

The quantitative discernibility relations are given here to show the value of x_i, x_j and x_k that do not meet the indiscernible pair of objects.

Method. The proposed method consists of five steps:

1. Given an information table S with $V_a \in \{0, 1\}$, assign the indiscernibility matrix ($|U| \times |U|$) to each element $ind(A)(x_i, x_j)$ defined as:

$$ind(A)(x_i, x_j) = \frac{|\{a \in A \mid I_a(x_i) = I_a(x_j)\}|}{|A|}.$$

   Each element of an indiscernibility matrix stores the attributes on which the corresponding two objects have the same values.

2. Compute the indiscernibility level for each of the objects according to the formula. Then construct an indiscernibility level matrix ($|U| \times |U|$) for each element $\gamma(x_i, x_j)$ defined as:

$$\gamma(x_i, x_j) = \frac{|\{a \in A \mid I_a(x_i) = I_a(x_j)\}|}{|\{a \in A \mid I_a(x_i) = I_a(x_j)\}| + |\{a \in A \mid I_a(x_i) = I_a(x_j) = I_a(x_k)\}|}$$

3. Combine two matrices, $ind(A)(x_i, x_j)$ and $\gamma(x_i, x_j)$.
4. For the combination value ($ind(A)(x_i, x_j) \mid \gamma(x_i, x_j)$), choose the maximum value for $ind(A)(x_i, x_j)$ first and then $\gamma(x_i, x_j)$ to merge two objects.
5. Construct a dendogram using hierarchical clustering based on step 4.

4 Comparing Cluster Techniques Using Synthetic Data Set

To evaluate and compare the dendrogram of traditional hierarchical clustering and clustering based on indiscernibility and indiscernibility level we use simulation data. Simulated binary data for 5 rows x 3 columns are generated by the *bindata* package in R software [26], which produces

```
      [,1]  [,2]  [,3]
[1,]    1     0     0
[2,]    1     1     0
[3,]    1     0     1
[4,]    0     0     1
[5,]    1     1     1
```

We use this result and changes it to, say, product A, B, C, D, and E. Product A takes the values from first row, product B takes the values from second row and so on. The columns represent, say, a features of each product. So, this dataset consists of three attributes (a, b, c) features for five products (A, B, C, D, E) as shown in table 3.

Table 3. Features of datasets

Objects	*a*	*b*	*c*
Product A	1	0	0
Product B	1	1	0
Product C	1	0	1
Product D	0	0	1
Product E	1	1	1

A *features* [25] with binary attribute value which is coded 1 to indicate the presence of an attribute and 0 to indicate its absence.

This synthetic dataset illustrates a businessman who would like to make an advertisement for 5 products and sale for 5 products with different brand and would like to give customer a discount price for buying two or three products together. Therefore he needs to determine the products group. This dataset also shows how different orders of objects may lead to different tree structure of the dendrogram.

For the simulated data set, we clustered the objects of features based on two clustering algorithm : the first were a hierarchical clustering methods, average linkage, complete linkage, single linkage and Ward's implemented by the R function *hclust* [21]. The other were, average linkage, complete linkage, single linkage and ward's implemented by the R function *agnes* [22] in the *cluster package.* Four clustering methods implemented by function *hclust* and *agnes* are presented graphically in Figure 3 and Figure 4, respectively. For the average, complete linkage and Ward's method result, we can examine from the dendrograms showing how the objects that are partitioned are equal for each R function but their height of objects merged were different. The single linkage method only gives one cluster for both R functions.

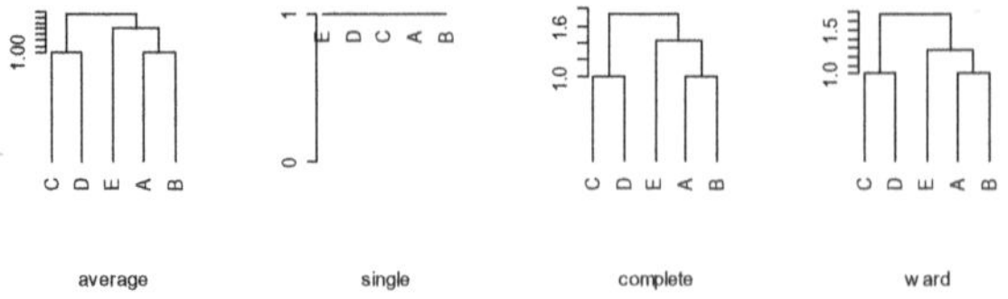

Fig. 3. Dendrograms for R function *hclust*

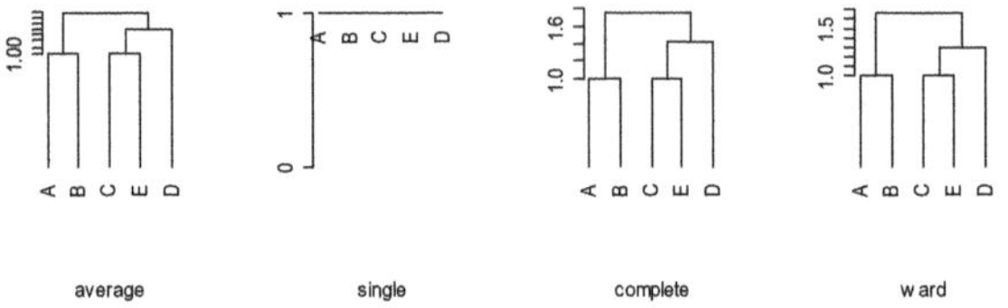

Fig. 4. Dendrograms for R function *agnes*

For the 2-cluster partitions result, the separation among clusters are different; using the method based on the function *hclust* gives (C, D) in one cluster and (E, A, B) in another cluster, while using the method based on the function *agnes*, (A , B) in one cluster and (C, E, D) in another cluster.

Before drawing final conclusion from the clustering result and due to rough set assumption that the data table represents a closed world [23], we should consider to get optimal and mining all possible information from the dataset for knowledge discovery by performing permutation to the order of objects. In general, if we have n objects, there are $n!$ different ways of putting them in order. Since there are five objects, the number of different orderings is 5! =120 ways. Table 4 shows all possible result of 120 order which is performed by R function *hclust* using complete linkage and shows different tree structures with initial position. Other methods (average, single and Ward) and R function *agnes* give same tree structure as Table 4. For the simulation of different orders from initial position, the dendrogram result shows different tree structures and heights.

Table 4. Dendrogram Result due to Data Input Order

No.	Dendrogram Result	Data input order	
1	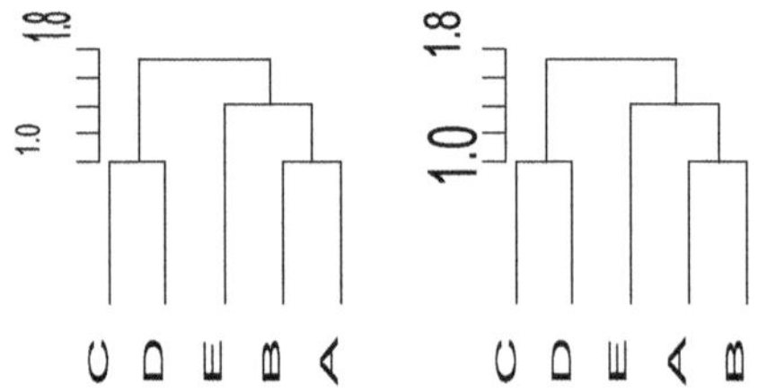	ABCDE ADBCE ADEBC ABDEC ADBEC ABDCE BACDE BCDAE	BCADE BDAEC BADEC BDACE BADEC BDCAE BADCE

Table 4. (*Continued*)

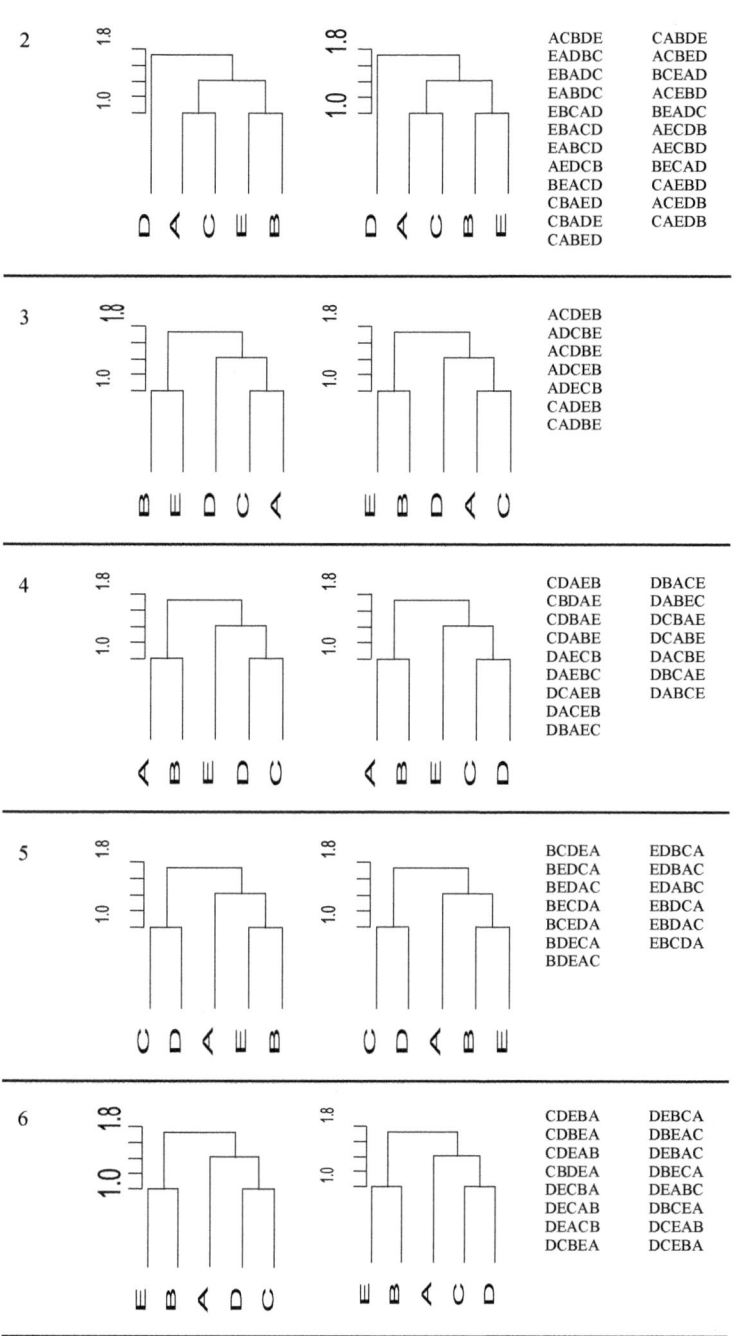

2	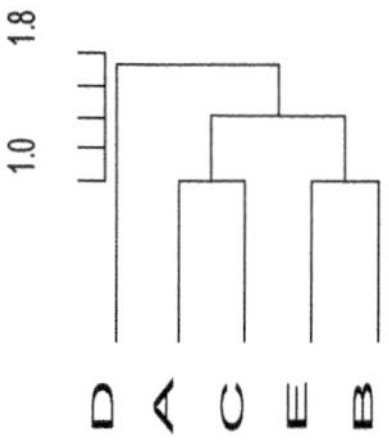	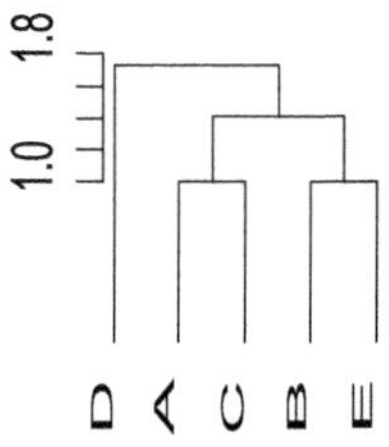	ACBDE EADBC EBADC EABDC EBCAD EBACD EABCD AEDCB BEACD CBAED CBADE CABED CABDE ACBED BCEAD ACEBD BEADC AECDB AECBD BECAD CAEBD ACEDB CAEDB
3	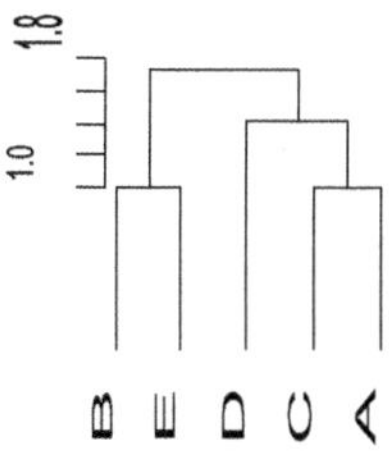	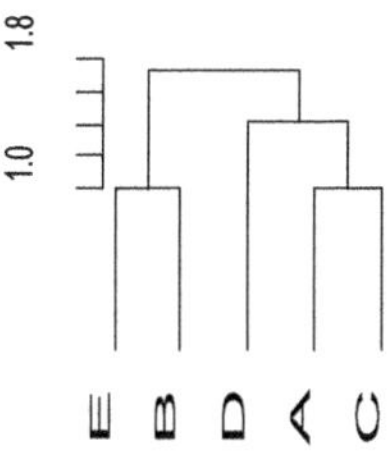	ACDEB ADCBE ACDBE ADCEB ADECB CADEB CADBE
4	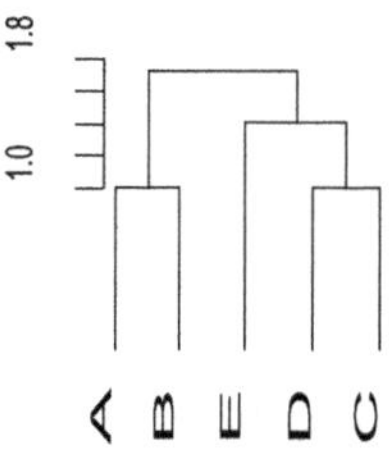	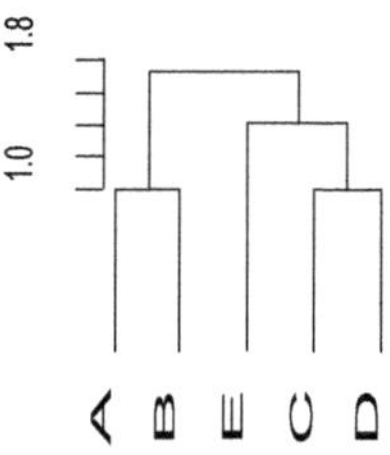	CDAEB CBDAE CDBAE CDABE DAECB DAEBC DCAEB DACEB DBAEC DBACE DABEC DCBAE DCABE DACBE DBCAE DABCE
5	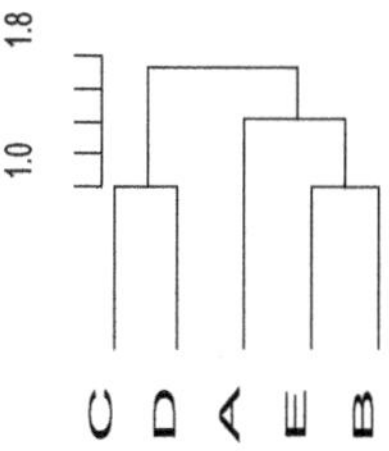	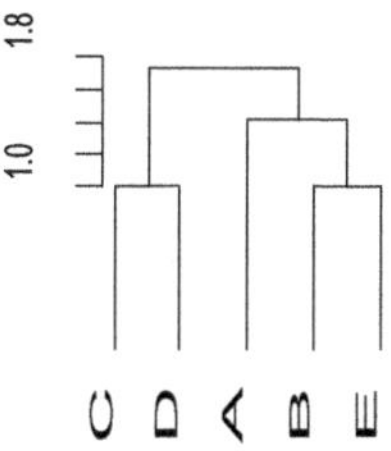	BCDEA BEDCA BEDAC BECDA BCEDA BDECA BDEAC EDBCA EDBAC EDABC EBDCA EBDAC EBCDA
6	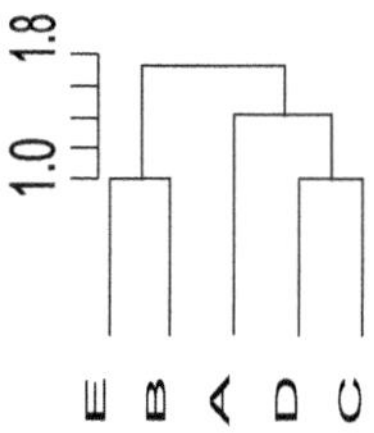	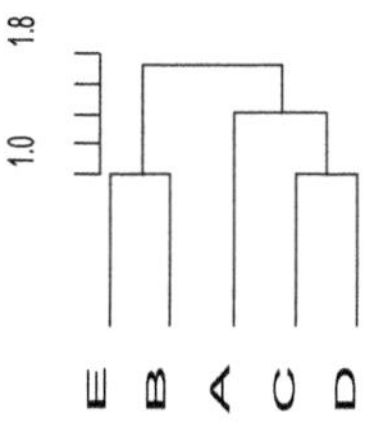	CDEBA CDBEA CDEAB CBDEA DECBA DECAB DEACB DCBEA DEBCA DBEAC DEBAC DBECA DEABC DBCEA DCEAB DCEBA

Table 4. (*Continued*)

7

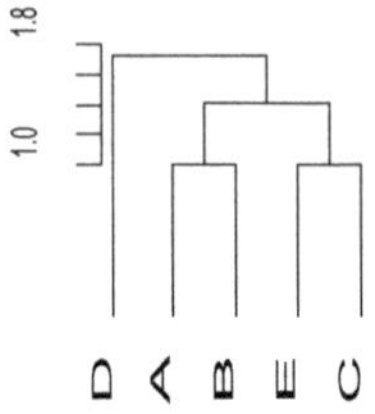

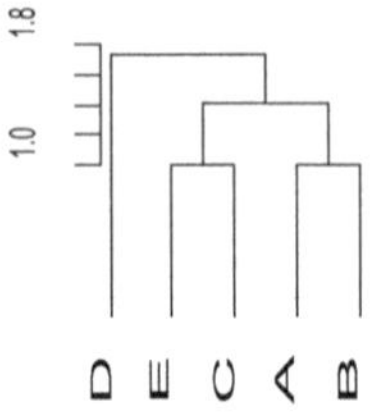

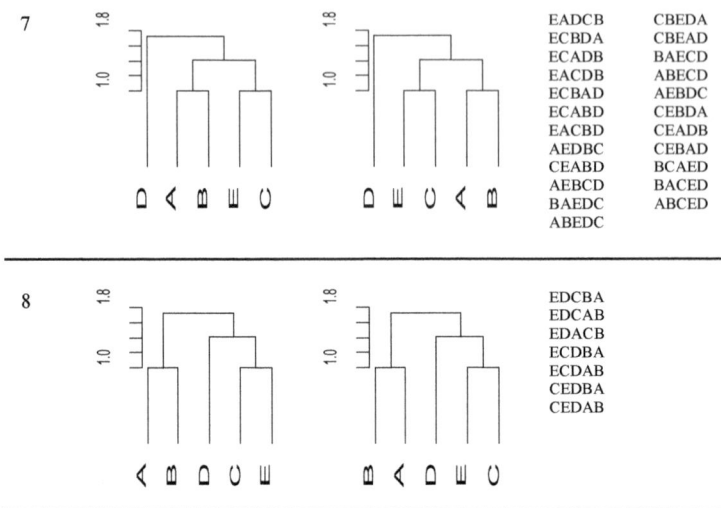

EADCB	CBEDA
ECBDA	CBEAD
ECADB	BAECD
EACDB	ABECD
ECBAD	AEBDC
ECABD	CEBDA
EACBD	CEADB
AEDBC	CEBAD
CEABD	BCAED
AEBCD	BACED
BAEDC	ABCED
ABEDC	

8

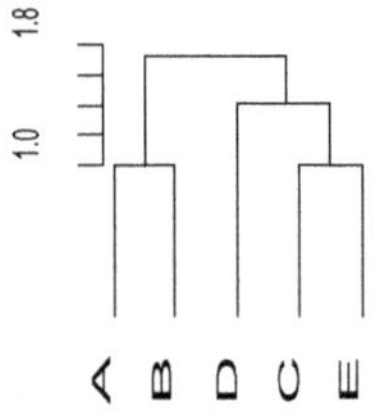

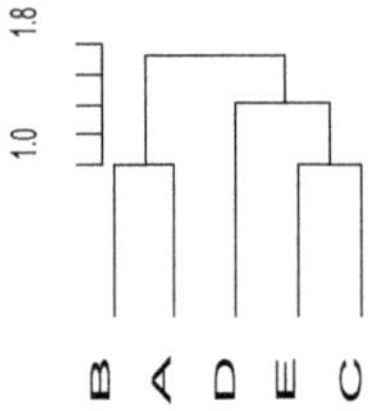

EDCBA
EDCAB
EDACB
ECDBA
ECDAB
CEDBA
CEDAB

Table 4 presents eight tree structures or dendrogram formed for all 120 order. The 2-cluster partitions from all possible different orders are presented in Table 5 and can be summarized to four different groups, as shown in Table 6.

Table 5. 2-cluster partitions yielded

No	2-cluster partitions
1	(C, D) and (A, B, E)
2	(D) and (A, C, B, E)
3	(B, E) and (A, C, D)
4	(A, B) and (C, D, E)
5	(C, D) and (A, B, E)
6	(B, E) and (A, C, D)
7	(D) and (A, B, C, E)
8	(A, B) and (C, D, E)

It shows that different orders of objects lead to different tree structure of the dendrogram. We could evaluate the results obtained by function *hclust* or *agnes* using the Rand statistic [24].

Table 6 . 2-cluster partitions summarized

No	2-cluster partitions
1	(C, D) and (A, B, E)
2	(D) and (A, C, B, E)
3	(B, E) and (A, C, D)
4	(A, B) and (C, D, E)

The Rand statistic is a value ranging from 0 to 1; if two different clustering algorithms are applied to the same set of data or one clustering algorithms are applied to the same set of data but with different order of data and as it approaches 1, it indicates that the clusters within each clustering method has perfect agreement.

Suppose we wish to evaluate the group results of function *hclust* with complete linkage method and consider the result of two different orders of data that gave 2-cluster partitions: the first group consists of two clusters (A, B) and (C, D, E) while the second group consist of two clusters (B, E) and (A, C, D), the resulting Rand statistic is 0.4. Another example, we wish to evaluate two different groups ((D), (A, B, C, E)) and ((A, B), (C, D, E)), Rand statistic is also 0.4. This low value of Rand statistic index indicates that traditional hierarchical clusterings show minimum agreement due to permutations of input objects order.

For comparison, we will use the hierarchical clustering based on indiscernibility and indiscernibility level. Representing Table 4 to information system table or features, we obtain five objects and three attributes, *a*, *b* and *c* as shown in Table 7. We will use this table for the rest of paper.

Table 7. Features from table 3

U	*At*		
	a	*b*	*c*
A	1	0	0
B	1	1	0
C	1	0	1
D	0	0	1
E	1	1	1

We can obtain quantitative indiscernibility relations for all pairs of objects in *U* and tabulated then in Table 8.

Table 8. Quantitative indiscernibility relations

	A	B	C	D	E
A	3/3=(1.00)	2/3=(0.67)	2/3=(0.67)	1/3=(0.33)	1/3=(0.33)
B		3/3=(1.00)	1/3=(0.67)	0/3=(0.00)	2/3=(0.67)
C			3/3=(1.00)	2/3=(0.67)	2/3=(0.67)
D				3/3=(1.00)	1/3=(0.67)
E					3/3=(1.00)

Their indiscernibility levels are tabulated in Table 9.

Table 9. Indiscernibility level table

	A	B	C	D	E
A	3/9=(0.33)	2/4=(0.50)	2/5=(0.40)	1/2=(0.50)	1/3=(0.33)
B		3/8=(0.38)	1/3=(0.33)	0/0=(0.00)	2/4=(0.50)
C			3/10=(0.30)	2/4=(0.50)	2/5=(0.40)
D				3/7=(0.43)	1/2=(0.50)
E					3/9=(0.33)

For example, the indiscernibility level $\gamma(A, B)$ of objects *A* and *B* in Table 5 is calculated as follows:

$$\gamma(A, B) = \frac{ind(At)(A, B)}{ind(At)(A, B) + ind(At)((A, B), U_i))}, \text{ where } U_i = C, D, E$$

$$= \frac{|\{a,b,c \in At \mid I_{a,c}(A) = I_{a,c}(B)\}|}{|\{a,b,c \in At \mid I_{a,c}(A) = I_{a,c}(B)\}| + |\{a,b,c \in At \mid I_{a,b,c}(A,B) = I_{a,b,c}(U_i)\}|} = \frac{2}{2+2} = 0.50.$$

Let us explain this example: since product *A* and *B* are indiscernible at attributes *a* and *c*, the value of the numerator ($|\{a,b,c \in At \mid I_{a,c}(A) = I_{a,c}(B)\}|$) is 2. To calculate $|\{a,b,c \in At \mid I_{a,b,c}(A,B) = I_{a,b,c}(U_i)\}|$, U_i = C, D, E, we see that ((A, B), C) and ((A, B), E) are indiscernible at attribute *a*, hence the value of $|\{a,b,c \in At \mid I_{a,b,c}(A,B) = I_{a,b,c}(U_i)\}|$ is 2. Indiscernibility levels matrix containing each element *ind*(*At*)(U_i, U_j) for all other pairs in *U* are tabulated in table 7. Thus the pairs of quantitative indiscernibility and indiscernibility level relations (*ind*(*At*)(U_i, U_j)| γ(U_i, U_j)) could be brought into one Table 10.

Table 10. Indiscernibility and indiscernibility level matrix (*ind*(*at*)(u_i, u_j) | γ(u_i, u_j))

	A	B	C	D	E
A	1.00 | 0.33	0.67 | 0.50	0.67 | 0.40	0.33 | 0.50	0.33 | 0.33
B		1.00 | 0.38	0.33 | 0.33	0.00 | 0.00	0.67 | 0.50
C			1.00 | 0.30	0.67 | 0.50	0.67 | 0.40
D				1.00 | 0.43	0.33 | 0.50
E					1.00 | 0.33

Note that, the clustering start using the highest value of dual (*ind*(*At*)(U_i, U_j) | γ(U_i, U_j)), $i \neq j$, and objects U_i have already been regarded forming their own cluster. There are three pairs of objects that give the same value: A and B, B and E, C and D *i.e.* (0.67 | 0.50). We can choose to start from A and B (or (B, E), or (C, D)) then the second grouping is the cluster (A, B) with another object that has the highest value of the (*ind*(*At*)(U_i, U_j)| γ(U_i, U_j)). For example:

((A, B), E) = *max* ((A, E), (B, E))
= *max* ((*ind*(*At*)(*A*, *E*)| γ(*A*, *E*), *ind*(*At*)(*B*, *E*)| γ(*B*, *E*))
= *max* ((0.33 | 0.33), (0.67|0.50))
= (0.67 | 0.50)

From Table 10 above, the first step clustering give,

	A, B	C	D	E
A, B	0.67 | 0.50	0.67 | 0.40	0.33 | 0.50	0.67 | 0.50
C		1.00 | 0.30	0.67 | 0.50	0.67 | 0.40
D			1.00 | 0.43	0.33 | 0.50
E				1.00 | 0.33

The second step clustering gives:

	A, B, E	C	D
A, B, E	0.67 | 0.50	0.67 | 0.40	0.33 | 0.50
C		1.00 | 0.30	0.67 | 0.50
D			1.00 | 0.43

Two objects that have another high value are C and D so C and D form new cluster. The third step of clustering gives:

	A, B, E	C, D
A, B, E	0.67 | 0.50	0.67 | 0.40
C, D		0.67 | 0.50

The final step is:

	A, B, E, C, D
A, B, E, C, D	0.67 | 0.40

Figure 5 shows the dendogram of the final grouping,

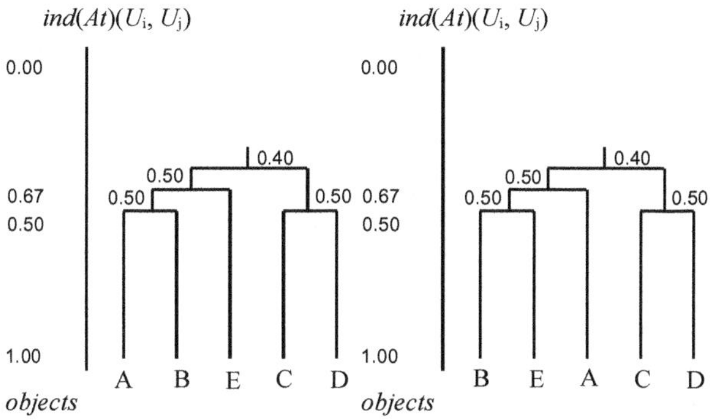

Fig. 5. Dendrogram of clustering based on indiscernibility and indiscernibility level. Only two possible tree structure occur for clustering based on this method. For 2-partitions the clusters are (A, B, E) and (C, D).

For the clustering based on indiscernibility and indiscernibility level result, we can examine the dendrograms showing how the objects were partitioned (Figure 5). There will be only two models of tree structure of the objects clustered and since we do not know a priori the correct number of clusters, for the 2-cluster partitions, the Rand statistic shows perfect agreement and the separation among clusters are (*A, B, E*) and (*C, D*). Altering or permuting the order of the objects would not affect this method; the value of quantitative indiscernibility relations and its indiscernibility level would not change. Even though it is impossible to say whether using the clustering based on indiscernibility and indiscernibility level produces a better partition of the data set, clearly at least this method is more consistent on the tree structure of the dendrogram and this method reduces the instability of the clustering result due to data input order compared to traditional clusterings. The result of the clustering based on indiscernibility and indiscernibility gives additional value to help the businessman to make decision for choosing two groups of the products.

5 Conclusion

In this paper, we compared traditional hierarchical clustering method with a new clustering method based on the combination of quantitative indiscernibility relations and indiscernibility level. Using a simple simulation dataset, we have demonstrated that the method is not influenced by the order of the objects and could produce a consistent tree structure which is reducing the dendrogram instability. The dendrogram does not only show the measure of indiscernibility but also gives the level of indiscernibility that represents the level of granularity of object pairs in the information system.

References

1. Hao, Y., Yao, Y., Luo, F.: Data Analysis Based on Discernibility and Indiscernibility. Journal of Information Science 177, 4959–4976 (2007)
2. Hakim, R.B., Fajriya, S.: Clustering Based on Indiscernibility and Indiscernibility Level. In: Proceeding 2009 IEEE International Conference on Granular Computing, pp. 192–196 (2009) ISBN : 978-1-4244-4830-2
3. Hirano, S., Sun, X., Tsumoto, S.: Comparison of Clustering Methods for Clinical Databases. Information Sciences 159, 155–165 (2004)
4. Mingoti, S.A., Lima, J.O.: Comparing SOM Neural Network with Fuzzy C-Means, K-Means and Traditional Hierarchical Clustering Algorithms. European Journal of Operational Research 174, 1742–1759 (2006)
5. Budayan, C., Dikmen, I., Talat Birgonul, M.: Comparing the Performance of Traditional Cluster Analysis, Self Organizing Maps and Fuzzy C-Means Method for Strategic Grouping. Expert System with Applications 36, 11772–11781 (2009)
6. Finch, H.: Comparison of Distance Measures in Cluster Analysis with Dichotomous Data. Journal of Data Science 3, 85–100 (2005)
7. Hitchcock, D.B., Chen, Z.: Smoothing Dissimilarities to Cluster Binary Data. Computational Statistics and Data Analysis 52, 4699–4711 (2008)
8. Van Der Kloot, W. A., Alexander M. J., Spaans, M.J., Heiser, W.J.: Instability of Hierarchical Cluster Analysis Due to Input Order of the Data: The PermuCluster Solution. Psychological Methods 10(4), 468–476 (2005)
9. Hardle, W., Simar, L.: Applied Multivariate Statistical Analysis. Springer, Heidelberg (2007)
10. Pawlak, Z.: Rough Sets: Theoretical Aspects of Reasoning About Data. Kluwer Academic Publishers, Boston (1991)
11. Pawlak, Z., Skowron, A.: Rudiments of rough sets. Information Sciences 177, 3–27 (2007)
12. Pawlak, Z., Skowron, A.: Rough Sets: Some Extensions. Information Sciences 177, 28–40 (2007)
13. Parmar, D., Wu, T., Blackhurst, J.: MMR: An Algorithm for clustering categorical data using Rough Set Theory. Data & Knowledge Engineering 63, 879–893 (2007)
14. Kumar, P., Radha Krishna, P., Bapi, R.S., De Kumar, S.: Rough Clustering of sequential data. Journal of Data & Knowledge Engineering 63, 183–199 (2007)
15. Upadhyaya, S., Arora, A., Jain, R.: Rough Set Theory: Approach for Similarity Measure in Cluster Analysis. In: Proceeding of the 2006 International Conference on Data Mining, DMIN 2006, Las Vegas, Nevada USA,(2006) ISBN 1 60132-004-3
16. Hirano, S., Tsumoto, S.: An indiscernibility-based clustering method with iterative refinement of equivalence relations –rough clustering. Advanced Computational Intelligence and Intelligent Informatics 7(2), 169–177 (2003)
17. Hirano, S., Tsumoto, S.: Indiscernibility-based clustering: Rough clustering. In: De Baets, B., Kaynak, O., Bilgiç, T. (eds.) IFSA 2003. LNCS, vol. 2715, pp. 378–386. Springer, Heidelberg (2003)
18. Hirano, S., Sun, X., Tsumoto, S.: Comparison of Clustering Methods for Clinical Databases. Information Sciences 159, 155–165 (2004)

19. Hirano, S., Tsumoto, S.: Hierarchical Clustering of Non-Euclidean Relational Data using Indiscernibility-Level. In: Wang, G., Li, T., Grzymala-Busse, J.W., Miao, D., Skowron, A., Yao, Y. (eds.) RSKT 2008. LNCS (LNAI), vol. 5009, pp. 332–339. Springer, Heidelberg (2008)
20. Hirano, S., Tsumoto, S.: Indiscernibility-based Clustering of Non- Euclidean Relational Data, `http://www.ecmlpkdd2007.org/CD/workshops/RSKD`
21. Kaufman, L., Rousseeuw, P.J.: Clustering by Means of Medoids. In: Dodge, Y. (ed.) Statistical Data Analysis Based on the L1 Norm, North Holland, Amsterdam, pp. 405–416 (1987)
22. R Development Core Team, R : A language and environment for statistical computing, R Foundation for Statistical Computing, Vienna, Austria (2006), `http://www.r-project.org/`
23. Pawlak, Z.: A Primer on Rough sets: A New Approach to Drawing Conclusions from Data. Cardozo Law Review, Vol 22(5 -6), 1407–1415 (2001)
24. Rand, W.M.: Objective criteria for the evaluation of clustering methods. J. Amer. Statist. Assoc. 66, 846–850 (1971)
25. Demri, S.P., Orlowska, A.E.S.: Incomplete Information: Structure, Inference, Complexity. Springer, Heidelberg (2002)
26. Leisch, F., Weingessel, A., Hornik, K.: Bindata: Generation of Artificial Binary Data, R package version 0.9-12 (2005)

Dual Quorum Replication (DQR): A Protocol for Data Replication in Data Grid

Rohaya Latip and Aminah Hidayah Mohamad Ariff

Faculty of Computer Science and Information Technology
Universiti Putra Malaysia
{rohaya}@fsktm.upm.edu.my

Abstract. Grid computing is a large dynamic network where data can be copied anywhere in Storage Element (SE) located somewhere in the network. Having SE at some number of location will increase the communication cost. Therefore having the best number of SE is one of the issues in grid. In this paper, we introduced a new protocol, named Dual Quorum Replication (DQR) for data replica control protocol for a large dynamic network by using quorum techniques to improve communication cost because quorum techniques reduce the number of copies involved in reading or writing data. The protocol of DQR replicates data for large dynamic network such as grid network by putting the protocol in two connected quorum and access consistent data by ensuring the quorum not to have a nonempty intersection quorum. To evaluate our protocol, we developed a simulation model in Java. Our result proved that DQR improves the performance of the communication cost compare to the latest replica control protocol named Enhanced Diagonal Data Replication using 2D Mesh (EDR2M).

Keywords: Data replication, Data Grid, Simulation, Communication cost, replica control protocol.

1 Introduction

A grid is a distributed network computing system, a virtual computer formed by a networked set of heterogeneous machines that agree to share their local resources with each other. A grid is a very large scale, generalized distributed network computing system that can scale to internet size environment with machines distributed across multiple organizations and administrative domains [1, 2, 3]. I. Foster [4] defines grid as technologies and infrastructure to support the sharing and coordinated use of diverse resources in dynamic, distributed virtual organizations which are the creation from geographically distributed components operated by distinct organization with differing policies of virtual organization that are sufficiently integrated to deliver the desired QoS.

Virtual organization Ensuring efficient access to such a huge network and widely distributed data is a challenge to those who design, maintain and manage the grid network. Replicating data at a minimum communication cost is one of the issues for dynamic network such as data grid. In our work, we investigate the use of replication on a grid network to improve its communication cost to access data efficiently.

Y. Zhang et al. (Eds.): DTA/BSBT 2010, CCIS 118, pp. 226–233, 2010.

We use Java to run this replication protocol. The paper is organized as follow. Section 2, overview the previous work on grid replication and distributed database. In section 3, we introduce our replication protocol and its algorithm. Section 4 describes the Simulation Design and Section 5 discusses the simulation results. Brief conclusions and future works are discussed in Section 6.

2 Related Work

2.1 Enhanced Diagonal Data Replication Using 2D Mesh Protocol (EDR2M)

EDR2M is a protocol for data replication protocol in Data Grid introduced by [5] in year 2009. In this protocol, all nodes are logically organized into two dimensional grid structure. In [5] assumed that the replica copies are in the form of data file and all nodes are operational, meaning that the copy at that node is available. The data file is replicated to only one node of the diagonal sites.

This protocol also uses quorum approach to put nodes in cluster. Quorum is grouping the nodes or databases as shown in Figure 1. Figure 1 illustrated how the quorums for network size of 7 x 7 are grouped by nodes of 5 x 5 in each quorum. Nodes which are formed in quorum intersect with other quorums. This is to make sure that they can communicate or read other data from other nodes which is in another quorum.

The replica which is the node that has the copy of the replicated data is located in the middle of diagonal nodes as illustrated in Figure 2. For example in Figure 2, where select the replica is node *s*(3,3).

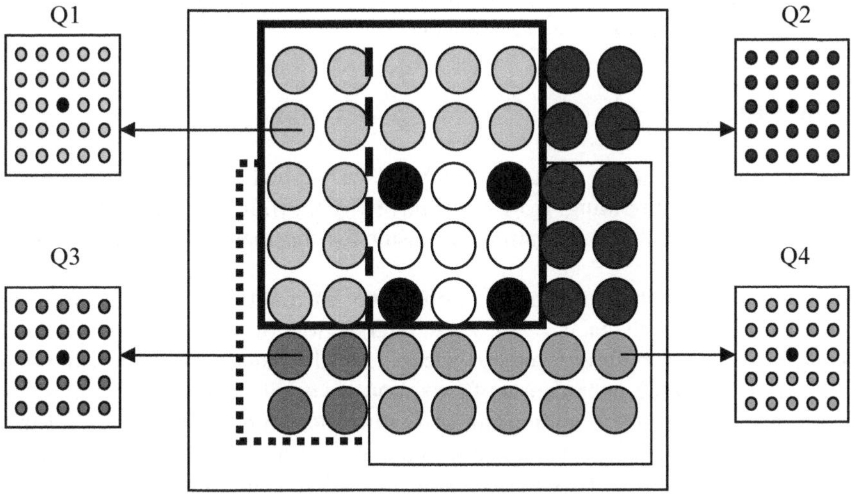

Fig. 1. Four quorums obtain in a 7 x 7 network size

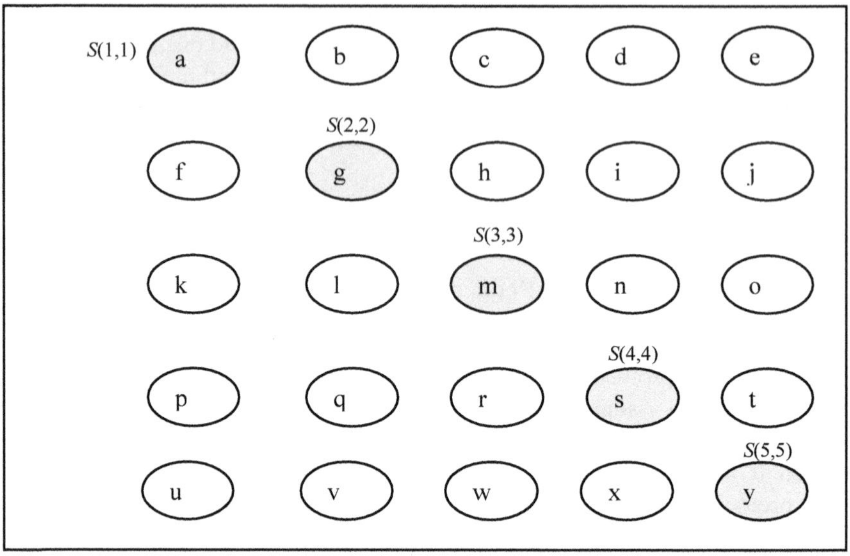

Fig. 2. A grid organization with 25 nodes, each of the node has a data file *a,b,…,* and *y* respectively

Definition 1: Assume that a database system consists of n x n nodes that are logically organized in the form of two dimensional grid structure. All sites are labeled s(i,j), $1 \leq i \leq n$, $1 \leq j \leq n$. The diagonal site to s(i,j) is {s(k,l)| k = i + 1, l = j + 1, and k,l $\leq$ n, if i = n initialized i = 0, if j = n, initialized j = 0}. A diagonal set, D(s), is a set of diagonal sites, D(s) = s(m,n), where m = n and m, n = 1, 2,…, 5 as in Figure 2. Figure 2 has 25 nodes where the size network is 5 x 5 nodes.

3 The Model

Dual Quorum Replication Protocol (DQR) is a replica control protocol for data replication in Grid computing. It was introduced by [5,6]. It uses quorum approach for putting the member of the operation in cluster for the member to perform read and write operation.

For DQR, the number of quorum is only two for a network location. Both of the quorums intersect with each other. The intersection of the quorum allocates the primary database for both quorums to access. As illustrated in Figure 3, the copy of the data is placed at the intersection of two quorums or clusters.

Let r and w denote the read quorum and write quorum respectively. To ensure that read operation always gets the updated data, $r + w$ must be greater than the total numbers of copies (votes) assigned to all sites. To make sure the consistency is obtained, the following conditions must be fulfilled [7].

i. $1 \leq r \leq L_B$, $1 \leq w \leq L_B$
ii. $r + w = L_B + 1$.

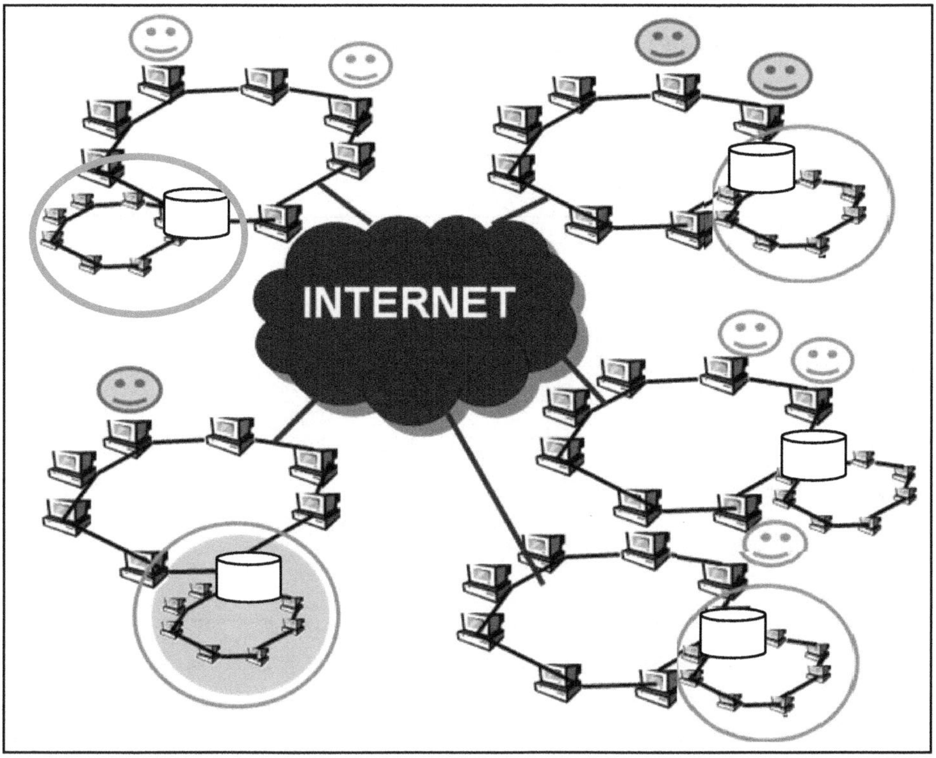

Fig. 3. Dual quorum intersect in DQR

These two conditions ensure that there is a nonempty intersection of copies between read and write quorum. Thus, the conditions ensure that a read operation can access the most recently updated copy of the replicated data.

Let *S*(*B*) be the set of sites at which replicated copies are stored corresponding to the assignment *B*. Then,

$$S(B) = \{s(i,j) \mid B(s(i,j)) = 1,\ 1 \le i \le n,\ 1 \le j \le n\}$$

This protocol is suitable for dynamic network such as grid network. For example 27 x 27 nodes, where n is 27. Number of quorum, Q is 5 as $Q = \lfloor \sqrt{27} \rfloor$ then R, the number of nodes in each quorum is 7, where $R = n/Q = 5.4$ and the next odd number is 7. All quorums will intersect as define in Definition 1. To replicate the data, the middle node of the diagonal node in *D*(*s*) will have the copy of data file as illustrated in Figure 2. The middle node acts as primary database because they assumed that all nodes and sources have the same requested priority and level. This is one of the behavior of grid which is its accessibility is peer to peer other than its dynamic number of nodes in the network.

DQR introduced Eq. (1) to compute the number of quorums needed in N size of network where n is the number of row and columns. The nodes are organized in 2D

Mesh and the total number of quorums, Q for the entire network, N is $q * q$ where q is the number of quorum in a row or column as shown in Eq. (1).

$$q = \left\lfloor \sqrt{n} - \frac{n}{10} \right\rfloor / 2 \tag{1}$$

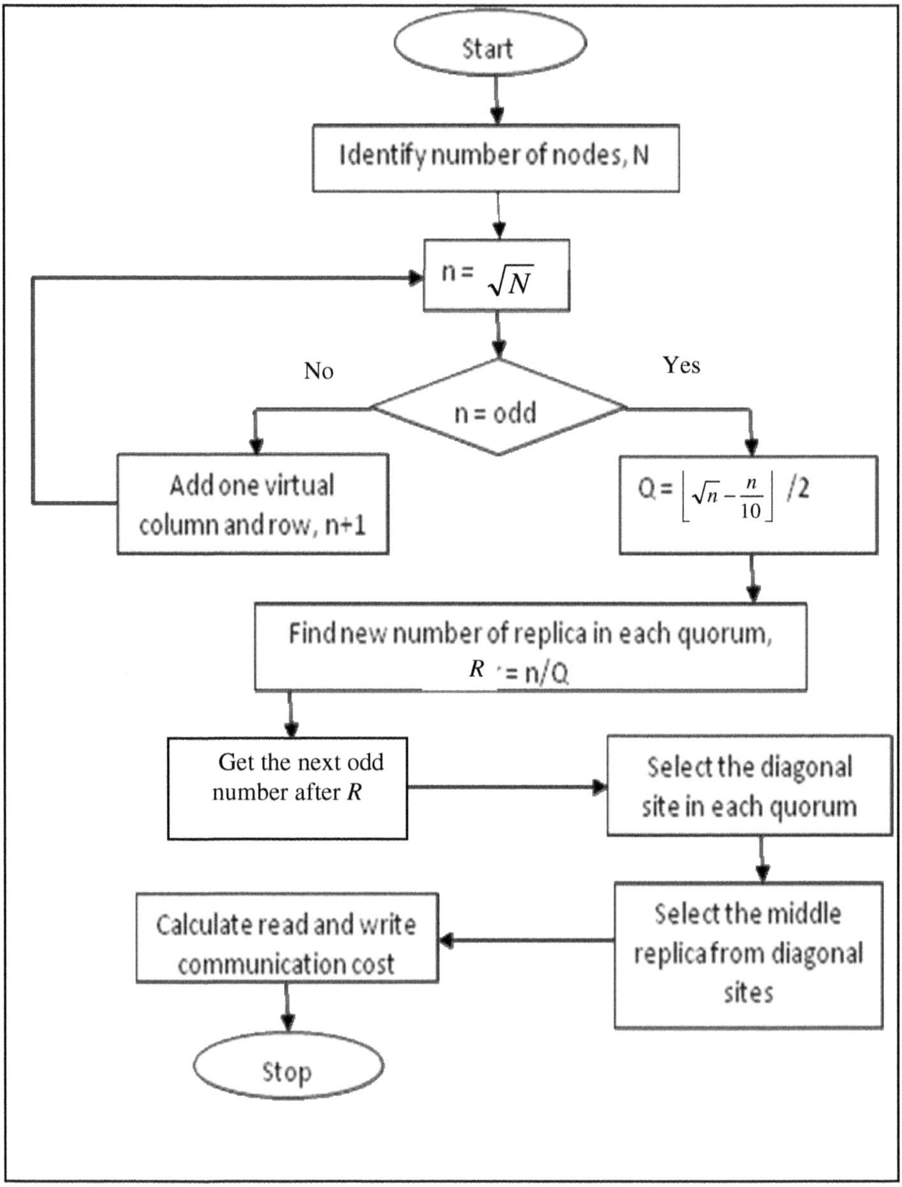

Fig. 4. Flow chart of simulation design

4 Simulation Design

The simulation model was developed using Java. The design of DQR model is as shown in Figure 4, where it shows the flow of the simulation. The simulation model must identify the number of nodes in the network for it to logically structure the network to *n* x *n* column. If the number of n is odd it will proceed to identifies the selected middle node. Otherwise, the simulation adds a column virtually.

DQR introduced Eq. (1) to compute the number of quorums needed in *N* size of network where *n* is the number of row or columns. The nodes are organized in two dimensional structures, it is *n* x *n* network then a virtual column and row are added logically. In DQR protocol, the total number of replica, *R* is identified for allocating a number of nodes in a quorum. The number of quorum will be identified using the Eq. (1) and then the number of nodes in each quorum is allocated using Eq. (2).

$$R = n/Q \tag{2}$$

5 Performance Analysis and Comparison

To analyze the performance of read and write communication cost, Eq. (3) and Eq. (4) is used from [8], where *n* is the grid column or row size, example *n* is 7, thus 7 x 7 nodes of two dimensional grid structure. Therefore the network size is 49.

For formula to calculate communication cost of read operation is as in Eq.(3) and to calculate communication cost of write operation is as in Eq(4).

$$C_{\text{read}} = \lceil n/2 \rceil \tag{3}$$

$$C_{\text{write}} = \lceil (n+1)/2 \rceil \tag{4}$$

To get the number of quorums which are suitable for the size of the grid network. A bigger grid network will have more quorums. The number of columns and rows in each quorum must be odd, to get the middle replica. By selecting only one middle replica in each quorum will minimize the communication cost.

Figure 5, presents the communication cost for read operation of nodes from 25 to 729 nodes. Figure 5 shows that the DQR protocol has reduced the number of communication cost for read operation compare to EDR2M from node 25 to 729 nodes. The protocol manages to select a small number of replicas to make a copy of a data. The result shows that the average for DQR is one (1) while for EDR2M is 1.333. DQR decreased 25% of communication cost compare to EDR2M.

Fig. 6 shows the results for write operations for number of nodes from 25 to 729 nodes, where the communication cost for write in the DQR protocol for the number of nodes such as 25, 49, 81 and 121 is one (1) while the other number of nodes are two (2). This is because the number of selected replicas for DQR protocol is lower than EDR2M. The average of communication cost for DQR protocol is 1.667 compare to

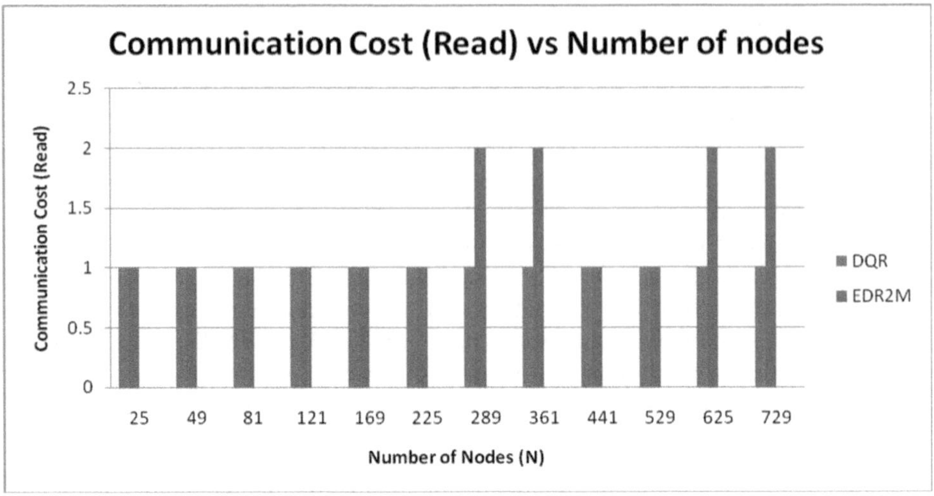

Fig. 5. Result of Communication Cost for Read Operation

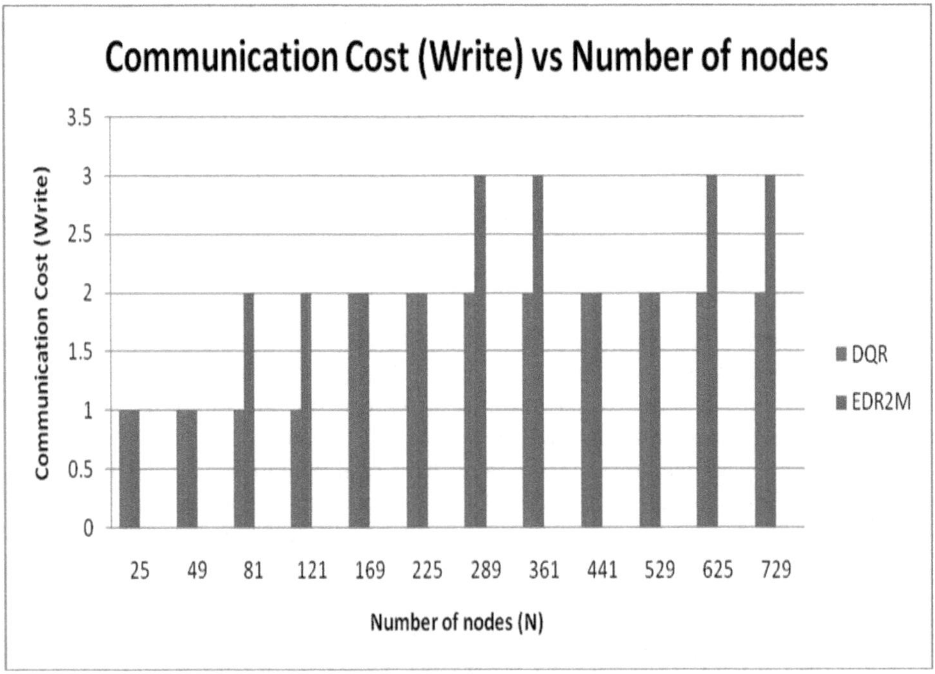

Fig. 6. Result of Communication Cost for Write Operation

EDR2M that is two (2). DQR protocol decreased 23% of communication cost for write operation compare to EDR2M. Therefore, DQR protocol is suitable for read and writes operations for any number of network size.

6 Conclusions

In this paper, a enhanced data replication protocol named Dual Quorum Replication (DQR) protocol is introduced to reduce the communication cost in grid network which is known as dynamic and more to peer to peer environment. DQR protocol puts the selected replica in a diagonal way in a grid logical structured by formulating new equations introduced in this paper. By having two quorum intersect and selecting replica, based on the middle location of the diagonal nodes proves that the data consistency is achieved in a low communication cost.

The simulator was created to study the performance of the DQR protocol. The results demonstrate that the DQR protocol has reduce both the read and write communication cost compared to the previous data replication protocol EDR2M.

In the future, we will investigate the data availability in the grid environment and this investigation will be used to evaluate the performance of our DQR protocol. Currently we have improved the protocol to a larger network and putting a certain number of nodes into tiers.

References

1. Krauter, K., Buyya, R., Maheswaran, M.: A Taxanomy and Survey of Grid Resource Management Systems for Distributed Computing. International Journal of Software Practice and Experience 32(2), 135–164 (2002)
2. Ranganathan, K., Foster, I.: Identifying Dynamic Replication Strategies for a High Performance Data Grid. In: Proceedings of International Workshop on Grid Computing, Denver (November 2001)
3. Lamehamedi, H.: Decentralized Data Management Framework for Data Grids. Ph.D. Thesis, Rensselaer Polytechnic Institute Troy, New York (2005)
4. Foster, I., Kesselman, C., Nick, J., Tuecke, S.: Grid Services for Distributed System Integration. Computer 35(6), 37–46 (2002)
5. Rohaya, L., Ibrahim, H., Othman, M., Abdullah, A., Sulaiman, M.N.: Quorum-based Data Replication in Grid Environment. International Journal of Computational Intelligence Systems (IJCIS) 2(4), 386–397 (2009)
6. Rohaya, L., Ibrahim, H., Othman, M., Abdullah, A., Sulaiman, M.N.: High Availability with Diagonal Replication in 2D Mesh (DR2M) Protocol For Grid Environment. Journal of Computer and Information Science by Canadian Center of Science and Education 1(2) (2008)
7. Mat Deris, M., Bakar, N., Rabiei, M., Suzuri, H.M.: Diagonal Replication on Grid for Efficient Access of Data in Distributed Database Systems. In: Bubak, M., van Albada, G.D., Sloot, P.M.A., Dongarra, J. (eds.) ICCS 2004. LNCS, vol. 3038, pp. 379–387. Springer, Heidelberg (2004)
8. Mat Deris, M., Bakar, N., Suzuri, H.M., Abu Osman, M.T.: Improving Data Availability Using Hybrid Replication Technique in Peer-to-Peer Environments. AINA (1), 593–598 (2004)

Fingerprint Area Detection in Fingerprint Images Based on Enhanced Gabor Filtering

Michal Dolezel[1], Dana Hejtmankova[1], Christoph Busch[2], and Martin Drahansky[1]

[1] Department of Intelligent Systems, Faculty of Information Technology, Brno University of Technology, Bozetechova 2, 612 66 Brno, Czech Republic
xdolez40@stud.fit.vutbr.cz, {hejtmanka,drahan}@fit.vutbr.cz
[2] Hochschule Darmstadt - CASED, Mornewegstr. 32, 64293 Darmstadt, Germany
christoph.busch@h-da.de

Abstract. This paper describes a new approach to fingerprint area detection in a digital fingerprint image. This approach was evaluated for real-world scenario, specifically for fingerprints scanned from dactyloscopic fingerprint cards. These images, which compose widely used fingerprint databases, have their specific problems and properties such as handwritten or printed characters, drawings or specific noise in the background or spreaded over the fingerprint itself. Our approach was compared with three other methods and yields significantly better results than the best of the benchmarked methods.

Keywords: Fingerprint, fingerprint area detection, semantic conformance testing, image segmentation, Gabor filters.

1 Introduction

Nowadays, semantic conformance testing for finger minutiae data is developed [6] for the purpose of validating the compliance of a minutia extractor with the ISO/IEC interchange standard 19794-2 [5]. In [2] a new methodology for conformance testing was presented. This methodology proposed three conformance rates, which describe to which extend a finger minutiae record is indeed a faithful representation of the physiological characteristic captured in the input image. One of the main objectives is the assessment whether an algorithm under test did or did not find false minutiae at the border of the fingerprint area or in the image background.

For the purposes of semantic conformance testing a special database (GTD – Ground Truth Database [2]) of fingerprints was prepared. This database consists of fingerprint images selected from the NIST special databases SD 14 and SD 29. Fingerprint images were thoroughly selected so, that the resultant GTD have a balanced ratio of pattern types, position codes (instance type) etc. However the majority of fingerprints in SD 14 and SD 29 databases are scanned from dactyloscopic fingerprint cards. These images have their specific properties and thus it is not possible to use standard algorithms for their processing. An example of such a fingerprint can be found in Fig. 1. Typical problems of fingerprint area extraction for these images are handwritten or printed characters, drawings, and the printed border of a cell of the fingerprint card or the dirt (noise) in the background. All these problems can occur in the image background or can interfere in the fingerprint area (e.g. right cell border of dactyloscopic fingerprint card in Fig. 1.), which represents a challenge for every algorithm.

Y. Zhang et al. (Eds.): DTA/BSBT 2010, CCIS 118, pp. 234–240, 2010.

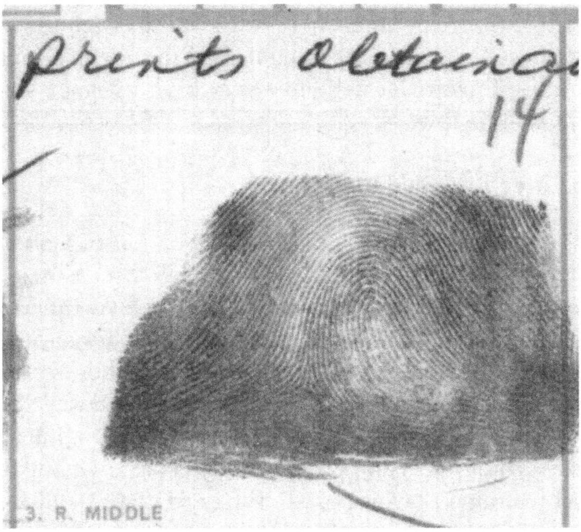

Fig. 1. Example of a fingerprint image from the Ground Truth Database (GTD)

2 Existing Methods

The detection of the fingerprint area is a relevant preprocessing step in many fingerprint analysis pipelines. However none of the pipelines requires a high precision as it is required for a conformance testing suite. In this Section we provide a survey of published concepts for fingerprint are detection and investigate their exactness. None of the surveyed methods was appropriate for our purpose, nevertheless they inspired our approach and all of them provide a baseline for benchmarking, as reported in Section 4.

2.1 NIST Algorithms

The National Institute of Standards and Technology provides implemented algorithms that can be used for fingerprint segmentation. For example, the NBIS (NIST Biometric Image Software) package contains the Segmentor routine [7], which deals with fingerprint segmentation for fingerprint classification purposes. By using special thresholding based on global and local pixel intensity minimums and maximums, massive erosion and edge detection, the Segmentor routine computes the most suitable fixed-size rectangle in input fingerprint and declares it as a segmentation result.

Second example of segmentation method using by NIST algorithm is segmentation based on NFIQ (NIST Fingerprint Image Quality). First the input fingerprint image quality map is computed using NFIQ algorithm. Then the result image is created by special thresholding, where areas with quality equal or better than the specific threshold are considered as fingerprint area whereas other areas are marked as background.

2.2 Ratha Algorithm

An interesting approach was chosen by Ratha et. al.[3]. They proposed a method that exploits the fingerprint orientations field. The orientation field is used to compute the optimal dominant ridge direction in each 16 x 16 block. Then they compute the variance of gray level in a direction perpendicular to the local orientation field. Foreground areas containing fingerprint will have very high variance whereas the variance of background areas will be low.

2.3 Basic Gabor Filter Based Algorithm

Allonso-Fernandez et al. [1] introduced a new application of Gabor filters for fingerprint segmentation, originally used for fingerprint quality measures [4]. By the help of several different orientated Gabor filters responses the so-called magnitude Gabor features are computed. Then it is possible to segment the fingerprint using thresholding, where the standard deviation of the magnitude Gabor features represents the threshold for each block. Allonso-Fernandez et al. also proposed some enhancements, for example, half block overlapping, ridge frequency computation etc. which can help with foreground/background decision problems. This basic Gabor filter based method provides quite good results on "well-posed" fingerprints but still has many disadvantages and fails in "ill-posed" cases. The segmented area is very jagged, and the method has problems with any kind of otherwise oriented patterns like edge lines in dactyloscopic fingerprint cards, descriptions, hand drawings, white scars inside fingerprint area etc.

3 Proposed Segmentation Pipeline

For the processing of the NIST special databases and the similar purposes, a more complex method is needed, as none of the methods described in Section 2 was able to produce sufficiently good results and distinguish reliably the fingerprint area from the drawing and noise in the background. Therefore, we further developed the method of Allonso-Fernandez Gabor Filter-Based segmentation and propose a fingerprint area segmentation pipeline, which consists of three phases. The first phase is a direct enhancement of the Allonso-Fernandez Gabor Filter-Based method. In a second phase artifacts are removed and in the third phases elements composing the entire fingerprint are selected.

3.1 Enhanced Gabor Filter Based Area Segmentation

A major enhancement can be achieved, if the overlap of blocks is not fixed to half the block size, as originally suggested by Allonso-Fernandez [1]. In proposed algorithm it will be possible to set up the size of overlap in horizontal and vertical direction in pixels. With maximal set overlap in both directions (blocks of 6x6 pixels overleaping in 5 pixels), the segmented area will be smooth enough to precise interpolation of fingerprint ridge endings, while sufficiently big blocks have a good standard deviation of magnitude features value for foreground/background thresholding. In the basic method proposed by Allonzo-Fernandez one threshold for each block was computed in a way that all pixels in each block had the same value after thresholding. In our revised method the average value of standard deviations for every pixel is computed during the Gabor filtering process. The average value of standard deviations for one image pixel is computed as a sum of deviations (based on 8 Gabor features) for all blocks containing that pixel divided by the number of such blocks.

As a result, the segmented image is very smooth. Since the segmented area is slightly larger in size than it is appropriate for our purposes, the minimal omnidirectional morphological erosion (square 6x6 px) is used.

3.2 Removing Artifacts

After the main segmentation phase, it is necessary to tackle unwanted artifacts like lines in the dactyloscopic fingerprint cards, annotated descriptions, hand drawings etc. All these objects are likely to be marked as foreground by the Gabor-Filter. In most image processing applications a morphological operation called binary opening is used for background noise removal. An opening is defined as binary erosion followed by a binary dilatation. We use the same structural element for both operations and intend to remove background noise. However such morphological opening may damage some fine details along the detected fingerprint area edge. Therefore, some more sophisticated variation of this method is needed. First, a temporary image is created by copying the input image (the status after the Gabor thresholding). This temporary image is eroded (by the use of square structure element 15x15 px) such as all unwanted entities are eliminated. After that, the temporary image is dilated, but with a structural element that is slightly larger in size (17x17 px), than the one used for erosion. Now we have two intermediate binary images: the temporal image without artifacts containing the main fingerprint area slightly enlarged with respect to the input image and the original input image. By using a logical conjunction operation we get the resulting thresholded image without lines, drawings and other artifacts.

3.3 Composition of the Entire Fingerprint Area

After removing unwanted artifacts, the pipeline has to address a further challenge in the third phase. After main segmentation and artifacts removal, the segmented image may contain more than just one separated foreground areas and each foreground area may contain one or more holes (inside "background" areas) caused by scars, noise etc. Therefore, we propose as third phase an algorithm for eliminating of the holes and insignificant foreground areas.

We start with an algorithm removing holes. First we extend the binary image by one line/row (background padding) and thus adding to the input image one white (background) row on the top, bottom and left and right side. Thus the binary fingerprint is despite all artifacts in the two preceding processing phases bordered as background area. This is essential in a situation where foreground detecting phase may split the background area into several parts. Next we detect all background (white) areas using flood seed fill, where every new detected background area is filled with a gray (temporary) color and a starting point as well as a number of filled pixels for every area is stored. Next step is filling the biggest detected area with white color (color of background) and other detected areas with black color (color of fingerprint area). Finally, we remove columns and rows added in the first step.

Removing insignificant foreground objects is a similar task. We detect all black areas and their sizes and then we eliminate insignificant areas by white filling. Decision which areas are insignificant is controlled by a detection policy. Our detection policy keeps always the larges area and other areas are removed if their size is less than ten percent of the input image. After these two steps, we get in the final phase of the pipeline a segmented image without any artifacts as described above.

4 Benchmarking Results

For the purpose of developing a semantic conformance testing methodology, we needed a reliable fingerprint area segmentation that is applicable for each fingerprint in our database. Thus we cannot rely on any automatic area segmentation and have for the sake of quality assurance implemented a program for manual extraction of the fingerprint area. We applied this parallel automated and manual processing to a set of 595 images in the Ground Truth Database (347 of them was originally from NIST SD 14 database and 248 was originally from NIST SD 29 database).

Table 1. Results of tests for selected part of database SD14

Method/Algorithm	Mean (%)	Median (%)
Our segmentation pipeline	4,129	2,618
NFIQ (quality map threshold = 2)	10,113	9,904
Gabor Filter-Based algorithm [1]	13,950	13,503
Gabor Filter-Based algorithm (enhancement proposed by Allonso-Fernandez) [1]	16,069	15,512

The results from manual extraction of the fingerprint area were compared with the automated approaches; NIST NFIQ quality map (only the best threshold is shown); basic Gabor Filter based algorithm and its second version with enhancements proposed by Allonso-Fernandez in [1]. The results from manual extraction were considered as baseline (100%). We report the difference between the baseline and the results of the benchmarked methods such that 0% indicates absolute overlap (consensus) between the manually extracted area and the automatically extracted area and 100% indicates absolute difference i.e. inverted selection. The results of our benchmark are reported in Table 1 and 2. An example of processed images and the associated results are displayed in Fig. 2.

Table 2. Results of tests for selected part of database SD 29

Method/Algorithm	Mean (%)	Median (%)
Our segmentation pipeline	4,396	2,742
NFIQ (quality map threshold = 1)	7,495	6,896
Gabor Filter-Based algorithm [1]	7,530	6,647
Gabor Filter-Based algorithm (enhancement proposed by Allonso-Fernandez) [1]	8,627	7,649

According to conducted benchmark, our segmentation pipeline was approximately two respectively three times better than the other methods. Our pipeline also produced in several cases the 100% correct area extraction, which was not achieved by other methods. Of course, a 100% correctness of area extraction is hard to justify, as the manual determined area may be different, if a second operator analyses the fingerprints. Unfortunately, manual extraction is very time consuming, but we plan to perform this test in the near future.

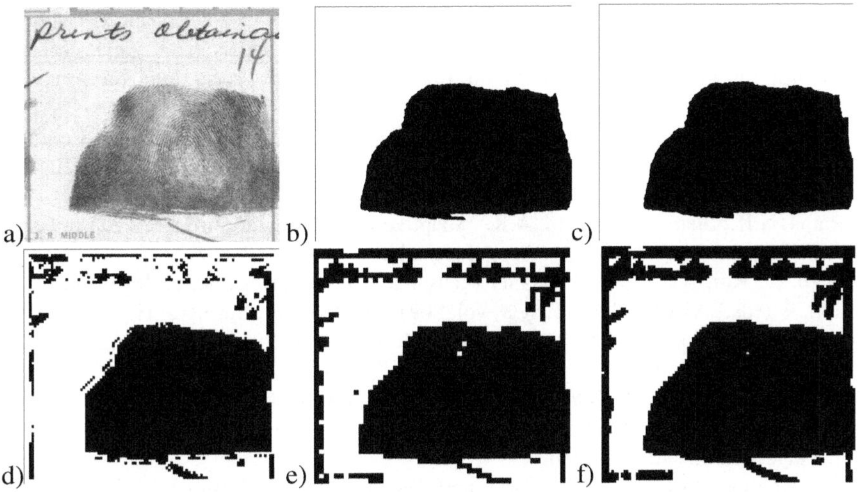

Fig. 2. a) Tested fingerprint; b) Fingerprint area extracted manually; fingerprint area extracted c) By our algorithm; d) By NIST NFIQ quality map with threshold 2; e) By Gabor Filter-Based algorithm [1]; and f) By Gabor Filter-Based algorithm (with enhancement proposed by Allonso-Fernandez) [1]

On the other side, our pipeline did not achieve so good results for fingerprints with low quality – fingerprints containing the large area(s) created by dotted papillary lines.

5 Conclusions

In this paper we have presented a new approach for fingerprint area segmentation, which was developed to process fingerprints scanned from dactyloscopic fingerprint cards. Our pipeline was benchmarked with other methods and achieved significantly better results than the other methods. The proposed pipeline is able to deal with the most drawing and characters, borderlines found on dactyloscopic fingerprint cards. Further the pipeline can well handle dirt in the background or interfering fingerprint areas. Nevertheless a problem with fingerprints with low quality papillary lines still remains.

Acknowledgments. This work is partially supported by the BUT FIT grants "Secured, reliable and adaptive computer systems", FIT-S-10-1, "Information Technology in Biomedical Engineering", GA102/09/H083 and "Support of education of Fundamentals of Artificial Intelligence and Soft-Computing courses", FR1613/2010/G1, and the research plan "Security-Oriented Research in Information Technology", MSM0021630528.

References

1. Alonso-Fernandez, F., Fierrez-Aguilar, J., Ortega-Garcia, J.: An Enhanced Gabor Filter-Based Segmentation Algorithm for Fingerprint Recognition Systems. In: Proceedings of the 4th International Symposium on Image and Signal Processing and Analysis, pp. 239–244 (2005)
2. Lodrova, D., Busch, C., Tabassi, E., Krodel, W., Drahansky, M.: Semantic Conformance Testing Methodology for Finger Minutiae Data. In: Proceedings of the Special Interest Group on Biometrics and Electronic Signatures, GI, Darmstadt, pp. 31–42 (2009)
3. Ratha, N.K., Shaoyun, C., Jain, A.K.: Adaptive Flow Orientation-Based Feature Extraction in Fingerprint Images. Pattern Recognition 28(11), 1657–1672 (1995)
4. Shen, L., Kot, A., Koo, W.: Quality Measures of Fingerprint Images. In: Bigun, J., Smeraldi, F. (eds.) AVBPA 2001. LNCS, vol. 2091, pp. 266–271. Springer, Heidelberg (2001)
5. International Standard ISO/IEC 19794-2 Information Technology - Biometric data interchange Formats – Part 2: Finger minutiae data (2005)
6. International Standard ISO/IEC 29109-2 AMD1 Information Technology - Conformance Testing Methodology for Biometric Interchange Formats defined in ISO/IEC 19794 – Part 2: Finger minutiae data (2010)
7. The National Institute of Standards and Technology: NIST Biometric Image Software, http://www.itl.nist.gov/iad/894.03/nigos/nbis.html

Portable Brain-Heart Monitoring System

Chih-Chung Fu, Chiu-Kuo Chen, Shao-Yen Tseng, Shih Kang, Ericson Chua, and Wai-Chi Fang

Department of Electronics Engineering and Institute of Electronics
National Chiao Tung University
1001 University Road, Hsinchu City, Taiwan (R.O.C.)
wfang@mail.nctu.edu.tw

Abstract. A portable brain-heart monitoring system is proposed to integrate and miniaturize those heavy equipments in the hospitals. The system comprises a 4-channel independent component analysis (ICA) engine for artifact removal from EEG, a heart-rate variability (HRV) analysis engine for on-line HRV analysis and a diffuse optical tomography (DOT) engine for reconstruction of the absorption coefficient image of the brain tissue. A lossless compression module achieves 2.5 compression ratio is also employed to reduce the power consumption of the wireless transmission. EEG, EKG and near-infrared signals acquired from the analog front-end IC are processed in real-time or bypassed according to user configurations. Processed data and raw data are compressed and sent to a remote science station by a commercial Bluetooth module for further analysis and 3-D visualization and remote diagnosis. The ICA and HRV engine are verified by real EEG and EKG signals while the DOT engine is verified by an experimental model. The system is implemented using UMC 65nm CMOS technology, and the core size is 680x680 um^2, and the estimated power consumption of the chip working at 24 MHz under full mode is 3.6 mW.

Keywords: Biomedical System, portable, Artifact Removal, wearable, ICA, DOT, HRV, Health-Care, Bluetooth, Signal Processing.

1 Introduction

Since the twenty-first century, the fast increment of an aged population is emerging as a preeminent worldwide phenomenon. Most of the elderly suffer from chronic ailments and illnesses related to central nervous system (CNS) in their later life. To ease the problems caused by insufficient nursing personnel, many health-care systems focusing on biomedical signal processing and monitoring have been developed. Traditional EEG measuring equipments require the patients to be confined to a small area due to their large size, bringing tremendous inconvenience to them. Therefore, integrated portable health-care systems have become an increasingly important topic.

Electroencephalogram (EEG) is a non-invasive tool for measuring the electrical activity in the brain. The EEG provides important information about the health of the central nervous system (CNS), particularly in the newborn [1]. However EEG signal is very sensitive, and very often may be contaminated by various disturbances, for

Y. Zhang et al. (Eds.): DTA/BSBT 2010, CCIS 118, pp. 241–250, 2010.

example ocular artifact, EMG and electrical noise from nearby instruments [2]. This problem can be alleviated by the independent component analysis (ICA) algorithm [3], which separates artifacts and noise from the measured EEG signals. Recent studies have shown that combined analysis of EEG together with heart rate variability or brain fNIR can aid in better diagnosis and treatment. For example EEG and HRV data were jointly analyzed for the automatic detection of seizures in newborns [4] and sleep apnea in hospital patients [5], while the advantage of combined analysis of EEG and fNIR data for cognitive rehabilitation and post traumatic stress syndrome was presented in [6]. Despite these studies indicating the need for joint monitoring of brain fNIR, EEG and HRV, an integrated brain-heart monitoring solution has not yet been developed. In this work, a portable brain-heart monitoring system comprising NIR-DOT, HRV and ICA is proposed. In addition, a lossless biomedical signal compressor is employed to reduce the power consumption of wireless transmission.

In Section 2, a brief description of the system from a top level view and its intended applications in portable brain-heart monitoring are given. Section 3 introduces the end-to-end system design including the performance analysis of the signal processing engines. Chip tape-out summary and a graphical user interface (GUI) on the science station that receives and displays the signal are shown in Section 4. Finally a conclusion is given in Section 5.

2 System Overview and Application Scenario

The proposed system in Fig.2 is interfaced with analog front-end circuits and a commercial Bluetooth module in Fig.1 (b). NIR, EEG and ECG signal are acquired, processed, compressed and transmitted via the Bluetooth wireless link. Biomedical data received at the science station are decompressed, displayed in real-time on the screen of the science station, and finally stored into non-volatile storage media for further off-line processing, analysis and diagnosis. In addition, the data can be forwarded from the science station to a remote workstation for online monitoring and remote diagnosis by doctors in hospitals. The application scenario is shown in Fig.1 (a).

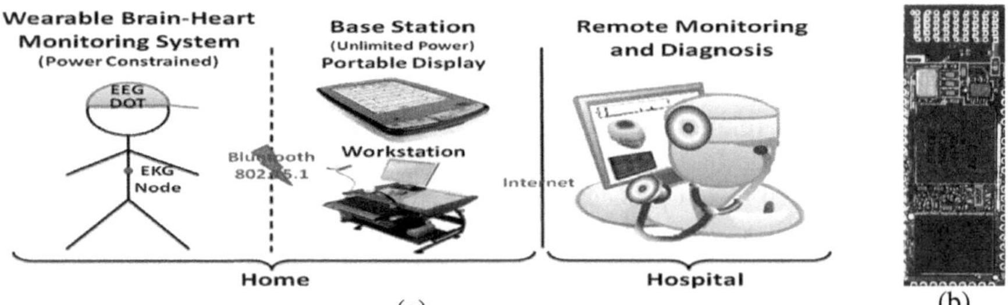

(a) (b)

Fig. 1. (a) Application scenario for the proposed wearable brain monitoring system (b) The employed commercial Bluetooth module

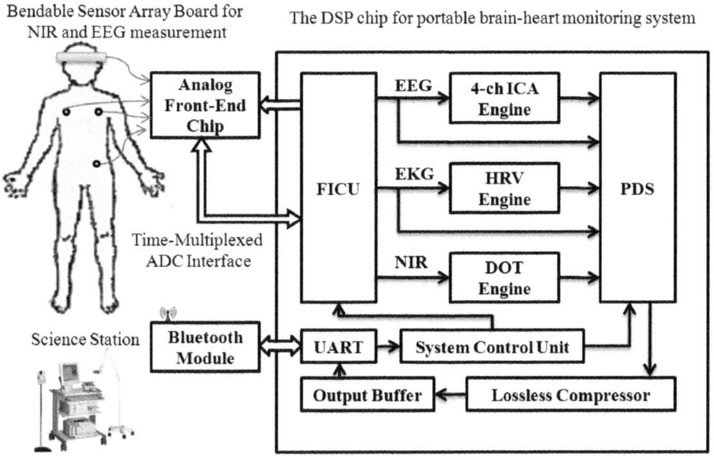

Fig. 2. Architecture for the portable brain-heart monitoring system

3 System Design and the Processing Engines

The overall system architecture is shown in Fig.2. First, the science station sends a trigger command including the working mode and compression mode to activate the system. The command is decoded and evaluated by the system control unit (SCU), and after the reset process, an internal trigger signal is then sent to the front-end interface control unit (FICU). Afterwards, time multiplexed data acquisitions are continuously triggered by two system counters in FICU.

The specifications and working modes are listed in Table 1.

Table 1. Specifications of the front-end circuits and working modes for the system

Parameter	DOT	EEG	EKG
Sample Rate(Hz)	1 frame (24 sensor values)	128	256
#Sensor(Channel)	12	4	3
#NIR LED	6	-	-
Filter BW	-	50	100
Gain (dB)	-	250	5000
Output Range (V)	0~2.2	0~2.5	0~2.5
Available Modes	DOT / off	raw / ICA / off	raw / raw+HRV / off
Compression	on / off	on / off	Only for raw data

Digitized raw bio-signals are sent to both the prioritized data selector (PDS) and the corresponding processing engines, that is, the ICA, HRV and DOT engine. When the processed data is available from the processing engines, PDS will check if the lossless compressor is busy or not, when the output buffer and the compressor are available, and then the queued data is sent by a fixed priority. Compressed data is packed and sent through the Bluetooth wireless link between the system and science station, so further display and signal processing can be performed on the science station.

3.1 Near-Infrared Diffuse Optical Tomography Engine (DOT)

To make sure the near-infrared doesn't leak through the space between the LED and the measuring surface, we designed the DOT sensor array board using bendable printed circuit board (PCB) as shown in Fig.3 (a). Aside from the NIR sensors and the bi-wavelength LEDs, there is an analog multiplexer chip for selecting the channel to be digitized and a decoder chip for selecting the LED to emit near-infrared.

The DOT engine is designed using CW (Continuous Wave) algorithm [7], and is composed of sub-frame operation control circuit [8], DOT reconstructor and the image post-processor, and its overall scheme is shown in Fig.3 (b). The DOT algorithm can be divided into a forward model and its inverse problem. Forward model includes abundant optical parameters and mathematic equations. Indeed, when the depth of the surface to be calculated is fixed, the inverse matrix is always the same. Therefore, the pre-calculated inverse matrices are stored in a look-up table. The DOT reconstructor controlled by the sub-frame operation control circuit is mainly used to perform matrix operations. Pixels reconstructed represent absorption coefficient variance $\Delta\mu^{a}_{x,y}$. Normally, $\Delta\mu^{a}_{x,y}$ are too small to be observed, so an image post-processor is employed to perform linear mapping, contrast enhancement and color mapping so that we can observe clear images on the science station monitor.

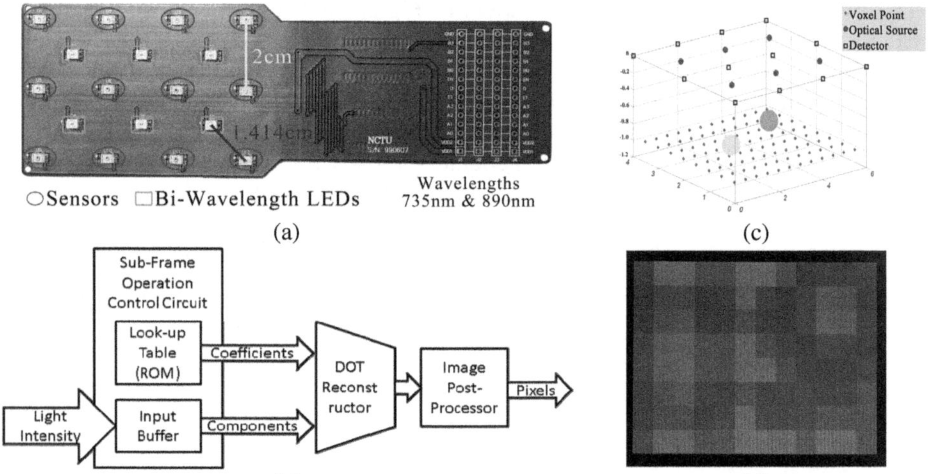

Fig. 3. (a) The bendable DOT sensor array board (b) Block diagram of the DOT engine (c) An experimental model (d) Reconstructed image on the LCD of the development platform

To verify the proposed DOT engine, we use the experimental model shown in Fig.3 (c). The frame area is $4x6cm^2$ and the volume of voxels is $(0.25cm)^3$. The background medium is homogenous with $\mu_a^{bg} = 0.05cm^{-1}$ and the reduced scattering coefficient $\mu_s^{'bg} = 10cm^{-1}$. Two kinds of inhomogeneous media were embedded at depth of 0.5cm below the surface. The absorption coefficients of the inhomogeneous mediums A and B (yellow) are $0.21cm^{-1}$ and $0.5cm^{-1}$, respectively. The reconstructed colored image using the system is shown in Fig.3 (d).

3.2 4-Channel Independent Component Analysis Engine (ICA)

The architecture of the ICA engine is shown in Fig.4 (a). The ICA engine can be divided into a pre-processor and ICA training unit. The pre-processor, including centering and whitening, is used to accelerate the convergence of ICA training. The ICA training unit is designed using Infomax ICA algorithm [9] capable of separating mixed sub-gaussian and super-gaussian sources. To reduce the output delay, we assume that after convergence, the unmixing matrix W won't change rapidly. Thus we can use the W of the previous window for the current input. This sliding window scheme is shown in Fig.4 (b).

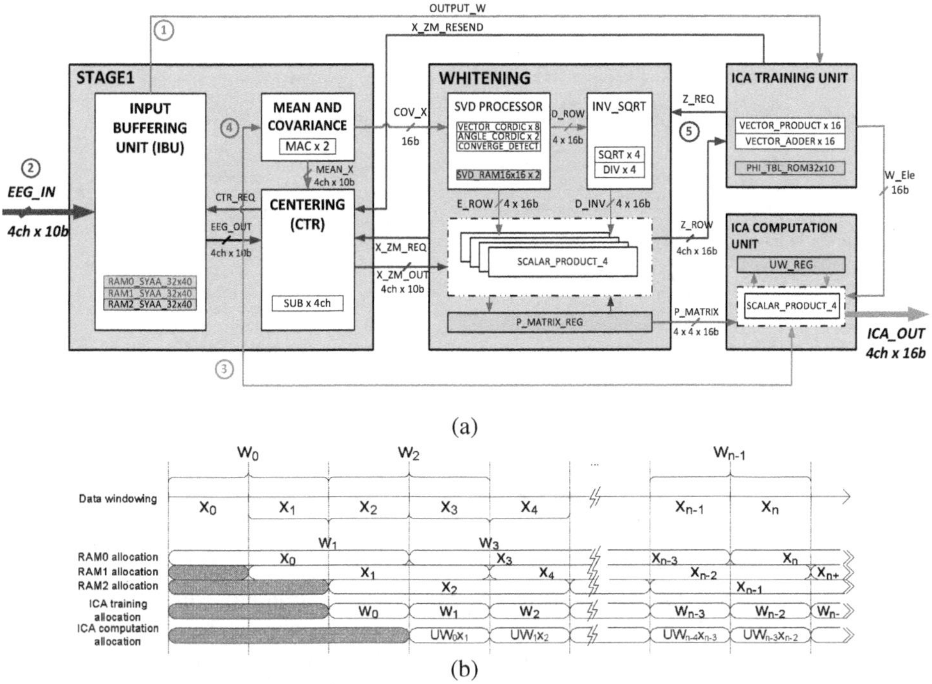

Fig. 4. (a) Hardware architecture of ICA engine (b) Data windowing and operation scheduling

An ICA demonstration using real EEG source recorded by NeuroScan system is shown below. In Fig.5 (a), the waveforms shows the four channel EEG signals recorded from the NeuroScan system and the 4-channel are from FP1, FPZ, FP2 and F1 in international 10-20 system of electrode placement standard. The signals are sampled at 128 Hz, and the signals are filtered using high-pass and low-pass filters with cut-off frequencies at 0.15 Hz and 55 Hz. In Fig.5 (b), the waveforms in the first row are the off-line ICA result using EEGLab; the waveforms in the second row are the real-time ICA result using the proposed ICA engine. The channel orders are shuffled through the algorithm, so the corresponding components are marked by the same component number, and the correlations between the results from on-line hardware calculation and off-line algorithm are also shown in the figure. Although the data are

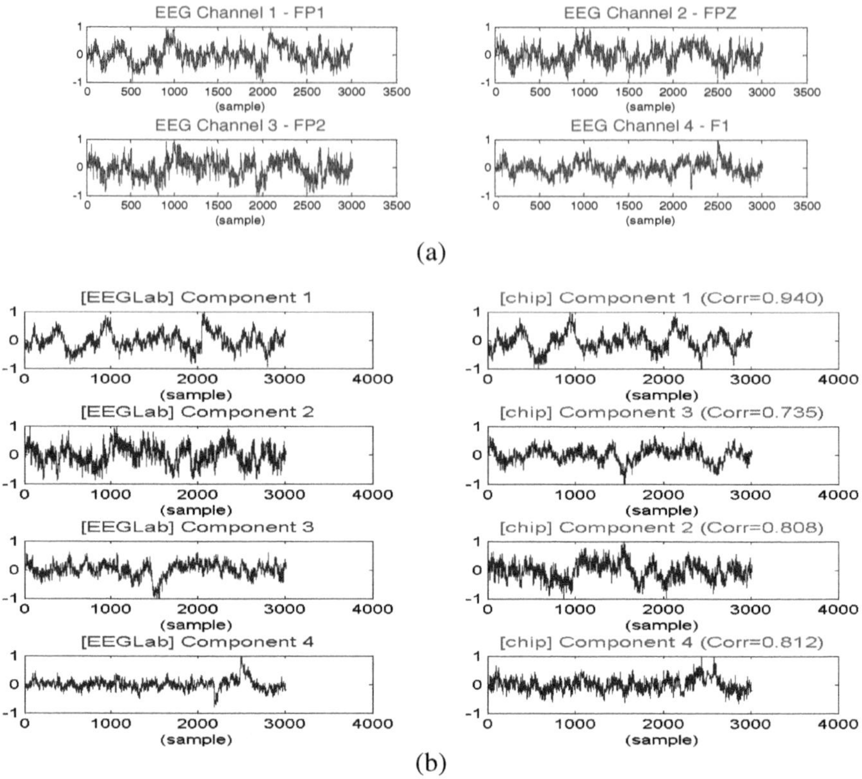

Fig. 5. (a) 4-channel EEG signals recorded using NeuroScan System (b) Waveform comparison of independent components analyzed by EEGLab and the designed engine

windowed, the results have shown that each statistical independent component is locked in the same channel for different window. To compare with other hardware design of the ICA algorithm, a Super-Gaussian random waveform generator is used. Four-channel Super-Gaussian patterns are generated and mixed with a fixed mixing matrix. The mixed signals are processed by the engine, and the result shown that an

Table 2. Comparisons of complexity and correlation

	Shyu [13]	Huang [14]	This Work
Application	EEG	Speech	EEG
Channel	4	2	4
Pre-Processing	No	Yes	Yes
Memory (words)	4096	1200	2048
ADC sample rate	64 Hz	16 KHz	128 Hz
ADC resolution	8 bits	N/A	10 bits
Correlation	> 0.8	N/A	0.86
Data format	Floating	Floating	Fixed
Algorithm	Infomax	Fast	Infomax

average 0.86 of correlation is achieved. Table 2 shows comparison of our proposed engine design with other works on ICA. In our proposed design, the memory complexity is lower while the correlation is higher.

3.3 Heart Rate Variability Analysis Engine (HRV)

A novel HRV engine using a fast windowed Lomb periodogram [11] is proposed. The Lomb time-frequency distribution (TFD) is suited for spectral analysis of unevenly spaced data and has been applied to the analysis of heart rate variability.

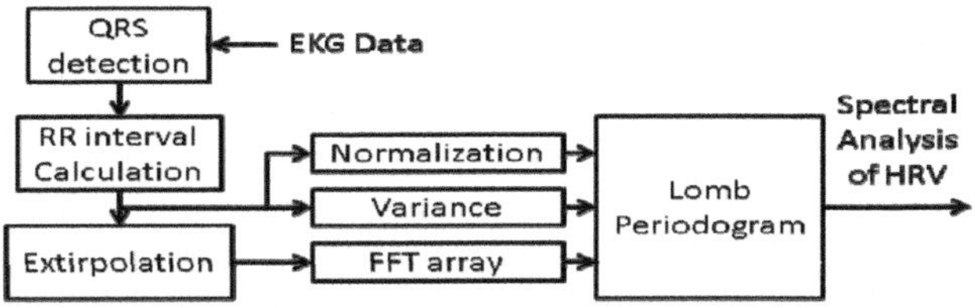

Fig. 6. Architecture of the HRV analysis engine

The HRV engine in Fig. 6 comprises the hardware implementation of the Lomb TFD as well as a simple RR interval calculation unit. In consideration of architecture simplicity and real-time properties, the classical derivative-based QRS detection algorithm introduced by Pan and Tompkins [10] was adopted as a baseline for the RR interval calculation unit. In the RR interval calculation unit, EKG signals first pass through a set of linear processes, including a band-pass filter comprising a cascaded low-pass and high-pass, and a derivative function. Non-linear transformation is then employed in form of a signal amplitude squaring function. Finally, a threshold is applied to detect the R-peaks of the QRS complexes. The RR intervals are then calculated from the detected peaks and HRV analysis is performed. Better time-frequency analysis of HRV is achieved through a de-normalized fast Lomb periodogram with a sliding window configuration.

The HRV engine is verified using the MIT/BIH database and results of the QRS detection algorithm were compared with offline simulations. The output of the QRS detection algorithm as well as the power spectrum of the RR intervals is shown in Fig. 7.

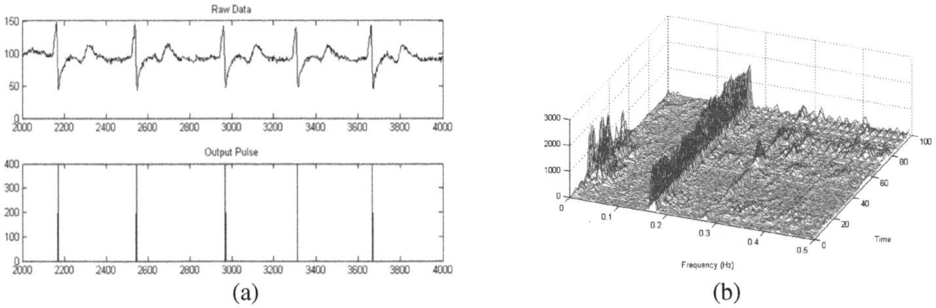

Fig. 7. (a) Input and the resulting output from the QRS detection unit (b) Time-frequency HRV analysis of EKG data from MIT-BIH arrhythmia database using Lomb TFD

3.4 Lossless Biomedical Signal Compressor

In recent years, many portable biomedical signal acquisition devices have been developed, wherein ECG or EEG data are gathered and transmitted wirelessly to a base station or PC for further analysis. In these devices, most of the power is consumed during the wireless transmission of the biomedical signal, and is directly proportional to the amount of data that needs to be transmitted. By performing data compression, much power can be saved as the extra power consumption due to the compression operation is much lower. We choose a lossless compression method because it should be left to the medical practitioner (e.g., doctor, medical researcher) to decide what part of the biomedical signal is to be considered as information and what is not.

The architecture of the compression module, shown in Fig.8, comprises a differential pulse code modulation (DPCM) predictor, a context-based k-parameter estimator, a prediction memory array (3-ch EKG, 4-ch EEG, 96 pixel DOT), a set of context variable upkeep modules, a Golomb-Rice entropy coder and a 40-bit output packaging unit. The 40-bit output packaging unit multiplexes compressed multichannel biomedical data onto a single data stream. Simulation results demonstrate an average compression ratio (CR) of around 1.7 for EEG data, and 2.5 for ECG data, which translates into power savings during wireless transmission, accordingly.

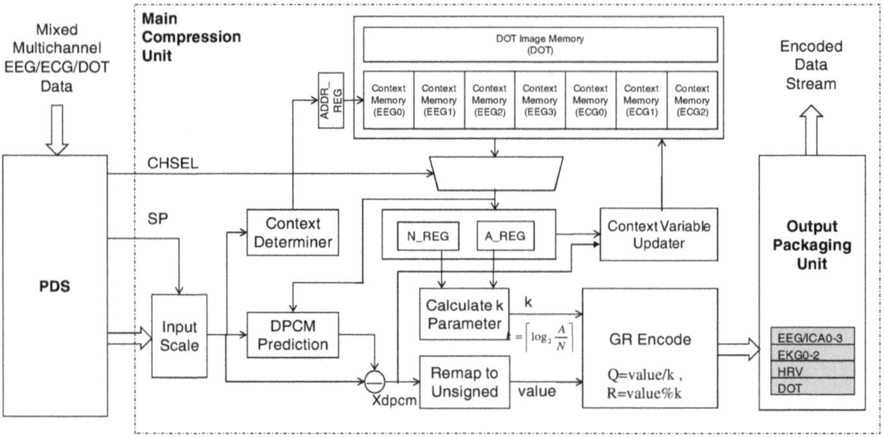

Fig. 8. Hardware architecture of the lossless bio-signal compressor

4 Result

The designed chip is currently under fabrication using UMC 65nm CMOS Technology. The die size is 1317x1317 um^2, and the core size is 680x680 um^2. We estimate the power consumption using Synopsys Prime Power, and the result shows total 3.6 mW is consumed, and around 64.7% of the total power is consumed by the 4-channel ICA engine. Due to the nature of the slow change of biomedical signals, we specified the system clock rate to be 24 MHz, and even lower frequency can be used to achieve lower power consumption. The partitioned chip layout generated by cadence SOC Encounter is shown in Fig.9.

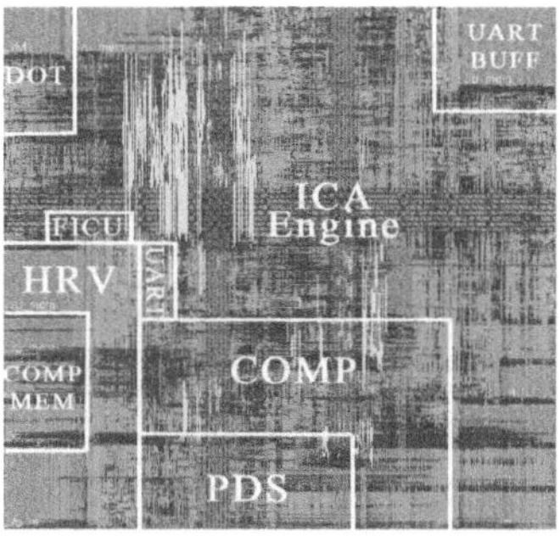

Fig. 9. The layout of the designed chip under fabrication

Table 3. Chip Specifications

	Chip Specifications
Technology	UMC 65nm CMOS 1P10M
Pad/Core Voltage	2.5 / 1 V
Die Size	1317 x 1317 um^2
Core Area	680 x 680 um^2
Equivalent Gate-Count	368314
System Operating Frequency	24 MHz

A graphical user interface (GUI) for PC-based science station is realized, while a portable 3D visualization system for our proposed monitoring system is being developed. The GUI is shown in Fig.10 (a) to (d) with views for DOT, EEG and EKG signals. In the GUI, each engine can be configured to sleep (bypass) mode including the biomedical signal compressor. In the chip implementation, we use clock-gating and power shut-off while the engine is in sleep mode so that the operation time can be prolonged.

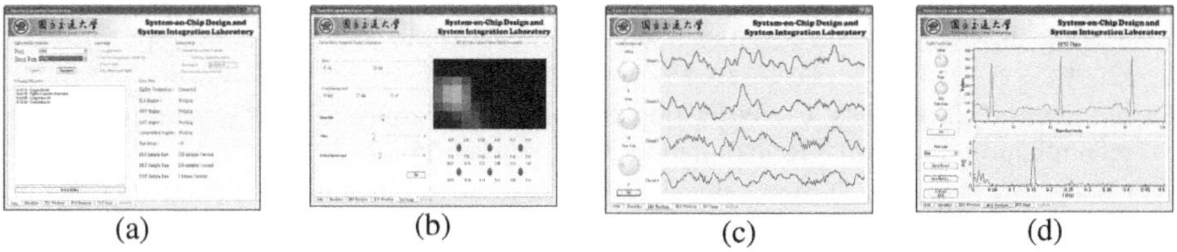

Fig. 10. (a) GUI for pc-based science station (b) DOT view in GUI (c) EEG view in GUI (d) HRV view in GUI

5 Conclusion

A portable brain-heart monitoring system comprises an fNIR-DOT engine, a 4-channel ICA engine that achieves 0.86 of correlation and an HRV analysis engine using Lomb periodogram is presented. Signals acquired from front-end sensor modules are processed in real-time or bypassed according to user configurations, and are then losslessly compressed by a biomedical signal compressor achieving an average 2.5 CR before being wirelessly sent to a base-station with a commercial Bluetooth module. By integrating three biomedical systems into a single chip, bulk associated with external circuitry is reduced. Improvements for the next stage would include raising the number of channels in the ICA engine, integration of the front-end analog chip and Bluetooth module using system-in-package (SiP) technology and adoption of more advanced low-power techniques suitable for this system to prolong the operation time.

Acknowledgement

This work was performed at NCTU System-on-Chip research center under the grants NSC99-2220-E-009-028 and NSC99-2220-E-009-030 sponsored by the National Science Council and the National Science and Technology Program for System-on-Chip, Taiwan. The authors would also like to thank the National Chip Implementation Center for CAD tool support. The chip fabrication was sponsored by UMC.

References

1. Rankine, L., Stevenson, N., Mesbah, M., Boashash, B.: A Nonstationary Model of Newborn EEG. IEEE Transactions on Biomedical Engineering 54(1), 19–28 (2007)
2. Vigario, R.: Extraction of ocular artifacts from EEG using independent component analysis. Electroencephalogr. Clin. Neurophysiol. 103, 395–404 (1997)
3. Vigario, R., Sarela, J., Jousmaki, V., Hamalainen, M., Oja, E.: Independent component approach to the analysis of EEG and MEG recordings. IEEE Trans. Biomed. Eng. 47(5), 589–593 (2000)
4. Malarvili, M.B., Mesbah, M.: Combining newborn EEG and HRV information for automatic seizure detection. In: 30th Annual International Conference of the IEEE, Engineering in Medicine and Biology Society, EMBS 2008, August 20-25, pp. 4756–4759 (2008)
5. Abdullah, H., Holland, G., Cosic, I., Cvetkovic, D.: Correlation of sleep EEG frequency bands and heart rate variability. In: Annual International Conference of the IEEE Engineering in Medicine and Biology Society, EMBC 2009, September 3-6, pp. 5014–5017 (2009)
6. Fidopiastis, C., Hughes, C.: Workshop 1: Use of psychophysiological measures in virtual rehabilitation. In: Virtual Rehabilitation 2008, August 25-27 (2008)
7. Cheng, X., Boas, D.: Diffuse optical reflection tomography with continuous-wave illumination. Med. Biol. 42, 841–854 (1997)
8. Hsu, Y.-H., Fu, C.-C., Fang, W.-C., Sang, T.-H.: A VLSI-inspired image reconstruction algorithm for continuous-wave diffuse optical tomography systems. In: IEEE/NIH Life Science Systems and Applications Workshop, LiSSA 2009, April 9-10, pp. 88–91 (2009)
9. Bell, A.J., Sejnowski, T.J.: An Information-Maximization Approach to Blind Separation and Blind Deconvolution. Neural Computing 7, 1129–1159 (1995)
10. Trahanias, P.E.: An approach to QRS complex detection using mathematical morphology. IEEE Transactions on Biomedical Engineering 40, 201–250 (1993)
11. Thong, T., McNames, J., Aboy, M.: Lomb-Wech periodogram for non-uniform sampling. In: Engineering in Medicine and Biology Society, IEMBS 2004, pp. 271–274 (2004)
12. Press, W.H., Rybicki, G.B.: Fast algorithm for spectral analysis of unevenly sampled data. Astrophysical Journal 338
13. [13] Shyu, K.-K., Li, M.-H.: FPGA Implementation of FastICA based on Floating-Point Arithmetic Design for Real-Time Blind Source Separation. In: International Joint Conference on Neural Networks, IJCNN 2006, pp. 2785–2792 (2006)
14. Huang, W.-C., Hung, S.-H., Chung, J.-F., Chang, M.-H., Van, L.-D., Lin, C.-T.: FPGA implementation of 4-channel ICA for on-line EEG signal separation. In: Biomedical Circuits and Systems Conference, BioCAS 2008, November 20-22, pp. 65–68. IEEE, Los Alamitos (2008)

Dermatologic Diseases and Fingerprint Recognition

Eva Brezinova[1], Martin Drahansky[2], and Filip Orsag[2]

[1] Masaryk University, Faculty of Medicine, Komenskeho nam. 2, CZ-662 43, Brno, Czech Republic
141896@mail.muni.cz
[2] Brno University of Technology, Faculty of Information Technology, Bozetechova 2, CZ-612 66, Brno, Czech Republic
{drahan,orsag}@fit.vutbr.cz

Abstract. This article discusses different dermatologic diseases which have the impact to the process of fingerprint acquirement. There are many people, who suffer under such skin diseases and are therefore excluded from the set of users of a biometric system. The classification of skin diseases is made from the medical point of view.

Keywords: dermatology, skin disease, fingerprint recognition, biometrics.

1 Introduction

Skin diseases represent a very important, but often neglected factor of the fingerprint acquirement. It is impossible to say in general how many people suffer from skin diseases, because there are so many various skin diseases – please refer e.g. to [1][2][3][4][5], but we must admit that such diseases are present in our society. When discussing whether the fingerprint recognition technology is a perfect solution capable to resolve all our security problems, we should always keep in mind those potential users who suffer from some skin disease.

In the following text, several skin diseases are introduced from the medical point of view, which attack hand palms and fingertips.

The situation after successful recovery of a potential user from such skin diseases is, however, very important for the possible further use of fingerprint recognition devices. If the disease has attacked and destroyed the structure of papillary lines in the epidermis layer of the skin, the papillary lines will not grow in the same form as before (if at all) and therefore such user could be restricted in his/her future life by being excluded from the use of fingerprint recognition systems, though his fingers don't have any symptoms of a skin disease any more.

2 Skin Diseases

There are a lot of skin diseases, which can affect palms and fingers. We find plenty of skin diseases including description of their influence on the structure and color of the skin in specialized medical literature, e.g. [4][5][6][7][8][9]. In following chapters we describe some of these diseases together with photographs. These clearly show that these diseases may cause many problems in automatic biometric systems.

Y. Zhang et al. (Eds.): DTA/BSBT 2010, CCIS 118, pp. 251–257, 2010.

2.1 Diseases Causing Histopathological Changes of Epidermis and Dermis

These diseases may cause problems for the most types of sensors.

Hand eczema [5][8] is an inflammatory non-infectious long-lasting disease with relapsing course. It is one of the most common problems encountered by the dermatologist. Hand dermatitis causes discomfort and embarrasment and, because of its locations, interferes significantly with normal daily activities. Hand dermatitis is common in industrial occupations. The prevalence of hand eczema was approximately 5.4% and was twice as common in females as in males. The most common type of hand eczema was irritant contact dermatitis (35%), followed by atopic eczema (22°), and allergic contact dermatitis (19%). The most common contact allergies were to nickel, cobalt, fragnance mix, balsam of Peru, and colophony. Hand eczema was more common among people reporting occupational exposure. The most harmful exposure was to chemicals, water and detergents, dust, and dry dirt.

Fingertip eczema [5] is very dry, chronic form of eczema of the palmar surface of the fingertips, it may be result of an allergic reaction or may occur in children and adults as an isolated phenomenon of unknown cause. One finger or several fingers may be involved. Initially the skin may be moist and then become dry, cracked, and scaly. The skin peels from the fingertips distally, exposing a very dry, red, cracked, fissured, tender, or painful surface without skin lines – see Figure 1.

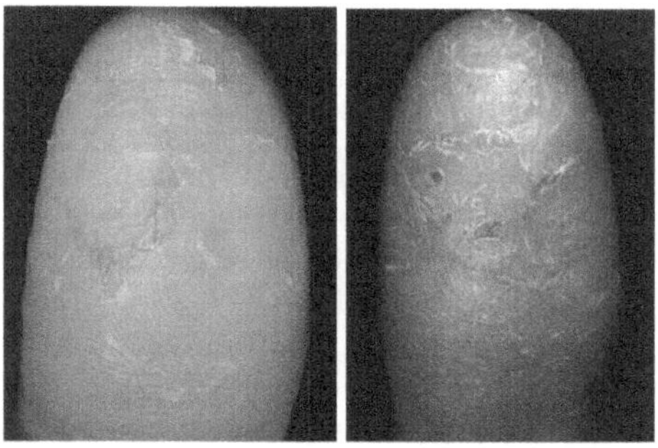

Fig. 1. Fingertip eczema [5]

Pomfolyx (*dishydrosis*) [4] is a distinctive reaction pattern of unknown etiology presenting as symmetric vesicular hand and foot dermatitis. Itching precedes the appearance of vesicles on the palms and sides of the fingers. The skin may be red and wet. The vesicles slowly resolve and are replaced by rings of scale. Chronic eczematous changes with erythema, scaling, and lichenification may follow.

Tinea of the palm [5][8] is dry, diffuse, keratotic form of tinea. The dry keratotic form may be asymptomatic and the patient may be unaware of the infection, attributing the dry, thick, scaly surface to hard physical labor. It is frequently seen in assotiation with tinea pedis which prevalence is 10 to 30%.

Pyoderma [8] is a sign of bacterial infection of the skin. It is caused by Staphylococcus aureus and Streptococcus pyogenes. Some people are more susceptible to these diseases (such as diabetics, alcoholics, etc.) - see Figure 2.

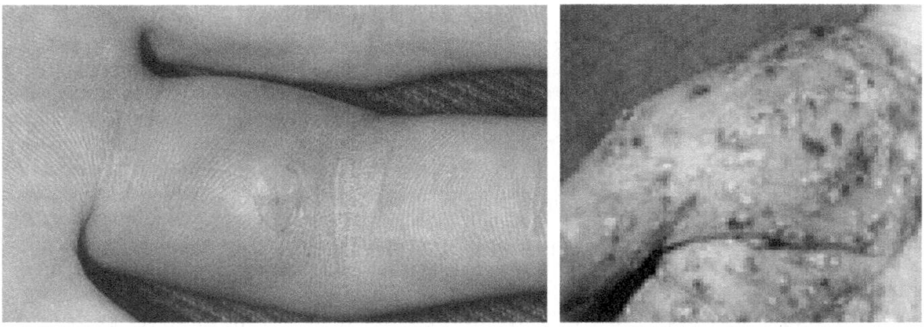

Fig. 2. Abscess on finger of patient with diabetes [4] and pyoderma [9]

Pitted keratolysis [5] is a disease mimicking tinea, especially for people who swelter and wear rubber gloves in the hot, humid environment. Hyperhydrosis is the most frequently observed symptom. The disease is bacterial in origin, characterized by many circular or longitudinal, punched out depressions in the skin surface. The eruption is limited to the stratum corneum.

Keratolysis exfoliativa [5] is a common, chronic, asymptomatic, noninflamatory, bilateral peeling of the palms of the hands. Its cause is unknown. The eruption is most common during the summer months and is often associated with sweaty palms and soles. It is characterized by scaling and peeling, the central area becomes slightly red and tender.

Lichen planus [8] is quite common, unique inflammatory cutaneous and mucous membrane reaction pattern of unknown etiology. LP of the palm and soles generally occurs as an isolated phenomenon. The lesions are papules aggregated into semitranslucent plaques with globular waxy surface, ulceration may occur.

Acanthosis nigricans [4][6] is non-specific reaction pattern that may accompany obesity, diabetes, tumors. AN is classified into benign and malignant forms. In all cases the disease presents with symmetric, brown thickening of the skin. During the process there is papillary hypertrophy, hyperkeratosis, and increased number of melanocytes in the epidermis.

Pyogenic granuloma [5] is a benign acquired vascular lesion of the skin that is common in children and young adults. It often appears as a response to an injury or hormonal factors. Lesions are small rapidly growing, yellow-to-bright red, dome-shaped.

Systemic sclerosis [6][8] is a chronic autoimmune disease characterized by sclerosis of the skin or other organs. Emergence of acrosclerosis is decisive for fingerprinting. Initially the skin is infused with edema mainly affecting hands. With the progressive edema stiff skin appears and necrosis of fingers may form. The disease leads to sclerodactyly with contractures of the fingers. For more than 90% of patients is typical Raynaud's phenomenon (see below). The typical patient is a woman over 50 years of age.

Raynaud's phenomenon [5][6][8] represents an episodic vasoconstriction of the digital arteries and arterioles that is precipitated by cold and stress. It is much more common in women. There are three stages during a single episode: pallor (white), cyanosis (blue), and hyperemia (red). Estimates of the prevalence of Raynaud's phenomenon ranged between 4.7 - 21% for women and 3.2-16% for men.

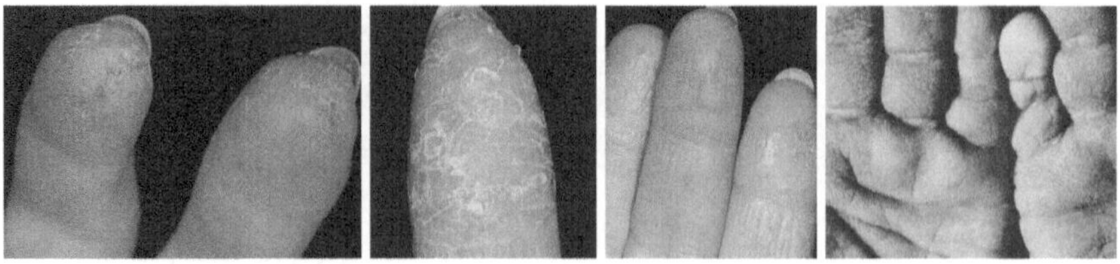

Fig. 3. Different types of eczema [5] (3× left) and acanthosis nigricans [4] (right)

Drug induced skin reactions [5] are among the most common adverse drug reactions. They occur in many forms and can mimic virtually any dermatosis. Occur in 2-3% of hospitalized patients. Sulfonamides, NSAIDs and anticonvulsants are most often applied in the etiology.

2.2 Diseases Causing Skin Discoloration

These diseases are focused mainly on optical sensors.

Hand, foot, and mouth disease (HFMD) [4][5] is contagious enteroviral infection occuring primarily in children and characterized by a vesicular palmoplantar eruption. The skin lesions begin as red macules that rapidly become pale, white, oval vesicles with red areola.

Xantomas [5][6][8] are lipid deposits in the skin and tendons that occur secondary to a lipid abnormality. These localized deposits are yellow and are frequently very firm.

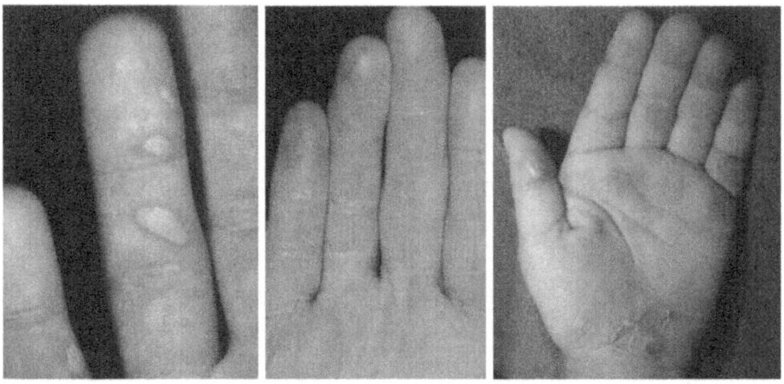

Fig. 4. Hand, foot and mouth syndrome [4]; xantomas [6]; epidermolysis bullosa [7]

Scarlet fever (scarlatina) [5][8] is contagious disease produced by streptococcal, erythrogenic toxin. It is most common in children (ages 1 to 10 years). In the ending stages of the disease large sheats of epidermis may be shed from the palms in glove-like cast, exposing new tender and red epidermis beneath.

Scabies [7][8] is highly contagious disease caused by the mite Sarcoptes scabiei. It is characterized by red papules, vesicles and crusts located usually on the areas with tender skin, palms and soles especially in infants included.

Secondary syphilis [6][8] is characterized by mucocutaneous lesions, which may assume a variety of shapes, including round, elliptic, or annular. The color is characteristic, resembling a „clean-cut ham" or having a copery tint.

2.3 Diseases Causing Histopatological Changes in Junction of Epidermis and Dermis

These disease are focused mainly on ultrasonic sensors, the diagnosis also belong to the first group.

Hand eczema – particularly chronic forms (see above).

Warts (verruca vulgaris) [8] are benign epidermal neoplasms that are caused by human papilloma viruses (HPVs). Warts commonly appear at sites of trauma, on the hand, in periungual regions. HPVs induce hyperplasia and hyperkeratosis.

Psoriasis [6][7][8] is characterized by scaly papules and plaques. IT occurs in 1% to 3% of the population. The disease is transmitted genetically, environmental factors are needed to precipitate the disease. The disease is lifelong and characterized by chronic, recurrent exacerbations and remissions that are emotionally and physically debilitating. Psoriasis of the palms and fingertips is characterized by red plaques with thick brown scale and may be indistinguishable from chronic eczema.

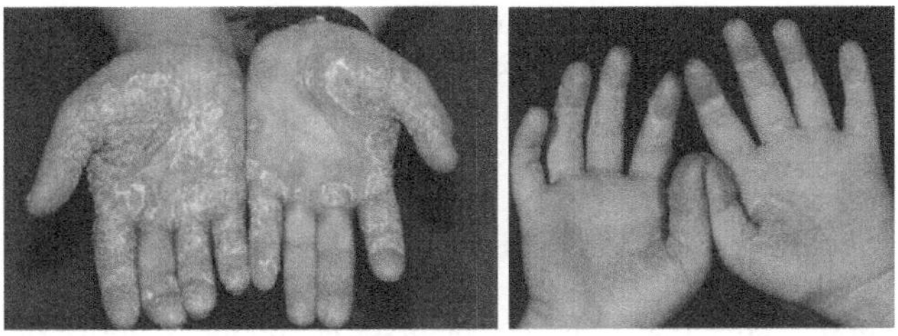

Fig. 5. Psoriasis [7]; scarlet fever [5]

Systemic lupus erytematosus (SLE) [5] is a multisystem disease of unknown origin characterized by production of numerous diverse of antibodies that cause several combinations of clinical signs, symptoms and laboratory abnormalities. The prevalence of LE in North America and nothern Europe is about 40 per 100,000 population. In the case of acute cutaneous LE indurated erythematous lesions may be presented on palms.

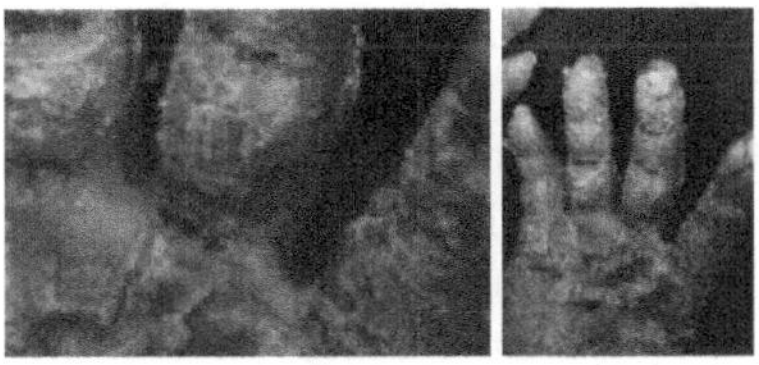

Fig. 6. Psoriasis vulgaris [9]

Epidermolysis bullosa [6][7] is a term given to groups of genetic diseases in which minor trauma causes noninflammatory blistering (mechanobullosus diseases). Repetitive trauma may lead to a mitten-like deformity with digits encased in an epidermal „cocoon". These diseases are classified as scarring and nonscarring and histologically by the level of blister formation. Approximately 50 epidermolysis cases occur per million live births in the United States.

3 Conclusions

It is clear from each subsection that either the color of the skin or the structure of papillary lines on the fingertip could be influenced. If only the color has changed, some of optical fingerprint scanners might be influenced and so this change is not crucial. On the other hand, the change of skin structure is very significant, because if papillary lines are damaged, it is impossible to find the minutiae and therefore to recognize the person. If we are unable to recognize/enroll a person, then such person cannot use the biometric system based on the fingerprint recognition technology, and therefore the implementing company has a big problem – how to authorize such person, if they don't want to use PINs (Personal Identification Numbers) or other authorization methods.

Some of these diseases are only temporary, i.e. after the healing of such disease, the papillary line structure or color is restored and the user is again able to use his/her fingers for the fingerprint recognition in authorization tasks in security systems. However, some diseases leave irrecoverable finger damage restraining a new growth of papillary lines and respective user is then unable to use his/her fingerprints for appropriate recognition tasks in automated fingerprint security systems.

Acknowledgement. This research has been done under support of the following grants: "Secured, reliable and adaptive computer systems", FIT-S-10-1 (CZ), "Security-Oriented Research in Information Technology", MSM0021630528 (CZ) and by the project "Information Technology in Biomedical Engineering", GD102/09/H083 (CZ).

References

[1] Evaluation of Fingerprint Recognition Technologies - BioFinger, Public Final Report, version 1.1, Bundesamt für Sicherheit in der Informationstechnik, p. 122 (2004)

[2] Jain, A.K., Flynn, P., Ross, A.A.: Handbook of Biometrics, p. 556. Springer, Heidelberg (2008) ISBN 978-0-387-71040-2

[3] Bolle, R.M., Connell, J.H., Pankanti, S., Ratha, N.K., Senior, A.W.: Guide to Biometrics, p. 364. Springer, Heidelberg (2004) ISBN 0-387-40089-3

[4] James, W.D., Berger, T.G., Elston, D.M.: Andrew's Diseases of the Skin – Clinical Dermatology, 10th edn., p. 961. Saunders Elsevier, Canada (2006) ISBN 0-8089-2351-X

[5] Habif, T.P.: Clinical Dermatology, 4th edn., Mosby, China, p. 1004 (2004)

[6] Wolff, K., Johnson, R.A., Suurmond, D.: Color Atlas and Synopsis of Clinical Dermatology, 5th edn., p. 1085. McGraw-Hill, USA (2005)
[7] Weston, W.L., Lane, A.T., Morelli, J.G.: Color Textbook of Pediatric Derma-tology, p. 446. Mosby Elsevier, China (2007)
[8] Štork, J., et al.: Dermatovenerologie, Galén, Praha, p. 502 (2008) ISBN 978-80-7262-371-6
[9] Niedner, R., Adler, Y.: Kožní choroby - kapesní obrazový atlas, Triton, Praha, p. 359 (2005) ISBN 80-7254-734-8

Changes in Cognitive Performance Due to Three Types of Emotional Tension

Mi-Hyun Choi[1], Su-Jeong Lee[1], Jae-Woong Yang[1], Ji-Hye Kim[1], Jin-Seung Choi[1], Jang-Yeon Park[1], Jae-Hoon Jun[1], Gye-Rae Tack[1], Dae-Woon Lim[2], and Soon-Cheol Chung[1,*]

[1] Department of Biomedical Engineering, Research Institute of Biomedical Engineering, College of Biomedical & Health Science, Konkuk University, Chungju, Korea
mhchoi0311@gmail.com, sj8618@kku.ac.kr, jwyang1@kku.ac.kr, yeppari@hanmail.net, jschoi98@gmail.com, jypark@kku.ac.kr, jjun81@kku.ac.kr, grtack@kku.ac.kr, scchung@kku.ac.kr

[2] Department of Information & Communication Engineering, Dongguk University, Seoul, South Korea
dwlim01@dgu.edu

Abstract. The purpose of this study was to investigate how three types of emotional tension levels affect performance of a cognitive task. Ten university male (age 25.7 ±1.5) and ten female (age 24.5 ±1.8) students participated in this experiment. We used a 3-back task as a cognitive task. Using pictures selected from a group test, three types of tension levels, i.e. tensed, neutral, and relaxed emotions, were induced. The experimental design consisted of six phases; Rest 1 (2 min), Picture 1 (presenting emotion tensioning photos for 2 min), 3-back Task 1 (2 min), Picture 2 (presenting emotion tensioning photos for 2 min), 3-back Task 2 (2 min), and Rest 2 (2 min). Galvanic skin response (GSR) was also measured during all phases of the experiment. The accuracy rate of 3-back task performance was the highest at a neutral emotional state, followed by relaxed and then tensed emotional state. Through this study it could be inferred that tension, induced by stimuli unrelated to cognitive tasks, decreases the performance of cognitive tasks.

Keywords: Cognitive performance; Emotion; Tension level; Accuracy.

1 Introduction

Many researchers have examined diverse aspects of human behavior, including behavioral changes, subjective assessment, and physiological responses, to study emotion [1-4]. Psychologists who used the discrete state model to understand the inner states of human beings argue that emotion could be classified into several categories [2,4]. According to the dimensional model, the inner states can be placed in a two-dimensional space [5-6]. In other words, each emotion is positioned in a circular space with the two dimensions of valence (pleasantness-unpleasantness) and tension (tension-relaxation). The two-dimensional structure is considered a fairly stable and general structure for human emotion.

[*] Corresponding author.

Y. Zhang et al. (Eds.): DTA/BSBT 2010, CCIS 118, pp. 258–264, 2010.

Recent studies showed that one axis of two-dimensional space, pleasantness/ unpleasantness emotion might have an influence on cognitive processing abilities, especially memory. Kensinger, Garoff-Eaton, and Schacter [7] reported that people memorized pictures which induced negative (unpleasantness) emotions better than pictures which induced neutral emotions. The memory difference in event-related details and personal details was investigated for those who watched the same baseball game (winning team: positive emotion, losing team: negative emotion, those who not a fan: neutral emotion) [8]. The same memory test was performed immediate after and six months after the game, and the amount of memory, memory consistency, confidence and vividness was measured. Results showed that those who had negative emotions while watching the game scored the highest in the amount of memory, memory consistency, confidence and vividness, with positive and neutral emotions ranked in order.

Previously, it was widely held that the memory which was influenced by emotion was distorted or more easily forgotten [9], but recent studies reported that the memories related to emotion were kept more vividly and in greater detail [7-8,10]. Many studies concerning changes in cognitive ability due to emotion have been carried out, but have been confined to the valence aspect (positive/negative) [7-10].

Therefore, this study investigated how the tension axis of human emotion affects cognitive performance ability. After inducing three emotions, tension, neutral and relaxation via pictures, it was observed how these emotions influenced the accuracy of a 3-back task. To measure changes in cognitive performance ability, a 3-back task which can exclude the learning effect was used in this study. To check whether the proper tension level was achieved by using emotion inducing pictures, galvanic skin response (GSR) was measured and subjective evaluation was performed after experiment.

2 Methods

To induce the proper emotion for experimental purpose, pictures were selected using a group test. The group test was performed with 73 university students (age 26.0 ±1.6). 300 pictures from the international affective picture system (IAPS) were used during the group test [11]. Participants were advised to mark questionnaires on a 5 point scale (tensed/pleasant 5, neutral 3, and relaxed/unpleasant 1) based on the tensed-relaxed and pleasant-unpleasant degrees after looking at each picture. After the group test, the average of the tensed-relaxed and pleasant-unpleasant degrees of 300 pictures was calculated. 24 pictures for the tensed emotion were selected in the order of the highest average of tensed-relaxed degree while closest to point 3 in pleasant-unpleasant dimension. 24 pictures for the neutral emotion were selected in the order of the closest to point 3 in both dimensions, and 24 pictures for the relaxed emotion in the order of the lowest average of tensed-relaxed degree while closest to point 3 in pleasant-unpleasant dimension. The average of selected pictures was 3.9±0.7 for the tensed, 2.9±0.3 for the neutral, and 1.6±0.9 for the relaxed. Results of one-way repeated measures ANOVA using SPSS ver. 12.0 (SPSS Inc. Chicago, Illinois) showed that there was a significant different among three types of emotion induced pictures as shown in Fig. 1 (a) ($F=40.419$, $df=2$, $p<0.001$). In contrast, the differences in pleasant level were not significant among three types of pictures ($F=2.076$, $df=2$, $p=.131$).

Ten healthy male (25.7 ± 1.5 years old) and ten healthy female (24.5 ± 1.8 years old) university students participated in the study. None of the participants reported having a history of psychiatric or neurological disorders. The overall experimental procedure was explained to all subjects who released consent for the procedure. All examinations were performed under the regulations of our Institutional Review Committee.

Biopac MP30 and acqknowledge 3.5 (Biopac System, Inc. USA) were used to measure the average amplitude of galvanic skin response (GSR) from the index and middle fingers on the left hand. The sampling rate of the physiological data was 500 samples/sec.

The experiment consisted of three runs of a 3-back cognition test. Each run consisted of six phases: Rest 1 (2 min), Picture 1 (presenting emotion tensioning photos for 2 min), 3-back Task 1 (2 min), Picture 2 (presenting emotion tensioning photos for 2 min), 3-back Task 2 (2 min), and Rest 2 (2 min). Rest 1 was a 2-min stabilization period. During Picture 1 phase, 12 emotion inducing pictures selected from group test were presented randomly at a 10 seconds interval. During Picture 1 & 2 phases total 24 emotion induced pictures were presented. During the 3-back Task 1 phase, one of 40 alphabetical characters was presented at a 3 seconds interval. The subject was asked to press the answer button if the currently presented character was same to the third previous presented character. Among them the number of correct answer was ten. During 3-back Task 1 & 2 phases, a total of 80 alphabetical characters were presented and total correct answer was 20. Six sets of 3-back tasks were made. The six 3-back task types were counterbalanced across each run. The pictures and 3-back tasks were presented using SuperLab 1.07 (Cedrus Co. San Pedro, USA). Rest 2 was a 2-min rest period. Each subject was run through the procedure three times, once for each tension level. The experimental order was randomized. Subjects rested for 1 hour between runs. After each run, the subjects estimated their tension level induced by the pictures using a 5 point scale (tensed 5, neutral 3, and relaxed 1). GSR was measured in all phases.

The amplitude of GSR of each subject was normalized by the Rest 1 value after calculating the average value of each phase. To investigate if there was any statistical difference under the three tension levels and between each phase for amplitude of GSR, two-way repeated measures ANOVA was employed with tension level (the tensed, neutral, and relaxed) and phase as independent variables. The accuracy rate on the 3-back test was calculated. Significance in the difference in accuracy rate based on tension levels was determined using one-way repeated measures ANOVA.

3 Result

Fig. 1 (b) shows the self estimated results of the induced emotion. The tension level was 3.9±0.6 for the tensed pictures, 2.7±0.7 for the neutral pictures, and 1.5±0.5 for the relaxed pictures. This result was similar to the tension scores (Fig. 1 (a)) from the group test. One-way repeated measures ANOVA showed a statistical difference among tension levels of 3 types of emotion induced pictures ($F=41.511$, $df=2$, $p<0.001$).

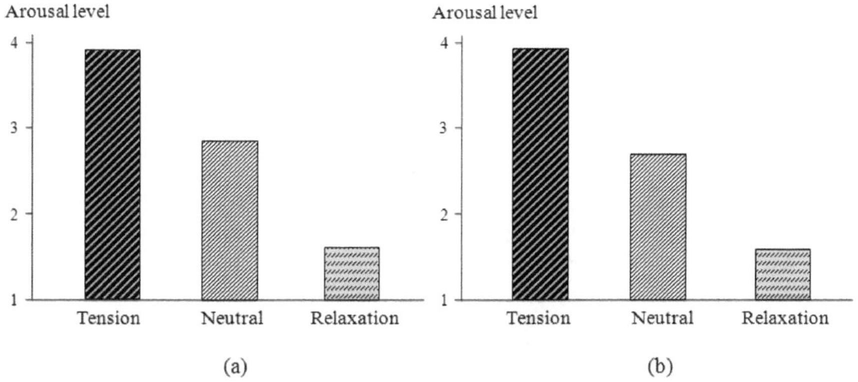

Fig. 1. (a) Magnitude of tension level for tensed, neutral and relaxed pictures from group test (b) Results of subjective evaluation for 3 types of emotion induced pictures after each experiment

Fig. 2 showed the mean amplitude of GSR at each phase at presenting 3 types of emotion inducing pictures. As shown in Fig. 2, the mean amplitude of GSR at each phase was the greatest during presentation of the tensed pictures, then neutral and relaxed pictures in order. Two-way repeated measures ANOVA showed a significant difference in the tension level (F=4.393, df=2, p=0.019) and phase (F=9.493, df=4, p<0.001). Since there was an interaction effect (F=2.016, df=8, p=0.048) between the tension level and phase, there was a difference in changes of GSR based on the types of tension levels.

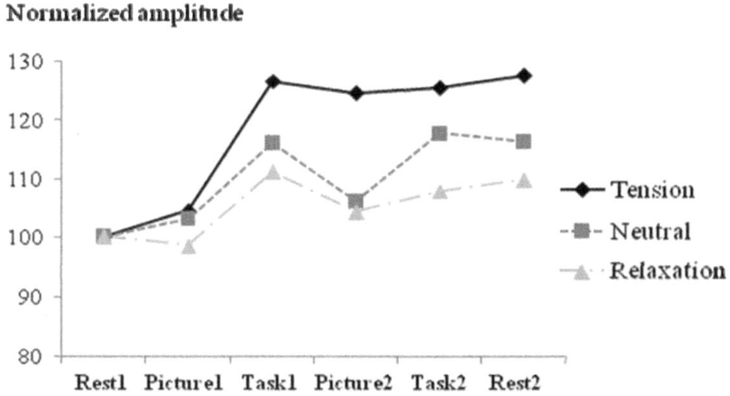

Fig. 2. The normalized amplitude of GSR in each phase by 3 types of emotion induced pictures

As shown in Fig. 3, the accuracy rate of the 3-back task was the highest when presented with neutral pictures (78.0 ± 16.4 [%]), then relaxed pictures (74.0 ± 15.7 [%]), and then tensed pictures (70.3 ± 14.4 [%]), in order (F=4.938, df=2, p=0.012).

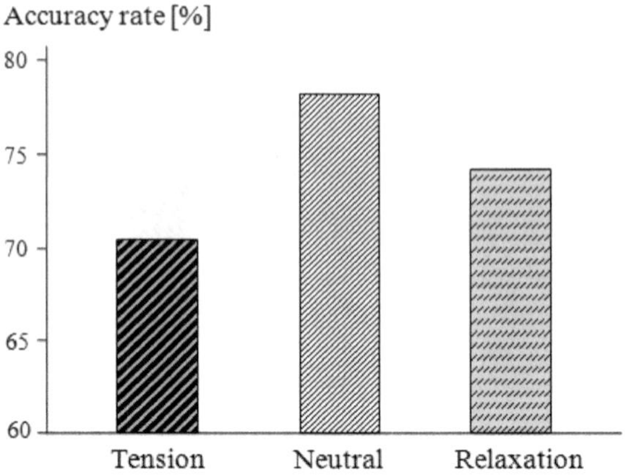

Fig. 3. Accuracy rate of the 3-back task by 3 types of emotion induced pictures

4 Discussion

This study investigated the effect of tension level on the performance of a 3-back task. It was well known that GSR reflected tension levels sensitively [12-13]. As shown in self estimated results in Figure 1 and in the GSR in Figure 2, the 3 types of pictures used in this study induced 3 types of tension level effectively.

Results showed that the accuracy rate of the 3-back task was the highest at a neutral emotional state, then a relaxed state and a tensed state, in order. Therefore, it could be concluded that the performance of a 3-back task increases when in a neutral emotional state compared with tensed and relaxed.

It has been reported that the change in feeling due to external stimuli had an effect on cognitive processing, judgment, and evaluation through indiscreet ignition stage unconsciously without conscious intervention process [14-15]. Physiological stimulation especially had a negative effect on information processing of succeeding input [16-17], and it was reported that the increase in physiological stimulus had a negative effect on cognitive processing [18]. From these studies, it could be inferred that the increase in the physiologically stimulated state (tensed state), which is irrespective of cognitive processing, had a negative effect on information processing of succeeding input. From this, it would be expected that the ability of performing a 3-back task would decrease, as shown in this study.

It was well known that cognitive processing activated the sympathetic nervous system [19-20]. Attention and concentration on a given task was accompanied by the activation of the sympathetic nervous system. Inducing a relaxed state of this study can activate the parasympathetic nervous system, and by disturbing the concentration for a given task, it can be expected that the ability to perform a given task decreases.

The tension (tension/relaxation) axis of human emotion is closely related to physiological awakening, unlike the valence (pleasantness/unpleasantness) axis. From this study, it can be inferred that the increase or decrease in the physiologically stimulated state, which is irrespective of cognitive processing, decreases the ability of performing tasks by disturbing proper concentration for given tasks.

Preceding studies about the valence (pleasantness/unpleasantness) axis showed that the memory ability with emotion was higher, whether or not the memory was related with tasks [7-8,10]. This study on the tension (tension /relaxation) axis showed that the memory ability without emotion was higher. Therefore, it is necessary to clarify how the human emotional state affects cognitive processing ability by considering both axes of human emotion at the same time.

Acknowledgments

This work was supported by Mid-career Researcher Program through NRF grant funded by the MEST (No. 2009-0084784).

References

1. Chung, S.C., Min, Y.K., Lee, B., Tack, G.R., Yi, J.H., You, J.H., Jun, J.H.: Development of the real-time subjective emotionality assessment (RTSEA) system. Behav. Res. Meth. 39, 144–150 (2007)
2. Ekman, P.: Universals and cultural differences in facial expressions of emotion. In: Cole, J. (ed.) Nebraska symposium on motivation (1971), Lincoln, vol. 19, pp. 207–282. University of Nebraska Press, NE (1972)
3. Min, Y.K., Chung, S.C., Min, B.C.: Physiological evaluation on emotional change induced by imagination. Appl. Psychophysiol Biofeedback 30, 137–150 (2005)
4. Tomkins, S.: Affect theory. In: Scherer, K., Ekman, P. (eds.) Approaches to emotion, pp. 163–196. Erlbaum, Hillsdale (1984)
5. Russell, J.: A circumplex model of affect. J. Personal. Soc. Psychol. 39, 1161–1178 (1980)
6. Russell, J., Lewicka, M., Nitt, T.: A cross-cultural study of a circumplex model of affect. J. Personal. Soc. Psychol. 57, 848–856 (1989)
7. Kensinger, E.A., Garoff-Eaton, R.J., Schacter, D.L.: Memory for specific visual details can be enhanced by negative arousing content. J. Mem. Lang. 54, 99–112 (2006)
8. Kensinger, E.A., Schacter, D.L.: When the Red Sox shocked the Yankees: Comparing negative and positive memories. Psychonomic Bulletin and Review 13, 757–763 (2006)
9. Brown, R., Kulik, J.: Flashbulb memories. Cognition 5, 73–99 (1977)
10. Kensinger, E.A.: Negative emotion enhances memory accuracy: Behavioral and neuroimaging evidence. Psychol. Sci. 16, 213–218 (2007)
11. Lang, P.J.: International affective picture system (IAPS): Technical manual and affective ratings. NIMH center for the study of Emotion and Attention, Gainsville (1997)
12. Andreassi, J.: Psychophysiology: Human behavior and physiological response, 4th edn. Lawrence Erlbaum Associates, Mahwah (2000)
13. Kak, A.V.: Stress: an analysis of physiological assessment devices. In: Salvendy, G., Smith, M.J. (eds.) Machine Pacing and Occupational Stress, Taylor and Francis, London (1987)
14. Bargh, J.A.: Conditional automaticity: Varieties of automatic influence in social perception and cognition. In: Uleman, J.S., Bargh, J.A. (eds.) Unintended thought, pp. 3–50. Guilford Press, New York (1989)
15. Kitayama, S., Howard, S.: Affective regulation of perception and comprehension: Amplification and semantic priming. In: Niedenthal, P.M., Kitayama, S. (eds.) The heart's Eye: Emotional Influences in Perception and Attention, pp. 41–65. Academic Press, New York (1994)

16. Bower, G.H.: Mood and memory. Am. Psychologist 36, 129–148 (1981)
17. Bower, G.H.: Mood congruity of social judgments. In: Forgas, J.P. (ed.) Emotion and social judgments, pp. 31–53. Pergamon Press, Oxford (1991)
18. Pennebaker, J.W.: Confession, inhibition, and disease. In: Berkowitz, L. (ed.) Advances in Experimental Social Psychology, vol. 22, pp. 211–244. Academic Press, New York (1989)
19. Backs, R.W., Seljos, K.A.: Metabolic and cardiorespiratory measures of mental effort: The effects of level of difficulty in a working memory task. Int. J. Psychophysiology 16, 57–68 (1994)
20. Chung, S.C., Lee, B., Tack, G.R., Yi, J.H., Lee, H.W., Kwon, J.H., Choi, M.H., Eom, J.S., Sohn, J.H.: Physiological mechanism underlying the improvement in visuospatial performance due to 30% oxygen inhalation. Appl. Ergon. 39, 166–170 (2008)

Amino Acid Substitution Positions on the VP7 and NSP4 Molecules of Rotavirus Circulating in China

Xiaofeng Song[1,*], Yan Hao[1], and Ping Han[2]

[1] Department of Biomedical Engineering, Nanjing University of Aeronautics & Astronautics, Nanjing 210016, P.R. China
Tel.: +86 25 84891938
xfsong@nuaa.edu.cn
[2] The First Hospital affiliated to Nanjing Medical University, Nanjing 210029 China

Abstract. Rotavirus is the leading cause of severe diarrhea among infants and young children, and its outer capsid structural protein, VP7, stimulates the production of distinct neutralizing antibodies in the host and also determines the serotype of the virus strain. NSP4, the rotavirus nonstructural protein, plays a role in viral assembly and is becoming an attractive candidate for vaccine development. By using the Maximum likelihood method with codon-substitution models, amino acid substitutions on the VP7 and NSP4 protein-coding sequences of G1~G3 serotype rotavirus circulating in China are analyzed, and no amino acid site under positive selection in VP7 is found, but three amino acid sites under positive selection in NSP4 are detected. Since these sites are located in different functional sequence segments, we may draw a conclusion that these sites are crucial to related virus biological function, and should be paid much attention in developing vaccine.

Keywords: Rotavirus; VP7 protein; NSP4 protein; amino acid substitution.

1 Introduction

Rotavirus has been regarded as a major cause of severe diarrhea among infants and young children worldwide [1], since the discovery of human rotavirus (HRV) 25 years ago. The rotavirus comprises a single genus within the Reoviridae family with a genome of 11 segments of double-stranded RNA (dsRNA). The rotavirus genome encodes six structural (VP1-VP4, VP6 and VP7) and five non-structural (NSP1-NSP5) proteins. A mature infectious rotavirus (RV) particle has three concentric layers: the protein core, an inner protein capsid, and an outer protein capsid [2]. Among 11 proteins of rotavirus, outer capsid structural protein VP7 constitutes the outer shell of the virus, and is also known to be involved in the neutralization of the viral particle, in which the predominant reactivity is directed against it. This protein is encoded by the seventh, eighth, or ninth gene segment of the genome, and contains nine regions (VR1-VR9) which are conserved within the same serotype but highly divergent among different serotypes [3]. As one of the non-structural proteins encoded by rotavirus gene segment 10 or 11, NSP4 is a multifunctional protein.

* Corresponding author.

Y. Zhang et al. (Eds.): DTA/BSBT 2010, CCIS 118, pp. 265–272, 2010.

Previous researches indicate that NSP4 may represent a viral enterotoxin, in which NSP4 protein itself or a peptide fragment of NSP4 is released or secreted from cells infected with the virus, thereby affecting uninfected tissue and ultimately causing a fluid secretary response[4][5].

Because rotavirus genes have a very high mutation rate which is crucial for them to evade the response of the host's immune system, molecular adaptation evolution of rotavirus should be deeply investigated. Detecting amino acid substitution under diversifying selective pressure may help identify potential targets of the immune response and/or the most virulent strains contributing to vaccine design [6]. Rotavirus affects a wide variety of mammal such as Homo sapiens, swine, Bos tarurs and so on, therefore it is important to investigate what factors affect the evolution of immune-related genes such as VP7 and NSP4.

The basic process of adaptive evolution by natural selection is the replacement of one allele gene by another with higher adaptability in the population. Gene evolution is commonly controlled by positive selection, negative selection, or neutral mutation, which implies that detecting the adaptive evolution would be helpful to better understand the bio-evolutionary mechanism and corresponding variation in structure and function of a protein [7]. Negative selection may help detect regions or residues of functional importance [8]. If gene suffers strong negative selection, they might not be free from functional constraints. Some important functional alterations may arise from the nonsynonymous mutations in the key protein functional region. As for positive selection, it plays a significant role in immune-related genes which are major positively selected gene groups [9].

The objective of this paper is to analyze the amino acid substitution on the VP7 and NSP4 molecules of Rotavirus circulating in China using Maximum Likelihood method (ML) with codon-substitution models. The datasets are retrieved from the Genbank. With the premise of different codon substitution models, this research is not only on the selection of the best phylogenic tree that fit the real data, but also on the detection of the selecting pressure on sites, especially amino acid sites undergoing positive selection.

2 Materials and Methods

2.1 Sequence Data

All the VP7 and NSP4 protein sequences of rotavirus circulating in China are retrieved from the GenBank, with VP7 Dataset consisting of 17 coding sequences and NSP4 Dataset consisting of 25 coding sequences. We used BLAST to remove the redundant proteins using 90% sequence identity as the cutoff, thus we obtained 10 sequences in VP7 Dataset and 19 sequences in NSP4 Dataset (see Table 1).

2.2 Data Analysis

2.2.1 Phylogenetic Analysis, Tree Construction

After selection of all the sequences, sequences in VP7 Dataset are aligned together by using CLUSTALW (Ver.1.83) and NSP4 Dataset alignment are implemented in the same way [11].

The phylogenetic trees of VP7 and NSP4 are constructed by the use of NJ (neighbor-joining) (MEGA 4.0) [12][13] and Clade robustness is evaluated by the bootstrap method with 1000 replicates.

Table 1. The dataset of VP7 and NSP4 proteins

protein	serotype	AC number	tag
VP7	G1	DQ512997	Chi-84
		DQ512996	Chi-83
	G2	DQ904518	CH-146
		DQ904517	CH-86
		DQ904519	CH-188
	G3	DQ904501	Chi-182
		DQ904501	Chi-157
		DQ904499	Chi-10
		DQ904499	Chi-4
		EF495118	Chi-3
NSP4	G1	AY159640	97B11
		AY159639	97B6
		AY159644	97H12
		AY159641	97B55
		AY159647	97S36
		AY159646	97S35
		AY159645	97S34
		AY159638	98G34
		AY159634	98SH49
		AY159637	98G7
		AY159631	98B39
		AY159630	98B35
		AY159632	98B36
		AY159648	97SHRV
	G2	AY159649	97SZ8
	G3	AB008274	CH-32
		AB008272	CH-927A
		AB008276	CH-55
		AB008275	CH-C17

2.3 ML Models

The non-synonymous/synonymous rate ratio (ω=dN/dS) provides a straightforward measurement of the selective pressure acting on a protein-coding sequence. Assuming that synonymous mutations are subjected to almost strictly neutral selection, the ω values of <1, 1, >1 represent negative (purifying) selection, random drift and positive (diversifying) selection, respectively. An ω significantly greater than 1 means that the non-synonymous mutation is fixed in a higher rate than the synonymous mutation, and the evolution of this gene is driven by positive selection. The model with maximum likelihood ratio is considered as the best model to fit the data. The Llikelihood-rate Test (LRT) is used to compare twice the log-likelihood differences between two nested models, and to identify the statistic significance with a χ^2 distribution. The degrees of freedom (df.) used in LRT are equal to the difference in the number of parameters between the two models [14]. We employ three pairs of models in PAML[15] to establish three LRTs: M0 (one-ratio) and M3 (discrete), M1a(nearly neutral) and M2a (positive selection), and M7(β) and M8(β&ω). The simplest model, M0, assumes one identical ω value for all sites. M1a allows two different selective pressure acting on the sites with $0<\omega 0<1$ and $\omega 1=1$, corresponding to probability p0 for the conserved sites and p1=1-p0 for the neutral sites, respectively. On the basis of M1a model, an additional class of sites with ω2 estimated from the data is added in M2a model with the probability p2=1-p1-p0. For the M3 model, an unconstrained discrete distribution is used to model heterogeneous ω ratios among sites. It is assumed that a β(p,q) distribution is used in M7 model for $0\leq\omega\leq 1$. Following M7 model, one additional substitution rate ω1 with probability p1 acting on sites is taken into account in M8 model, while the rest of sites (at probability p0=1-p1) undergone substitution rate ω between 0 and 1 in the condition of the β(p,q)-distribution. This M8 model can be compared with M7 to test the presence of positive sites by using LRT. In this study, site-specific models are used by Codeml in the PAML 3.14b package. Positive selection is tested over sites of coding sequences by comparing twice the log-likelihood differences between M1a vs. M2a and M7 vs.M8 with a χ^2 distribution in the LRT.

Table 2. Likelihood values and parameter estimation for VP7 and NSP4

	Modle	lnL	Parameter estimate	2Δl	Positive selected site
VP7	M0	-2663.34	ω=0.04		\
	M3	-2638.77	p0=0.43638 p1=0.50621 p2=0.05742 ω0=0.00000 ω1=0.05539 ω2= 0.61060	49.14	\
	M1a	-2643.34	p0=0.91994 p1=0.08006		\
	M2a	-2643.34	p0=0.91993 p1=0.02539 p2=0.05467 ω2=1.00000	0	216 I
	M7	-2640.19	p=0.34538 q=5.79593		
	M8	-2639.45	p0=0.96416 p=0.49946 (p1=0.03584) q=12.27754 ω=1.00000	1.48	125 T 145 N 216 I
NSP4	M0	-1543.87	ω=0.14		\
	M3	-1534.14	p0=0.41220 p1=0.49713 p2=0.09066 ω0= 0.08489 ω1= 0.08489 ω2= 1.08422	19.46	**75I** 138A 139N **140V** 141L 168S
	M1a	-1534.07	p0=0.90186 p1=0.09814		\
	M2a	-1534.06	p0=0.90801 p1=0.00000 p2=0.09199 ω2=1.00000	0.01	**75I** 138A **140V** 141L 144S **168S**
	M7	-1535.67	p= 0.36144 q=1.77859		\
	M8	-1534.07	p0=0.91166 p=9.48340 (p1= 0.08834) q=99.00000 ω=1.09136	3.2	**75I** **140V** 168S

***positive selected site are in bold**

3 Results

3.1 Selective Pressure Detection

Fig.1 and Fig.2 show the phylogenetic trees with each serotype belonging to one branch, and Table 2 displays the ML results.

No site in VP7dataset is detected as significant positive selection site, indicating that synonymous substitution occurs more frequently than non-synonymous substitutions on average.

In NSP4 dataset, three sites, 75I, 140V, and 168S, are detected to be positive selective sites.

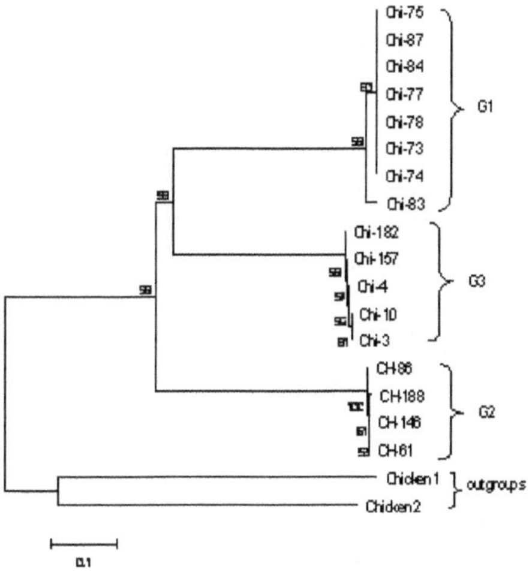

Fig. 1. Phylogenetic tree for 10 VP7 sequences from G1~G3 serotype with bootstrap support values, two chicken sequences being used as outgroup

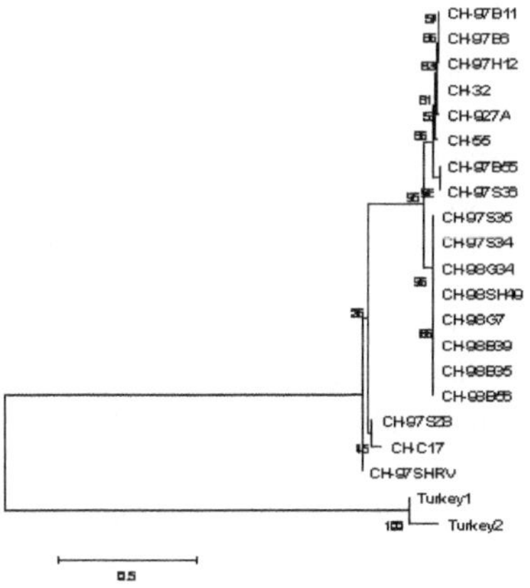

Fig. 2. Phylogenetic tree for 19 NSP4 sequences from G1~G3 serotype with bootstrap support values, two chicken sequences being used as outgroup

4 Discussion

It is well known that different amino acid sites have different biological functions and even endure different evolution selective pressures. The ratio of non-synonymous to synonymous substitution (ω=dN/dS) is an important index to evaluate selective pressure based on the sequences. An ω significantly greater than one means that non-synonymous mutation predominates in the sequence evolution, and the lineage or the critical amino acid sites are regarded as positive selection. However, most amino acids in a functional protein are with structural and functional constraints, and the adaptive evolution of this functional protein only occurs at times. Thus only a few amino acid sites are under positive selection [16]. An average ω might lack power in detecting site positive selection. Many robust statistic tools which can detect variation of substitution ratio among lineages or amino acids sites are developed based on several hypothesis models [17]. Although immune-related genes are of one of the important pools for adaptive evolution [18][19], the results demonstrate that VP7 and NSP4 proteins have suffered strong purifying selection among rotaviruses circulating in China. However, several positively selected sites in VP7 and NSP4 protein sequences are detected during Rotavirus VP7 and NSP4 evolution.

In VP7 Dataset, no site was identified as the significant positive selection signal based on ML methods. The mutation rate in RNA viruses is higher by several orders of magnitude than that in DNA viruses. There are several reasons why most amino acid sites undergo purifying selection of genetic constraints. Firstly, by limiting genome size and content, RNA virus genome can evade harmful mutation accumulation, which causes virus genomes to be inclined to concerted evolution, parallel evolution and owning similar codon substitution patterns [20]. The substitution ratios in different sites tend to be consistent. Secondly, the accuracy and sensitivity of current analysis methods are inadequate. Given the reasons above, it should not be difficult to understand our experimental results.

VP7 is well known to be involved in the neutralization of the viral particle. Protection from disease appears to be effected by secretary antibodies directed against determinants on the protein. Rotavirus VP7 amino acid sequence has 9 regions known to be hyper variable regions (VP1~VR9)[21]. Among these regions, antigenic sites on the rotavirus primary outer-shell glycoprotein are mainly located in VR5, VR7 and VR8 [22]. The results of our analysis suggest that no positive selective sites exist in these regions, and positive selection is not the main cause for mutations in the hyper variable regions. NSP4, one of the non-structural proteins encoded by rotavirus gene segment 10 or 11, has been considered as a viral enterotoxin and possess membrane destabilization activity [23], especially within amino acid site region aa48~91 [24], indicating that the site 75I may contribute to destroying the integrality of small intestine epithelia, and causing diarrhea by affecting small intestine absorption to water, inorganic salt and other substance. As for 140V, located in aa131~140, and 168S, located in aa156~175, they are important to the Rotavirus infection ability and for inner capsid particle assembly, and may play important roles in rotavirus biological functions.

There has been a steady accumulation of evidence suggesting that several important disease associated virus genes have shown to evolve in a neutral pattern or to be dominated by purifying selection, such as HIV-I resistance mutant CC5-△32[25], and prion protein gene (PRNP) in bovine[26]. Our results also revealed the same evidence for rotavirus.

VP7 and NSP4 proteins have suffered strong purifying selection in rotaviruses circulating in China. We found no amino acid site under positive selection in VP7, but three amino acid sites under positive selection in NSP4. Since these sites are located in different functional sequence segments, they are crucial to related functions. Meanwhile, the results in this paper also prove that ML methods can be used as important tools to investigate the gene adaptive evolution of RNA virus such as rotavirus.

Acknowledgments. The study was supported by grants from Natural Science Foundation of Jiangsu Province in China (BK2010500).

References

1. Mirazimi, A., von Bonsdorff, C.H., Svensson, L.: Effect of brefeldin A on rotavirus assemble and oligosaccharide processing. J. Virol. 217(2), 554–563 (1996)
2. Prasad, B.V., Wang, G.J., Clerx, J.P., Chiu, W.: Three-dimensional structure of rotavirus. J. Mol. Biol. 199, 269–275 (1988)
3. Green, K.Y., Hosshio, Y., Ikegami, N.: Sequence analysis of the gene encoding the serotype-specific glycoprotein (VP7) of two new human rotavirus serotypes. J. Virol. 168, 429–433 (1989)
4. Lundgren, O., Svensson, L.: Pathogenesis of rotavirus diarrhea. Microbes Infect. 3, 1145–1156 (2001)
5. Zhang, M., Zeng, C.Q., Morris, A.P., Estes, M.K.: A functional NSP4 enterotoxin peptidesecreted from rotavirus-infected cells. J. Virol. 74, 11663–11670 (2000)
6. Yang, Z.: Molecular Evolution of the Hepatitis Delta Virus Antigen Gene: Recombination or Positive Selection? J. Mol. Biol. 59, 815–826 (2004)
7. Nei, M., Kumar, S.: Molecular Evolution and Phylogenetics. Oxford University Press Inc., New York (2000)
8. Nielsen, R.: Molecular signatures of natural selection. Annu. Rev. Genet. 39, 197–218 (2005)
9. Yang, Z.: The power of phylogenetic comparison in revealing protein function. Proc. Natl. Acad. Sci. 102, 3179–3180 (2005)
10. Suzuki, Y., Gojobori, T.: A method for detecting positive selection at single amino acid sites. Mol. Biol. Evol. 16(9), 1315–1328 (1999)
11. Higgins, D.G., Sharp, P.M.: CLUSTAL: A package for performing multiple sequence alignment on a microcomputer. Gene 73, 237–244 (1998)
12. Saitou, N., Nei, M.: The neighbor-joining method: A new method for reconstructing phylogenetic trees. Mol. Biol. Evol. 4, 406–425 (1987)
13. Tamura, K., Dudley, J., Nei, M., Kumar, S.: MEGA4: Molecular Evolutionary Genetics Analysis (MEGA) software version 4.0. Molecular Biology and Evolution 24(8), 1596 (2007)
14. Yang, Z.: Likelihood ratio tests for detecting positive selection and application to primate lysozyme evolution. J. Mol. Biol. 15, 568–573 (1998)
15. Yang, Z.: PAML: a program package for phylogenetic analysis by maximum likelihood. Comput. Appl. Biosci. 13, 555–556 (1997)

16. Anisimova, M., Bielawski, J.P., Yang, Z.: The accuracy and power of likelihood ratio tests to detect positive selection at amino acid sites. J. Mol. Biol. 18, 1585–1592 (2001)
17. Yang, Z.: Likelihood ratio tests for detecting positive selection and application to primate lysozyme evolution. J. Mol. Biol. 15, 568–573 (1998)
18. Massingham, T.: Detecting Amino Acid Sites Under Positive Selection and Purifying Selection. Genetics 169, 1753–1762 (2005)
19. Yang, Z.: The power of phylogenetic comparison in revealing protein function. Proc. Natl. Acad. Sci. 102, 3179–3180 (2005)
20. Holmes, E.C.: Error thresholds and the constraints to RNA virus evolution. Trends. Microbiol. 11, 543–546 (2003)
21. Green, K., Sears, J.F.: Prediction of human rotavirus serotype by nucleotide sequence analysis of VP7 protein gene. J. Virol. 62, 1819–1823 (1988)
22. Dyall Smith, M.L., Lazdins, I., Tregear, G.W., et al.: Location of the major antigenic sites involved in rotavirus serotype2specific neutralization. Proc. Natl. Acad. Sci. 83, 3465–34681 (1986)
23. Tian, P., Ball, J.M., Zeng, C.Q., et al.: The rotavirus nonstructural glycoprotein NSP4 possesses membrane destabilization activity. J. Virol. 70(10), 6973–6981 (1996)
24. Browne, E.P., Bellamy, A.R., Taylor, J.A., et al.: Membrane destabilizing activity of rotavirus NSP4 is mediated by a membrane2proximal amphipathic domain. J. Gen. Virol. 81(Pt8), 1955–1959 (2000)
25. Sabeti, P.C.: The case for selection at CCR5-Delta32. PLoS Biology 3(11), e378 (2005)
26. Seabury, C.M., Honeycutt, R.L., Rooney, A.P., Halbert, N.D., Derr, J.N.: Prion protein gene (PRNP) variants and evidence for strong purifying selection in functionally important regions of bovine exon 3. Proc. Natl. Acad. Sci. 101, 15142–15147 (2004)

A Time-Frequency HRV Processor Using Windowed Lomb Periodogram

Shao-Yen Tseng and Wai-Chi Fang

Department of Electronics Engineering and Institute of Electronics
National Chiao Tung University
1001 University Road, Hsinchu City, Taiwan (R.O.C.)
wfang@mail.nctu.edu.tw

Abstract. In this paper, a system for time-frequency analysis of heart rate variability (HRV) using windowed Lomb periodogram is proposed. The system is designed with considerations in SOC implementation for portable applications. Time-frequency analysis of HRV is achieved through a de-normalized Lomb periodogram with a sliding window configuration. The Lomb time-frequency distribution (TFD) is suited for power spectral density (PSD) analysis of unevenly spaced data and has been applied to the analysis of heart rate variability. The system has been implemented in hardware as an HRV processor and verified on FPGA. Artificial heart rate was used to evaluate the system as well as data from the MIT-BIH arrhythmia database and real EKG data. Simulations show that the proposed Lomb TFD is able to achieve better frequency resolution than short-time Fourier transform of the same hardware size.

Keywords: Biomedical system, portable, healthcare, remote monitoring, ECG, HRV, TFD, Lomb.

1 Introduction

In recent years, the study of HRV has become a significant topic in biomedical signal processing of physiological signals. HRV refers to the study of variations in the time interval between successive heartbeats and has been shown to be an important indicator of cardiovascular health [1]. As the regulation mechanism of the heart is closely governed by the sympathetic and parasympathetic nervous systems, HRV is often used as a quantitative marker of the autonomic nervous system. Studies have shown that the HRV is an important indicator in many diseases and may contribute to a better treatment [2]. Applications of HRV have been applied to many forms of medical researches including studies in sleep apnea, patient monitoring after cardiac arrest[3], and use in intensive care units [4].

Traditional methods such as FFT, autoregressive models, or wavelet analysis are often used in frequency and time-frequency HRV analysis. In these processes, as the RR intervals are not evenly spaced data, interpolation is needed to produce the evenly spaced data required by the spectral analysis algorithms. However, there is no common approach to the choice of re-sampling frequency. Whilst there is a recommended rule of thumb for selection of re-sampling frequency [5], it has been

Y. Zhang et al. (Eds.): DTA/BSBT 2010, CCIS 118, pp. 273–282, 2010.

shown that re-sampling of unevenly sampled data may lead to quantifiable errors in the final analysis [6]. The Lomb periodogram [7] is a method for deriving the power spectral density of unevenly spaced data sets without the need for interpolation. This periodogram has been applied to the study of heart spectra in many works [6, 8].

For portable biomedical applications, such as devices in healthcare monitoring, it is crucial to decrease device size through an efficient algorithm and hardware architecture. In this study, a system for time-frequency analysis of HRV using the Lomb periodogram is proposed. The corresponding hardware architecture of the HRV analysis system is also developed. The proposed system has been implemented as an HRV processor in hardware and verified on FPGA. Section 2 describes the methods of analysis and architecture of the proposed HRV processor. Section 3 shows the simulation results of our HRV system for various types of input signals. Finally, a conclusion is given in Section 4.

2 Methods

An ECG system for portable ECG monitoring including a processor for time-frequency analysis of HRV has been developed. The system acquires ECG data from front-end circuits through an analog-to-digital converter (ADC). The ECG data is then passed to our HRV processor for time-frequency HRV analysis. Analyzed HRV data as well as raw ECG data is then transmitted to a remote station for display or further analysis.

2.1 HRV Using Windowed Lomb Periodogram

HRV is the analysis of the variations between the time intervals of successive heart beats. Therefore, before we can perform HRV analysis, we must first obtain the time series corresponding to the intervals between heart beats.

The ECG contains a trace of the electrical activity produced by the depolarization and repolarization of heart muscles during each heart beat. We can define the time interval between successive heart beats as the interval between R-peaks in the ECG [9], as shown in Fig. 1(a). To detect the R-peaks from the ECG raw data, we have implemented a modified Pan and Tompkins algorithm [10] into our system for real-time beat detection.

The time intervals between R-peaks, or RR intervals, formulate a time series with the intervals as the magnitudes as well as the distance to the next sample, as shown in Fig. 1(b). In other words, it is an unequally sampled time series where the time distance to the next sample depends on the magnitude of the next sample. This results in complications in performing HRV analysis as traditional power spectral analysis methods require the data to be equidistantly sampled.

In order to perform spectral analysis using Fast Fourier Transform (FFT), the time series is required to be re-sampled using interpolation techniques. However, re-sampling as well as variation in usage of different interpolation schemes may cause errors to be introduced to the final HRV analysis [6]. The main source of this error is the assumption that the RR intervals can be described by some underlying model of oscillation, whereas, in truth, it is a physiological phenomenon. To address the errors caused by re-sampling without straying from an area efficient portable solution, the Lomb periodogram is used to perform HRV analysis in our system.

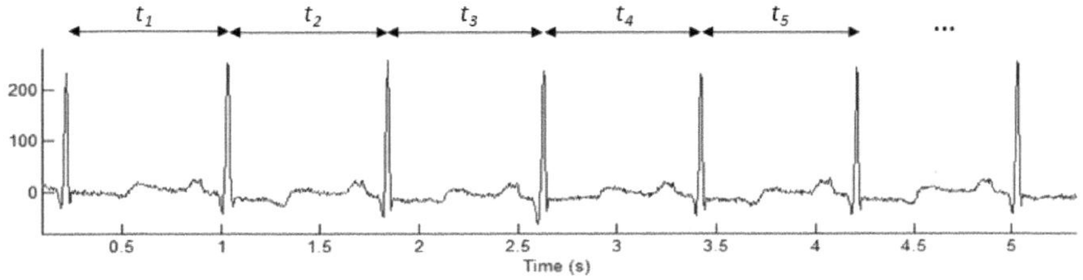

(a)

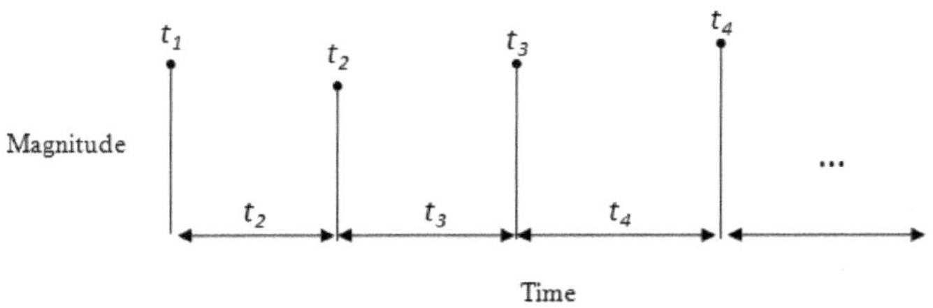

(b)

Fig. 1. (a) The R-peak to R-peak intervals of the ECG (b) The derived non-equidistant RR interval time series

Lomb developed a method for deriving the power spectral density of unevenly spaced data sets without the need for interpolation and re-sampling [7]. The Lomb method uses least squares fitting to estimate the amplitude of a given sinusoid with angular frequency ω_j over non-uniformly sampled data. In other words, the power of the given sinusoid, $P_N(\omega_j)$, for a set of data points of length N is computed using a least-squares fit to the model

$$x(t_i) = A\cos(\omega_j t_i) + B\sin(\omega_j t_i) + n(t_i) \tag{1}$$

for $i=0,1,\ldots,N$, where $n(t_i)$ is noise.

As the Lomb method weights the data on a "per point" basis rather than a "per time interval" basis [11], it is suitable for the analysis of non-uniform data. It is shown that in the case of equal spacing it reduces to the Fourier power spectrum [7]. In [12], Scargle improved the statistical performance of the Lomb periodogram. The Lomb-Scargle periodogram is given as

$$P_N(\omega) \equiv \frac{\left[\sum(x_j-\mu)\cos\omega(t_j-\tau)\right]^2}{2\sigma^2\sum\cos^2\omega(t_j-\tau)} + \frac{\left[\sum(x_j-\mu)\sin\omega(t_j-\tau)\right]^2}{2\sigma^2\sum\sin^2\omega(t_j-\tau)} \tag{2}$$

where μ and σ are the mean and variance, respectively, and τ is

$$\tau \equiv \frac{1}{2\omega}\tan^{-1}\left(\frac{\sum\sin 2\omega t_j}{\sum\cos 2\omega t_j}\right). \tag{3}$$

The constant τ is an offset for each angular frequency ω that makes the periodogram invariant to time-shifts.

As the HRV is a reflection of physiological phenomena, it is considered to be a non-stationary signal. To monitor the changes of heart rate over a course of time, time-frequency spectral analysis is used. In our study, the Lomb periodogram is modified to be implemented as a time-frequency distribution. Time-frequency analysis using Lomb periodogram is achieved by applying a window $w(t)$ to the data and evaluating each segment individually. Using a sliding window configuration $w(t_j\text{-}t)$ a time-frequency distribution $P_N(t,\omega)$ is obtained. The RR intervals are divided into windows of two minutes, in accordance with short-term frequency analysis of HRV [1]. A sliding window configuration with 50% overlap is adopted, as shown in Fig. 2.

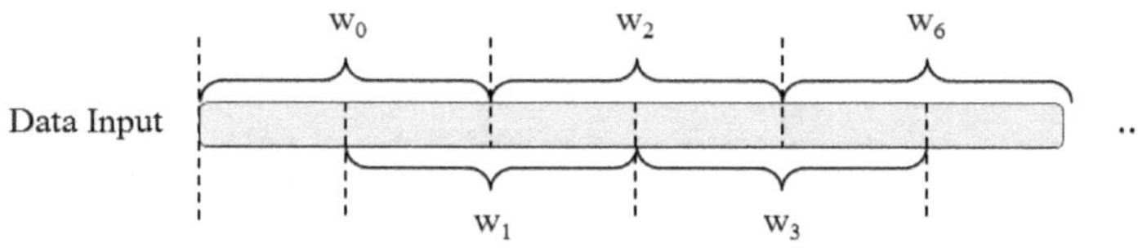

Fig. 2. The sliding window configuration. The length of windows w_i is two minutes with a 50% overlap to the next window.

Despite its ability to obtain a more accurate analysis of heart rate spectra [6], the Lomb method is not often adopted, especially in portable applications, due to its high computational complexity. A fast algorithm for evaluating Lomb using FFT transforms to reduce the order of operations from 10^2N^2 to $10^2N\log N$ was proposed in [11]. However, the fast algorithm yields only an approximation of the original Lomb whilst still requiring two FFT's to perform the calculation as well as addition preprocessing of the time series. Considering that the sliding window of our system is relatively long in regard to available system clocks, the Lomb periodogram is evaluated through hardware implementation of an iterative method in our HRV processor.

The Lomb periodogram $P_N(\omega)$ in discrete-time form can be re-written as

$$P_N(k) \equiv \frac{\left[\Sigma(x_j)\cos\left({kt_j}/{N}\right)\right]^2}{\Sigma\cos^2\left({kt_j}/{N}\right)} + \frac{\left[\Sigma(x_j)\sin({kt_j}/{N})\right]^2}{\Sigma\sin^2({kt_j}/{N})} . \tag{4}$$

The input x_j is assumed to be zero-mean and the time constant τ is set to zero as each window of data is independent with each set beginning at time zero. The time t_j of each point x_j can be calculated by adding the value of the current point to an accumulation of previous points. Thus, we can simplify the architecture for the Lomb periodogram to that shown in Fig. 3, where the trigonometric functions are evaluated within the *Lomb Calculation Unit*.

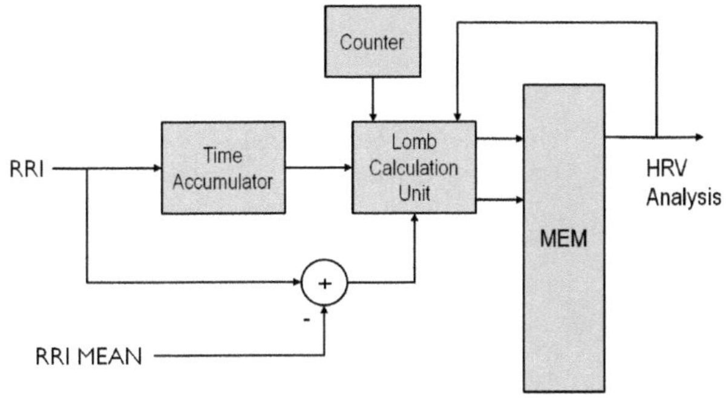

Fig. 3. Architecture for calculation of the Lomb periodogram

2.2 System Configuration

The hardware implementation of the Lomb TFD as well as a simple RR interval calculation unit is implemented as an HRV processor and verified with FPGA on an SOC development platform. An ARM926EJ-S processor is provided on the SOC development platform for control of the system. Various peripheral modules, including a Xilinx FPGA, are connected to the ARM processor through an Advanced Microcontroller Bus Architecture (AMBA) bus. Our HRV processor for time-frequency Lomb analysis is implemented on the FPGA and is accessible by the ARM processor through an AMBA wrapper. The architecture of the HRV processor is shown in Fig. 4.

The setup of the verification platform is shown in Fig. 5. Continuous EKG data is retrieved from the on-board ADC of the SOC development platform after front-end analog amplification and filtering and fed to the HRV processor. Processed HRV analysis data is transmitted to a PC via UART or displayed onto an LCD monitor on the platform.

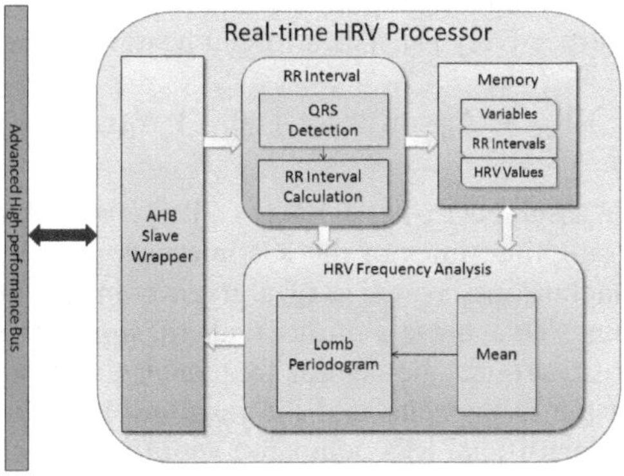

Fig. 4. Architecture of the HRV processor

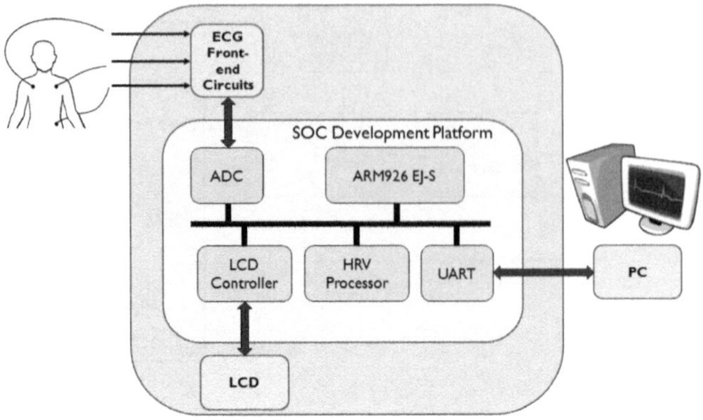

Fig. 5. Setup of FPGA system verification on an SOC development platform

3 Results

Simulations have been performed on the developed Lomb time-frequency distribution using computer generated artificial RR intervals, data from PhysioBank, and finally, real ECG data. Results were compared to HRV analysis using FFT with 4 Hz re-sampling.

3.1 Artificial Heart Rate Interval Data

As an understanding of the underlining oscillations within real RR intervals can rarely be obtained, artificial RR interval data was first used to verify the performance of the Lomb TFD in analyzing heart rate variability.

In generating our artificial RR signal, we assume that the RR intervals can be modelled as the sum of sine waves [13]. We consider that the peaks in the PSD of short-term RR intervals can be represented as 0.25 Hz and 0.1Hz corresponding to the HF and LF regions respectively [1]. The artificial heart rate function is then given as

$$HR(t) = HR_0 + A_L \sin(2\pi f_L t + \theta_L) + A_H \sin(2\pi f_H t + \theta_H) . \quad (5)$$

where θ_L and θ_H are assumed to be 0, $HR_0 = 70$, $A_L = 1$, $A_H = 1.5$, $f_L = 0.1$, and $f_H = 0.25$. As the sampling times of the RR intervals are the amplitudes rotated by 90 degrees, the instantaneous heart rate of a given point in time is used to calculate the next sampling time. To achieve a further realistic signal, the LF and HF frequency components are spread around the central frequencies with a Gaussian distribution. The distributions of the frequencies and the resulting RRI time series are shown in Fig. 6. The PSD of the artificial RRI compared with results using 1024 point FFT and the original Lomb-Scargle periodogram is shown in Fig. 7.

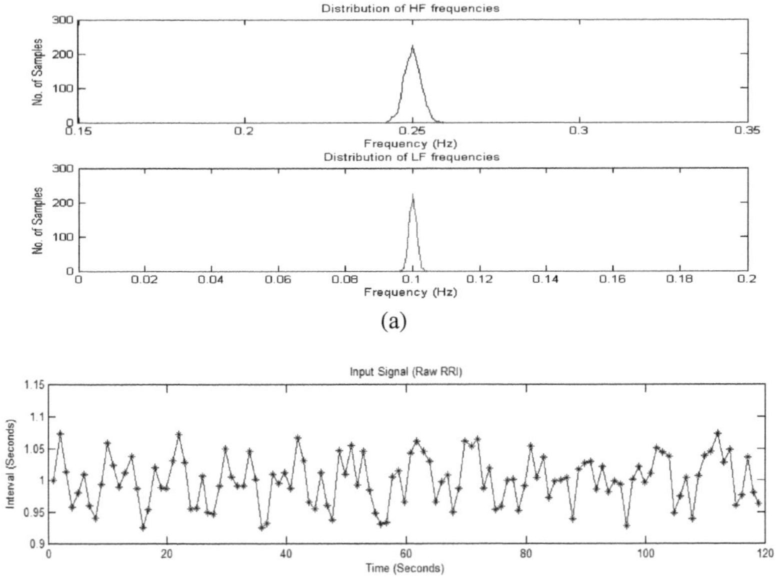

(a)

(b)

Fig. 6. (a) Distribution of the HF and LF frequencies (b) Artificial RRI time series

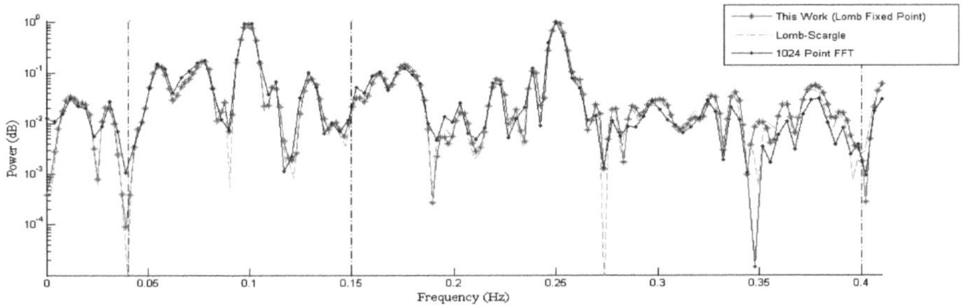

Fig. 7. Power spectral density of artificially generated RRI as calculated using 1024 point FFT, original Lomb-Scargle periodogram, and our Lomb method using fixed point. The RRI time series is re-sampled at 4 Hz before FFT analysis.

3.2 Results on Real Data

In addition to simulation using artificial RRI, data from PhysioBank's online ECG database has been used to verify the HRV processor design. Databases including the MIT-BIH Arrhythmia Database and the MIT-BIH Normal Sinus Rhythm Database have been used to compare results. Fig. 8 shows the PSD of a single window compared with results using 1024 point FFT and the original Lomb-Scargle periodogram. A time-frequency distribution of HRV is shown in Fig. 9.

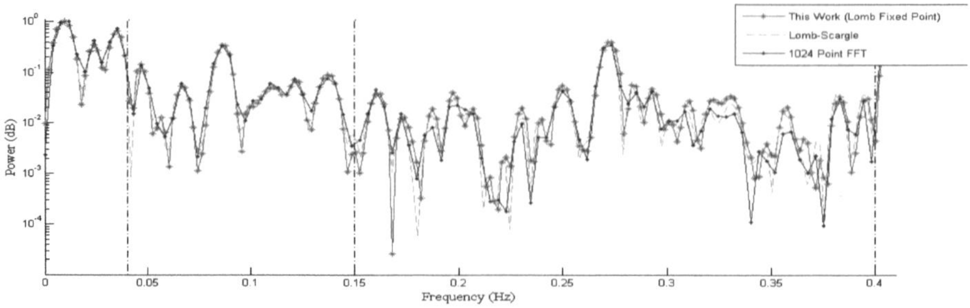

Fig. 8. Power spectral density of RRI data from the MIT-BIH Arrhythmia Database as calculated using 1024 point FFT, original Lomb-Scargle periodogram, and our Lomb method using fixed point. The RRI time series is re-sampled at 4 Hz before FFT analysis.

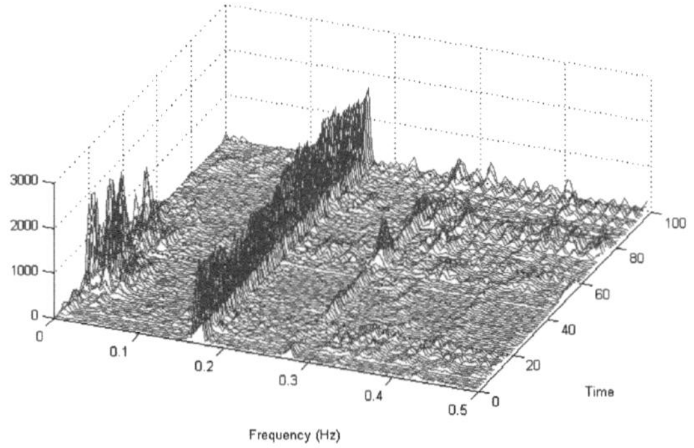

Fig. 9. Time-frequency HRV analysis using Lomb TFD of EKG data from the MIT-BIH arrhythmia database

4 Conclusion

The Lomb-Scargle periodogram has previously been demonstrated to be a good PSD estimator of unequally sampled data. In this paper, the Lomb-Scargle periodogram is applied to time-frequency analysis of HRV and modified to be implemented in hardware. An HRV processor has been developed to process ECG data for beat interval calculation as well HRV analysis using a sliding window configuration. The HRV processor has been implemented on FPGA as part of an SOC prototype system.

A time frequency distribution using Lomb periodogram for HRV analysis has been developed. The algorithm has been implemented in an HRV processor for processing raw EKG data and performing HRV analysis. The HRV processor has been implemented as an ECG system for portable ECG monitoring and HRV analysis on FPGA.

Although the errors due re-sampling are difficult to perceive using real bio-signals, as it is impossible to obtain the original components, it is important to note how efficient the Lomb periodogram is at performing PSD analysis of un-equally sampled data. When compared to traditional TFDs using FFTs of approximately the same area size (or total number of points), the Lomb periodogram yields higher frequency resolution within the frequencies of interest. Due to the need to re-sample RR intervals at higher frequencies for FFT analysis, only a portion of the FFT outputs are within frequencies of interest. The Lomb periodogram, however, maps all points to the frequencies of interest. This is due to the fact that the Nyquist rate is not well defined for unequally spaced data and we are able to obtain information from well above the Nyquist frequency [14]. With a higher re-sampling frequency, the number of required points for FFT becomes unrealistic for hardware implementation, whereas there is no need to consider re-sampling frequency when using Lomb periodogram. Therefore, the Lomb periodogram has been proven to be a hardware efficient method of performing PSD analysis of HRV. Our system, integrated with an HRV processor using windowed Lomb periodogram, is suitable for portable and low power medical devices for ECG monitoring and TFD HRV analysis.

Acknowledgments. This work was performed at the System-on-Chip Research Center at NCTU under the grants NSC98-2220-E-009-041 and NSC98-2220-E-009-044 sponsored by the National Science Council and the National Science and Technology Program for System-on-Chip, Taiwan. We would also like to thank the National Chip Implementation Center for CAD tool support.

References

1. Task Force of The European Society of Cardiology, The North American Society of Pacing and Electrophysiology: Heart Rate Variability: Standards of Measurement, Physiological Interpretation, and Clinical Use. Circulation 93, 1043–1065 (1996)
2. Rajendra Acharya, U., Paul Joseph, K., Kannathal, N., Lim, C., Suri, J.: Heart rate variability: a review. Medical and Biological Engineering and Computing 44, 1031–1051 (2006)
3. Tiainen, M., Parikka, H.J., Mäkijärvi, M.A., Takkunen, O.S., Sarna, S.J., Roine, R.O.: Arrhythmias and Heart Rate Variability During and After Therapeutic Hypothermia for Cardiac Arrest. Critical Care Medicine 37, 403–409 (2009)
4. Kasaoka, S., Nakahara, T., Kawamura, Y., Tsuruta, R., Maekawa, T.: Real-time monitoring of heart rate variability in critically ill patients. J. Crit. Care 25, 313–316 (2010)
5. Singh, D., Vinod, K., Saxena, S.C.: Sampling frequency of the RR interval time series for spectral analysis of heart rate variability. Journal of Medical Engineering & Technology 28, 263–272 (2004)
6. Clifford, G.D., Tarassenko, L.: Quantifying errors in spectral estimates of HRV due to beat replacement and resampling. IEEE Transactions on Biomedical Engineering 52, 630–638 (2005)
7. Lomb, N.R.: Least-squares frequency analysis of unequally spaced data. Astrophysics and Space Science 39, 447–462 (1976)
8. Moody, G.B.: Spectral analysis of heart rate without resampling. In: Conference Spectral Analysis of Heart Rate Without Resampling, pp. 715–718
9. Berntson, G.G., Thomas Bigger, J., Eckberg, D.L., Grossman, P., Kaufmann, P.G., Malik, M., Nagaraja, H.N., Porges, S.W., Saul, J.P., Stone, P.H., Van Der Molen, M.W.: Heart rate variability: Origins, methods, and interpretive caveats. Psychophysiology 34, 623–648 (1997)

10. Pan, J., Tompkins, W.J.: A Real-Time QRS Detection Algorithm. IEEE Transactions on Biomedical Engineering BME-32, 230–236 (1985)
11. Press, W.H., Rybicki, G.B.: Fast algorithm for spectral analysis of unevenly sampled data. Astrophysical Journal 338, 277–280 (1989)
12. Scargle, J.D.: Studies in astronomical time series analysis. II - Statistical aspects of spectral analysis of unevenly spaced data. The Astophysical Journal 263, 835–853 (1982)
13. McSharry, P.E., Clifford, G., Tarassenko, L., Smith, L.A.: Method for generating an artificial RR tachogram of a typical healthy human over 24-hours. In: Conference Method for generating an artificial RR tachogram of a typical healthy human over 24-hours, pp. 225–228 (2002)
14. Press, W., Teukolsky, S., Vetterling, W., Flannery, B.: Numerical recipes in C: the art of scientific computing, 2nd edn. Cambridge University Press, Cambridge (1992)

Molecular Markers Associated with Low Temperature Tolerance in Winter Wheat

Alireza Taleei[1,*], Reza Gholi Mirfakhraee[2], Mohsen Mardi[3], Abbas Ali Zali[1], and Cyrus Mahfouzi[3]

[1] Professors in the Department of Agronomy & Plant breeding, Faculty of Agricultural Sciences and Engineering, College of Agriculture and Natural Resources, University of Tehran, Karaj, P.O. Box 31587-11167, Iran
[2] PhD student in the Department of Agronomy & Plant breeding, Faculty of Agricultural Sciences and Engineering, College of Agriculture and Natural Resources, University of Tehran, Karaj, P.O. Box 31587-11167, Iran
[3] Seed and Plant Improvement Institute (SPII), 4119-31585, Karaj, Iran
ataleei@ut.ac.ir

Abstract. Low temperature tolerance (LT) is an important agronomic trait in winter wheat that determines the plants ability to cope with below freezing temperatures. To identify genomic regions, which determine the level of LT tolerance in hexaploid wheat, $F_{2:3}$ and $F_{2:4}$ populations produced from crossing between winter type tolerant parent Mirnovoskaya 808(LT_{50} =-20 °C) and spring type, susceptible parent, Pishtaz (LT_{50} =-7 °C) were analyzed. The levels of LT tolerance for these populations were evaluated using artificial freeze test $LT_{50,}$ the temperature at which 50%of plants were killed by LT stresses. The molecular analyses were assessed using 170 SSR primer pairs and 22 AFLP primer combinations. The result of phenotypic analysis showed continuous distribution of trait values (LT_{50} =-3 to -23 °C) which is in agreement with the distribution of trait expected for a polygenic and quantitatively inherited trait. The relationship between LT tolerance (LT_{50}) and genotypic data was analyzed using single marker analysis, interval mapping and composite interval mapping methods. Three detected QTLs for spring parent, Pishtaz, with partial dominant effects and three detected QTLs of winter parent, Mirnovoskaya 808, with over dominant effects. Because the detected QTLs located on the 5B and 7D chromosomes and other ones which were linked to AFLP markers were inherited in both parents*; therefore theses results do confirm the effectiveness of both parents for this characteristic.

Keywords: Low-temperature tolerance, LT_{50}, *Triticum aestivum*, QTL mapping.

1 Introduction

Freezing tolerance, the ability of plants to survive subfreezing temperatures, is the major component of winter survival and important characteristic necessary for

* Corresponding author.

Y. Zhang et al. (Eds.): DTA/BSBT 2010, CCIS 118, pp. 283 – 290, 2010.

optimum seed yield of winter wheat varieties [10 and 17]. Freezing tolerance may be assayed in field or in controlled conditions. The correlation between field survival and laboratory studies of freezing tolerance were reported between 0.77 and 0.92 [13]. Another important factor is acclimation ability, which is the ability of plants to increase its freezing tolerance, or survive at lower temperatures, after a period of cold-temperature treatment [10]. A strong correlation between cold acclimation and freezing tolerance in winter type wheat has been noticed [12]. Cold acclimation is associated with several physiological and biochemical alterations in the plants. Changes in plants during acclimation include increases in soluble sugars, proteins, amino acids and organic acids, accumulation of osmolytes and protective proteins as well as modification of membrane lipid composition and alterations in gene expression [6, 8, and 11]. In wheat cold induced genes have been isolated and characterized and there is a high correlation between the expression of some of these genes and the development of freezing tolerance, which appear to be up-regulated by low temperature [1, 4, 14, and 15]. Freezing tolerance is also strongly correlated with the capacity for maintaining high photosynthesis during cold acclimation, because it is indispensable to ensure an energy source during cold acclimation. Cessation of growth during cold acclimation is also necessary to reach the resistance [11and15]. During 1_{st} stage of cold acclimation, the water content of prehardened plants decreases and soluble sugars and free proline of leaves increase [10]. The genetic regulation of freezing tolerance and winter hardiness is complex in most or all crop species [5, 7, and 18]. The reported gene action for freezing tolerance has varied from recessive to partially dominant in winter wheat [3], largely additive [2], to partially dominant [9], in alfalfa and partially recessive in potato [16]. In recent years, molecular markers, as useful complementary tools for classical breeding methods, were used in selection programs for quantitative traits such as freezing tolerance. Various molecular markers were developed and applied for mapping QTLs. The objective of the present study was to identify SSR and RAPD markers linked to cold resistance genes in wheat.

2 Material and Methods

To identify genomic regions, which determine the level of LT tolerance in hexaploid wheat, $F_{2:3}$ and $F_{2:4}$ populations produced from crossing between winter type tolerant parent Mirnovoskaya 808(LT_{50} =-20 °C) and spring type, susceptible parent, Pishraz (LT_{50} =-7 °C) were analyzed. The levels of LT tolerance for these populations were evaluated using artificial freeze test $LT_{50,}$ the temperature at which 50%of plants were killed by LT stresses. The molecular analyses were assessed using 170 SSR primer pairs and 22 AFLP primer combinations. The result of phenotypic analysis showed continuous distribution of trait values (LT_{50} =-3 to -23 °C) which is in agreement with the distribution of trait expected for a polygenic and quantitatively inherited trait. The relationship between LT tolerance (LT_{50}) and genotypic data was analyzed by single marker analysis, interval mapping and composite interval mapping methods, using Win QTL Cartographer 2.5 [19] and LOD=2.5.

3 Resaults and Disccusion

Phenotypical Evaluation

LT_{50} values for parental lines along with 178 $F_{2:3}$ genotypes derived from a cross between them were shown in Figure 1 as a frequency distribution for 11 temperature levels. $F_{2:3}$ genotypes showed continuous distribution for this trait which revealed that there should be a polygenic inheritance for $LT_{50.}$ Mean value for LT_{50} was -14.52 °C. More than 5% of families (about 10 families) had LT_{50} values less than that of susceptible parent Pishtaz, and more than 24% of families (about 44 families) on the other hand showed LT_{50} values more than that of tolerant parent Mirnovoskaya 808.

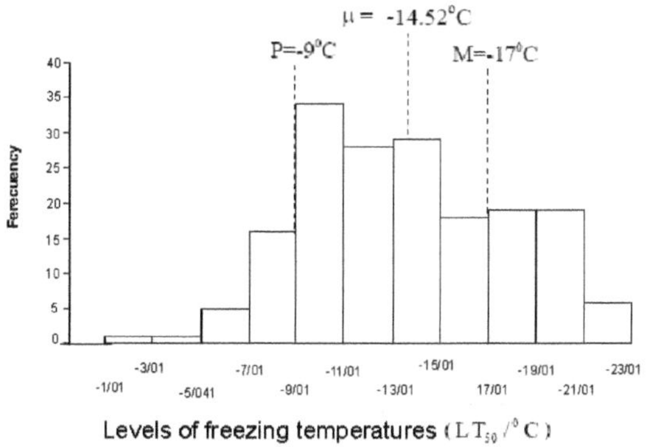

Fig. 1. The frequency distribution for $F_{2:3}$ populations at 11 freezing temperatures

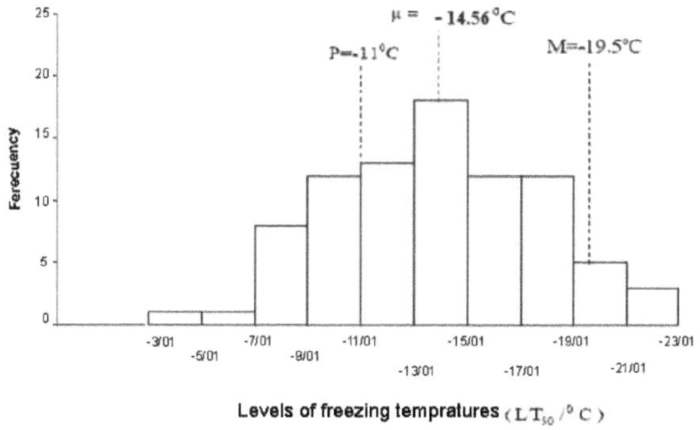

Fig. 2. The frequency distribution for $F_{2:4}$ populations at 11 freezing temperatures

The frequency distribution of LT_{50} values for parental lines along with 86 F2:4 genotypes derived from cross between them for 11 temperature levels as a frequency distribution was shown in Figure 2. As it shown in Figure 2, the distribution is continuous for this trait as well for the second year's experiment. Mean value for LT_{50} was -14.56°C. More than 14% of families (about 12 families) had LT_{50} values less

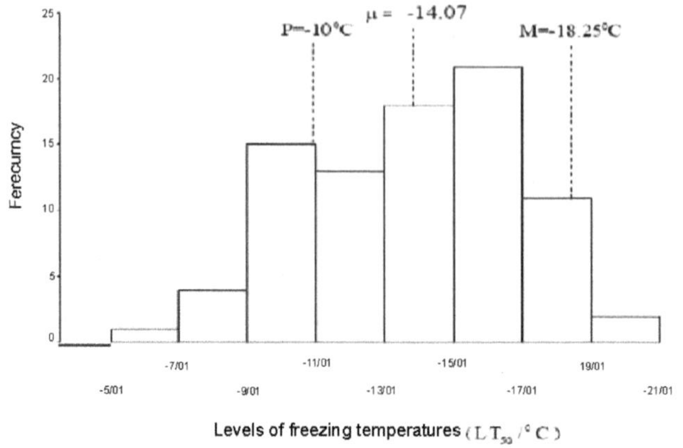

Fig. 3. The frequency distribution for the mean of $F_{2:3}$ and $F_{2:4}$ populations at 11 freezing temperatures

than that of susceptible parent Pishtaz, and more than 8% of families (about 7 families) on the other hand showed LT_{50} values more than that of the tolerant parent Mirnovoskaya 808. In Figure 3 the mean values of LT_{50} for parental lines and 85 pair of $F_{2:3}$ and $F_{2:4}$ genotypes derived from crosses between them shown as mean frequency distribution at 11 freezing temperatures. As it,s shown the mean value for this trait was -14.07 °C. More than 6% genotypes (about 8 genotypes) had LT_{50} values less than that of susceptible parent Pishtaz, and more than 4% of families (about 6 families) on the other hand showed LT_{50} values more than that of the tolerant parent Mirnovoskaya 808.

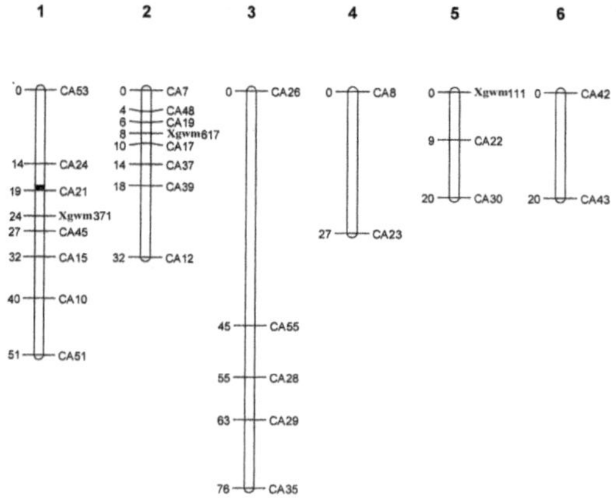

Fig. 4. Linkage groups of SSR and AFLP markers for wheat and the position of QTLs which controlling cold tolerance in the linkage group 1

Molecular Evaluation

The molecular analyses were assessed using 170 SSR primers pair and 22 AFLP primers combinations. Linkage groups of SSR and AFLP markers for wheat and the position of QTLs which controlling cold tolerance in the linkage group 1 shown in Figure 4.

Table 1. Molecular markers related to cold tolerance in a population derived from a cross between Mirnovoskaya 808 and Pishtaz

Marker	Chromosome	b_0	b_1	F(1,n-2)	P-value
CA24	5B	14/054	-1/633	12/876	0.000
CA21	5B	14/110	-1/866	16/749	0.000
Xgwm371	5B	14/140	-1/763	15/392	0.000
CA45	5B	14/102	-1/675	14/365	0.000
CA15	5B	14/150	-1/641	12/470	0.000
CA10	5B	14/086	-1/252	7/843	0.006
CA51	5B	14/252	-1/830	15/169	0.000
CA27	-	13/769	-1/364	6/692	0.010
Xgwm 397	4A	14/328	-1/800	14/203	0.000
Xgwm 174	5D	14/288	-1/356	5/601	0.019

The relationship between LT tolerance (LT_{50}) and genotypic data was analyzed using single marker analysis, interval mapping and composite interval mapping methods. Three detected QTLs for spring parent, Pishtaz, with partial dominant effects and three detected QTLs of winter parent, Mirnovoskaya 808, with over dominant effects. Because the detected QTLs located on the 5B and 7D chromosomes and other ones which were linked to AFLP markers were inherited in both parents; therefore theses results do confirm the effectiveness of both parents for this characteristic (Table1 and Figures 5-9).

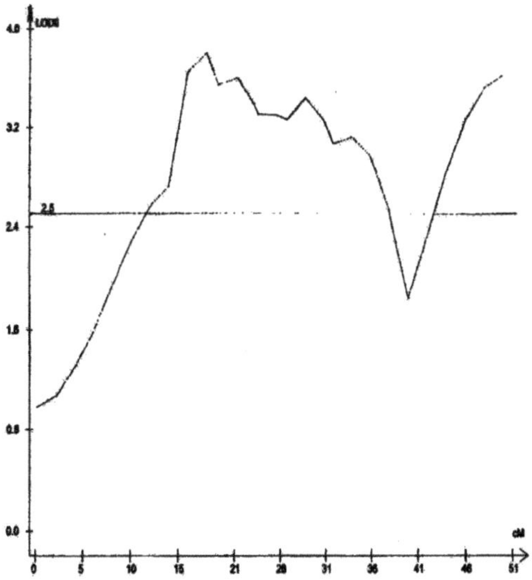

Fig. 5. Curve using single marker analysis for cold tolerance in the first year experiment

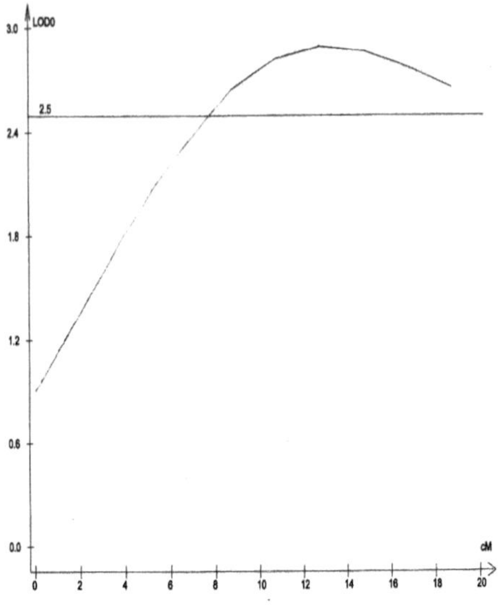

Fig. 6. Curve using single marker analysis for cold tolerance in the second year experiment

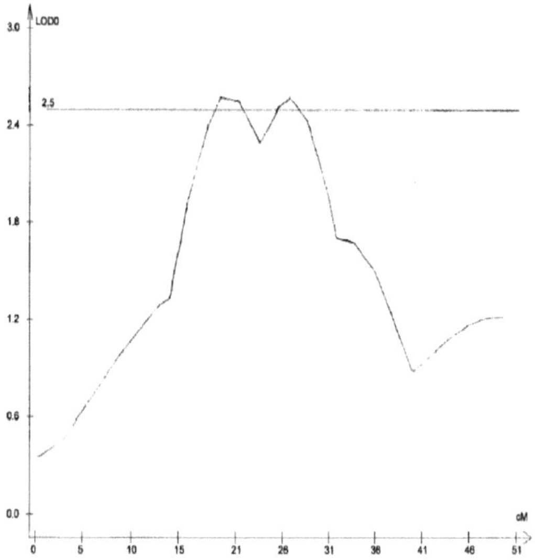

Fig. 7. Curve using single marker analysis for cold tolerance in wheat

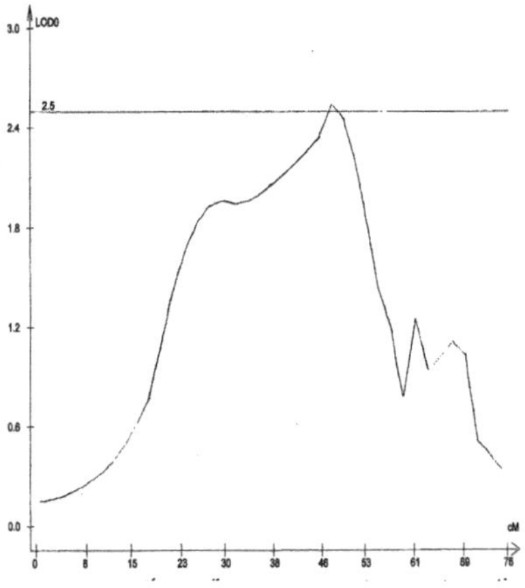

Fig. 8. Curve using single marker0020analysis for cold tolerance based on two years mean

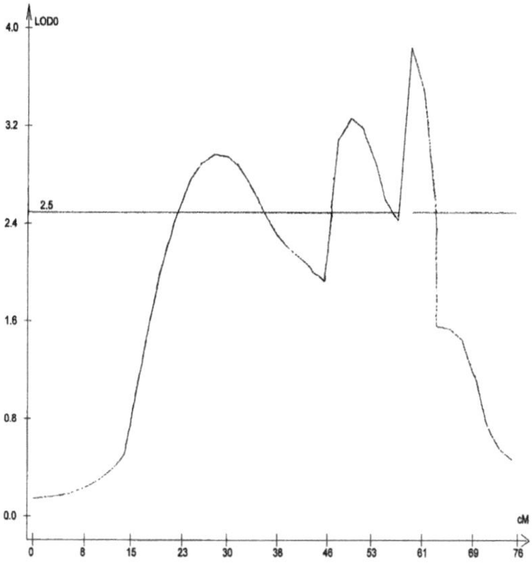

Fig. 9. Curve using composite interval mapping for cold tolerance based on two years mean

Acknowledgment

This project was supported jointly by the University of Tehran and Agricultural Research, Education and Extension Organization (AREEO), Iran.

References

1. Boothe, J.G., Beus, M.D., Johnson, F.A.M., De-Beus, M.D.: Plant physiology, vol. 108, pp. 795–803 (1995)
2. Daday, H., Greenham, C.G.: J. Hered, vol. 51, pp. 249–255 (1960)

3. Gullord, M., Olien, C.R., Everson, E.H.: Crop Sci., vol. 15, pp. 153–157 (1975)
4. Hawkins, G.P., Boothe, J.G., Nykiforuk, C.L., Johnson-Flanagan, A.M.: Genome, vol. 39, pp. 704–710 (1996)
5. Kole, C., Thormann, C.E., Karlsson, B.H., Palta, J.P., Gaffney, P., Yandell, B., Osborn, T.C.: Molecular Breeding, vol. 9, pp. 201–210 (2002)
6. Laroche, A., Geng, X.M., Singh, J.: Plant Cell and Environment, vol. 15, pp. 439–445
7. Low, C.N., Jenkins, G.: Genet. Res., vol. 15, pp. 197–208 (1992)
8. Palva, E.T., Welling, A., Tahtiharju, S., Tamminen, I., Puhakainen, T., Makela, P., Laitinen, R., Li, C., Helenius, E., Boije, M., Aspegren, K., Aalto, O., Heino, P.: Acta Hort., vol. 560, pp. 277–284 (2001)
9. Perry, M.C., Mcintosh, M.S., Wiebold, W.J., Welterlen, M.: Genome, vol. 29, pp. 144–149 (1987)
10. Rapacz, M.: Plant Sci., vol. 147, pp. 55–64 (1999)
11. Rapacz, M., Janowiak, F.: Crop Sci., vol. 182, pp. 57–63 (1999)
12. Rapacz, M., Markowski, F.: Crop Sci., vol. 183, pp. 243–253 (1999)
13. Salgado, J.P., Rife, C.L.: Cruciferae Newsletter, vol. 18, pp. 92–93 (1996)
14. Song, M.T., Copeland, L.O.: Korean Journal of Crop Science, vol. 40, pp. 69–76 (1995)
15. Song, M.T., Copeland, L.O., Song, M.T.: Korean Journal of Breeding, vol. 27, pp. 60–68 (1995)
16. Stone, J.M., Palta, J.P., Bamberg, J.B., Weiss, L.S., Harbage, J.F.: Proc. Nalt. Acad. Sci., vol. 90, pp. 7869–7873 (1993)
17. Teutonico, R.A., Palta, J.P., Osborrn, T.C.: Crop Sci., vol. 33, pp. 103–107 (1993)
18. Teutonico, R.A., Yandell, B., Satagopan, J.M., Ferreira, M.E., Palta, J.P., Osborn, T.C.: Molecular Breeding, vol. 1, pp. 329–339 (1995)
19. Wang, S., Basten, C.J. and Zeng, Z.B.: Windows QTL Cartographer 2.5, Department of Statistics, North Carolina State University, USA (2007)

Improving the Ability of Mining for Multi-dimensional Data

Yong Shi and Tyler Kling

Department of Computer Science and Information Systems
Kennesaw State University
1000 Chastain Road
Kennesaw, GA 30144

Abstract. In this paper, we present continuous research on data analysis based on our previous work on similarity search problems. *PanKNN*[13] is a novel technique which explores the meaning of K nearest neighbors from a new perspective, redefines the distances between data points and a given query point Q, and efficiently and effectively selects data points which are closest to Q. It can be applied in various data mining fields. In this paper, we present our approach to improving the scalability of the PanKNN algorithm. This proposed approach can assist to improve the performance of existing data analysis technologies, such as data mining approaches in Bioinformatics.

1 Introduction

With the advance of modern technology, the generation of multi-dimensional data has proceeded at an explosive rate in many disciplines. The similarity search problem has been studied in the last decade, and many algorithms have been proposed to solve the K nearest neighbor search[10,12,2,9,8]. *PanKNN*[13] is a novel technique which explores the meaning of K nearest neighbors from a new perspective, redefines the distances between data points and a given query point Q, and efficiently and effectively selects data points which are closest to Q. In this paper, we first give a brief introduction about our previous work on PanKNN, then propose a novel approach to improving the scalability of PanKNN, and demonstrate the experimental results.

2 Related Work

In traditional nearest neighbor problems, the similarity between two data points is based on a similarity function such as Euclidean distance which aggregates the difference between each dimension of the two data points. In other words, the nearest neighbor problems are solved based on the distance between the data point and the query point over a fixed set of dimensions (features). However, such approaches only focus on full similarities, i.e., the similarity in full data space of the data set. Also early methods [1,6,14] suffer from the "cure of dimensionality". In a high dimensional space the data are usually sparse, and widely used distance metric such as Euclidean distance may not work well as dimensionality goes higher. Recent research [7] shows that in high dimensions

Y. Zhang et al. (Eds.): DTA/BSBT 2010, CCIS 118, pp. 291–298, 2010.

nearest neighbor queries become unstable: the difference of the distances of farthest and nearest points to some query point does not increase as fast as the minimum of the two, thus the distance between two data points in high dimensionality is less meaningful. Some approaches [11,4,3] are proposed targeting partial similarities. However, they have limitations such as the requirement of the fixed subset of dimensions, or fixed number of dimensions as the input parameter(s) for the algorithms.

3 Solving Similarity Problems

In this section, we will briefly introduce our previous work on PanKNN[13]. PanKNN is a novel approach to nearest neighbor problems. We also analyze the nearest neighbor problems from a new perspective. We define the new meaning for the K nearest neighbors problem, and design algorithms accordingly. The similarity between a data point and a query point is not based on the difference aggregation on all the dimensions. We propose self-adaptive strategies to dynamically select dimensions based on the different situation of the comparison.

For a given data point X_i, and a given query point Q, we call the distance between X_i and Q as Pan-distance $PD(X_i, Q)$. $PD(X_i, Q)$ does not calculate the aggregated differences between X_i and Q on all dimensions. Instead, it only take into account those dimensions on which X_i is close enough to Q, and sum them up. This strategy not only avoids the negative impacts from those dimensions on which X_i is far to Q, but also eliminate the curse of dimensionality caused by similarity functions such as Euclidean distance which calculates the square root of the sum of squares of distances on each dimensions.

On more dimensions X_i is close (i.e., within the sets of K nearest neighbor) to Q, the smaller Pan-distance X_i has to Q. If we have two data points X_i and X_j, we judge which data point is closer to Q based on how many dimensions on which they are close enough (within dimension-wise K nearest neighbors) to Q, as well as their average distances to Q on such dimensions.

Given a data set DS, we first calculate the difference δ_{il} of each data point X_i to the query point Q on each dimension D_l. Then we sort the ids on each dimension D_l based on δ_{il}, and select the first K ids on each dimension D_l and put them into KS_l. We put all the ids in all KS_l to the set GS, and calculate the PD(X_i, Q) for each data point if its id is in GS. Finally, we sort the ids based on the Pan-distance and select the first K ids in the sorted list as the ids of K nearest neighbors of Q. We do not need to calculate the difference using different number of dimensions. The *number* of dimensions and the *subset* of dimensions associated with data point X_i are both dynamically decided depending on the values of X_i and their rankings on different dimensions.

4 A Scalable Approach to Finding Nearest Neighbors

PanKNN can efficiently and effectively find nearest neighbors when the size of the data set is small. However, the performance time increases dramatically with the increment of the data size. One of the reasons is that in PanKNN, given a query point, on each dimension, we need to sort the whole data set according their distances to the query point. It is well known that the average time complexity of most sorting algorithms is $O(n \log n)$. In this section we propose to design an algorithm which solves the scalability problem of PanKNN.

Let *n* denote the total number of data points and *d* be the dimensionality of the data space. Let D_l be the *l*th dimension, where l = 1, 2, ..., d. Let the input *d*-dimensional data set be **X**

$$\mathbf{X} = \{X_1, X_2, ..., X_n\},$$

which is normalized to be within the hypercube $[0, 1]^d \subset R^d$. Each data point X_i is a *d*-dimensional vector:

$$X_i = [x_{i1}, x_{i2}, ..., x_{id}]. \quad (1)$$

Data point X_i has the id number i. Let Q be the query point: $Q = [q_1, q_2, ..., q_d]$.

In PanKNN, we used $\Delta_i = [\delta_{i1}, \delta_{i2}, ..., \delta_{id}]$ as the array of differences between the data point X_i and the query point Q on each dimension. Given a data set DS of n data points $\mathbf{X} = \{X_1, X_2, ..., X_n\}$ with *d* dimensions D_1, D_2, ..., D_d, and a query point Q in the same data space, we first sort the data points on each dimension D_l, l=1, 2, ..., d, based on δ_{il} which is the difference between data point X_i and Q on dimension D_l. With the increment of the data size, this process can be very time consuming.

The purpose of sorting all data points in the data set on each dimension based on their distances to Q is to find a group of data points on each dimension which are closest to Q. Here we design an approach to acquire such groups without having to sort the whole data set on any dimension.

Given K as the number of data points we need to find which are closest to Q, on each dimension D_l, l=1, 2, ..., d, suppose $V_{l_{max}}$ and $V_{l_{min}}$ are the largest value and small value of data points in DS on D_l, respectively. Since we first normalize the data set, $V_{l_{min}}$ and $V_{l_{max}}$ will be 0 and 1, respectively. we divide [$V_{l_{min}}$,$V_{l_{max}}$] evenly by $\lfloor \frac{n}{K} \rfloor$, resulting in segments S_{l_1}, S_{l_2}, ..., $S_{l_{\lfloor \frac{n}{K} \rfloor}}$. Each segment will contain $\lceil K \rceil$ data points in average. However, some segments contain more than $\lceil K \rceil$ data points, and some contain less, due to the uneven distribution of data points on each dimension. It is easy to scan through the whole data set to calculate how many data points each segment S_{l_j} contains (denoted as $|S_{l_j}|$) and record the group of data point ids in S_{l_j}, for j=1, 2, ..., $\lfloor \frac{n}{K} \rfloor$.

We next locate the segment containing q_l which is the value of Q on D_l. Suppose segment S_{l_j} contains the value q_l, j=1, 2, ..., $\lfloor \frac{n}{K} \rfloor$. We design a process called *segment mergence* which is described as follows:

Process: segment mergence

Begin

1) In S_{l_j}, calculate the number of data points which are less than q_l, and denote it as N_l;

2) In S_{l_j}, calculate the number of data points which are larger than q_l, and denote it as N_r;

3) While N_l < K and there is a segment S_{left} on the left which is not merged

Begin

$S_{l_j} = S_{l_j} \bigcup S_{left}$;

$N_l = N_l + |S_{left}|$;

End;

4) While $N_l <$ K and there is a segment S_{right} on the left which is not merged
  Begin
    $S_{l_j} = S_{l_j} \bigcup S_{right}$;
    $N_l = N_l + |S_{right}|$;
  End;
5) return S_{l_j}
End.

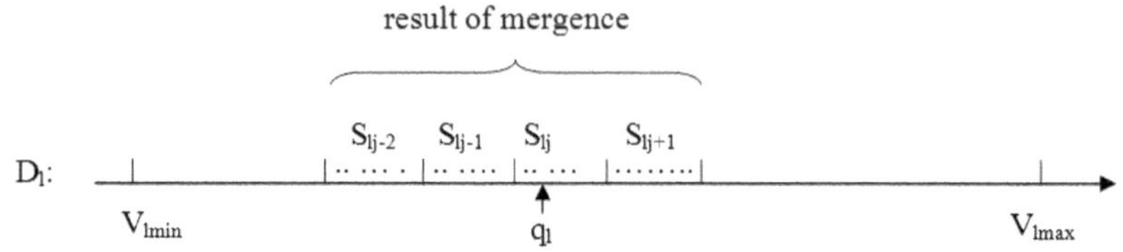

Fig. 1. An example of segment mergence on dimension D_l

Figure 1 shows an example of the segment mergence. Suppose K is set as 10, we divide the value range on D_l so each segment contains 10 data points in average. S_{l_j} contains q_l (the value of Q on D_l) and 5 data points, 2 of which are on the left side of q_l. So the initial value for N_l is 2, and the initial value for N_r is 3. We merge S_{l_j} with the segment on its left side ($S_{l_{j-1}}$), which contains 6 data points. N_l is updated as 8, which is still less than K which is 10. So we continue merging the segment on the left side ($S_{l_{j-2}}$) which contains 6 data points, and N_l is now updated as 14 which is larger than K. Next we merge the updated segment with the segment on its right side ($S_{l_{j+1}}$), which contains 8 data points, and N_r is updated as 11 which is larger than K, and we terminate the mergence process.

From the example we can see, after the segment mergence, the result segment will contain at least K data points or all the data points on the left side of q_l, and at least K data points or all the data points on the right side of q_r. The reason is that, in the extreme case, the K nearest neighbors of Q on D_l might all on the same side of Q. By having at least K data points on both sides, we assure that the K nearest data points are within the resulting segment S_{l_j} from the process of segment mergence.

By applying the process of segment mergence we do not need to sort the whole data set on each dimension in order to obtain the K nearest neighbors on each dimension.

4.1 Finding Nearest Neighbors with Scalability

Given a data set DS of n data points $\mathbf{X} = \{X_1, X_2, ..., X_n\}$ with D_l as the dimension l, l=1, 2, ..., d, a query point Q in the same data space, we try to find a set $PK_{scalable}$ which consists of k data points from DS so that for any data point $X_i \in PK_{scalable}$ and any data point $X_j \in DS - PK_{scalable}$, the PD(X_i, Q) is less than or equal to PD(X_j, Q). The set $PK_{scalable}$ is the Pan-K Nearest Neighbor set of Q in DS.

The PanKNN-scalable algorithm is described as follows:

1) For each $X_i \in DS$, we first calculate $\Delta_i = [\delta_{i1}, \delta_{i2}, ..., \delta_{id}]$ in which $\delta_{il} = |x_{il} - q_l|$;

2) On each dimension D_l, l=1, 2, ..., d, suppose its value range is $[V_{l_{min}}, V_{l_{max}}]$, we divide it evenly by $\lfloor \frac{n}{K} \rfloor$, resulting in segments S_{l_1}, S_{l_2}, ..., $S_{l_{\lfloor \frac{n}{K} \rfloor}}$.

3) Starting from S_{l_j} which contains the value q_l, we perform the *segment mergence process*. The resulting S_{l_j} will contain at least K data points.

We sort the ids of the data points in S_{l_j} instead of in DS, based on δ_{il} for X_i. Let G_l be the sorted list on D_l;

4) Let KS_l be the subset of G_l which contains the first K ids in G_l. For each data point X_i, i=1, 2, ... n, we generate $B_i = [b_{i1}, b_{i2}, ..., b_{id}]$ in which b_{il} = 1, if $i \in KS_l$; b_{il} = 0, if $i \notin KS_l$;

5) Let set $GS = \{i\}$ in which i $\in KS_l$, l=1, 2, ..., d. For each data point X_i, where $i \in GS$, we calculate $PD(X_i, Q)$.

Next we Sort $GS = \{i\}$ based on $PD(X_i, Q)$;

Let set $PK_{scalable}$ contain the first K $ids \in GS$, and we return $PK_{scalable}$.

4.2 Time and Space Analysis

Suppose the size of the data set is n. Throughout the process, we need to keep track of the information of all points, which collectively occupies $O(n)$ space. For one query point Q, we need to sort the data points in S_{l_j} after the segment mergence process on each dimension. The time required is $O(dKlogK)$.

5 Experiments

To assess the accuracy and efficiency of the proposed approach, comprehensive experiments on both synthetic and real data sets were conducted. Our experiments were run on Intel(R) Pentium(R) 4 with CPU of 3.39GHz and Ram of 0.99 GB.

5.1 Experiments on High-Dimensional Data Sets

To test the scalability of our algorithm over dimensionality, data size and K as the number of nearest neighbors required for the query points, we designed a synthetic data generator to produce data sets with normalized distributions.The sizes of the data sets

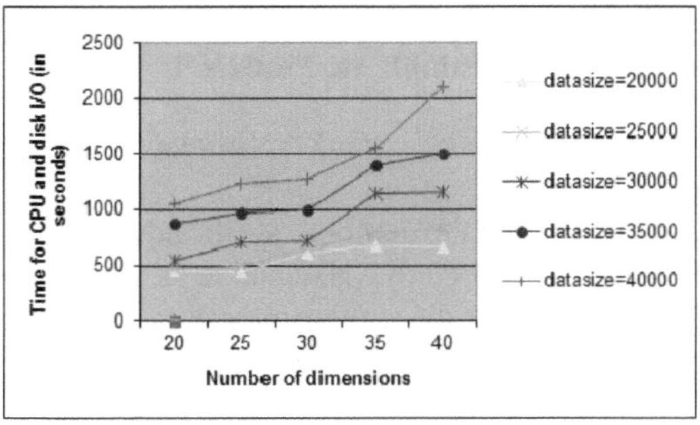

Fig. 2. Running time on one query point with increasing dimensions (K = 10)

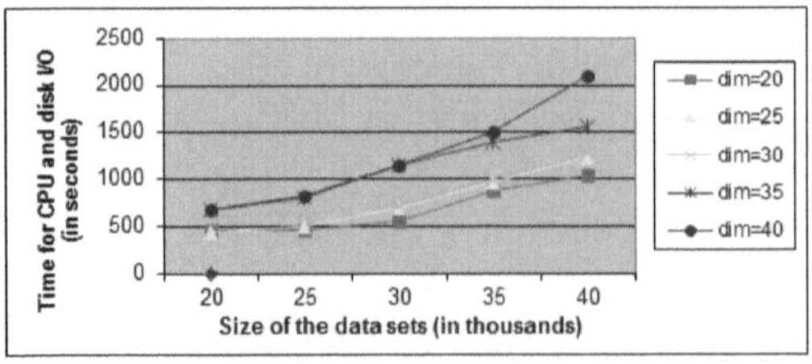

Fig. 3. Running time on one query point with increasing data set sizes (K = 10)

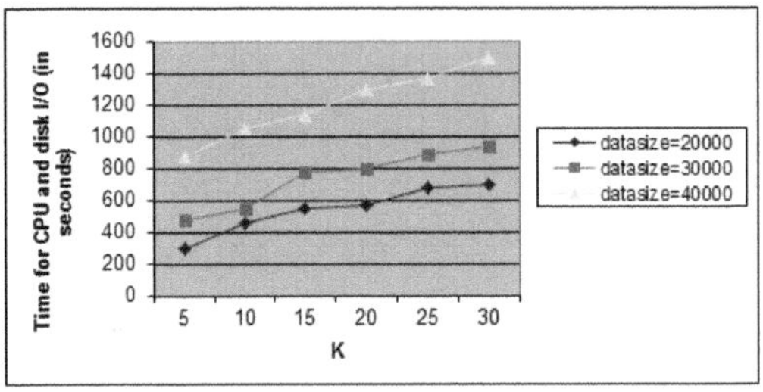

Fig. 4. Running time on one query point with increasing K values (dimensionality = 20)

vary from 20,000, 25,000, ... to 40,000, with the gap of 5,000 between each two adjacent data set sizes, and the dimensions of the data sets vary from 20, 25 ... to 40, with the gap of 5 between each two adjacent numbers of dimensions.

Figure 2 shows the running time of groups of data sets with dimensions increasing from 20 to 40. Each group has a fixed data size (from 20,000, 15,000, ... to 40,000). And we set K as 10.

Figure 3 shows the running time of groups of data sets on one query with sizes increasing from 20,000 to 40,000. Each group has fixed number of dimensions (from 20, 25, ... to 40). And we set K as 10. The two figures indicate that our algorithm is scalable over dimensionality and data size.

Figure 4 shows the running time of 3 groups of data sets with the size of 20000, 30000 and 40000 on one query with K increasing from 5,10,... to 30. And we set dimension as 20.

5.2 Experiments of PanKNN-Scalable vs. PanKNN

In this section we will demonstrate how PanKNN-scalable improves the performance compared to the original PanKNN.

We first use the synthetic data sets to demonstrate the advantage of PanKNN-scalable. Figure 5 shows the running time of PanKNN-scalable vs. PanKNN with sizes increasing from 20,000 to 40,000. We set the dimensionality as 20 and K as 10. From this picture we can see that PanKNN-scalable performs better than PanKNN when the data size increases.

We also use real data sets from UCI Machine Learning Repository [5] to demonstrate the performance difference of PanKNN-scalable vs. PanKNN.

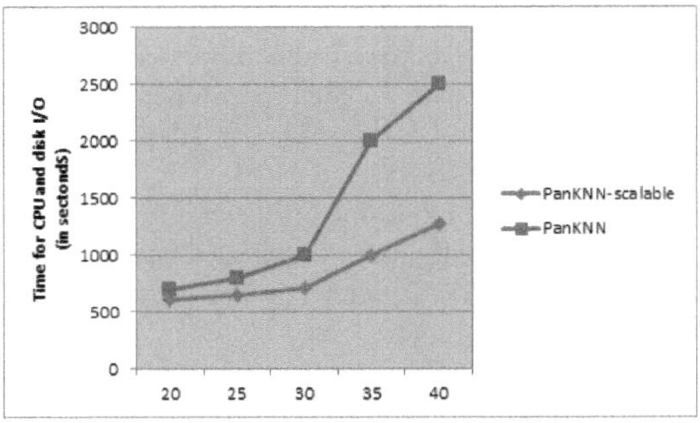

Fig. 5. Running time on one query point of PanKNN-scalable vs. PanKNN on increasing data sizes (dimensionality = 20 and K = 10)

Due to space limitation, here we demonstrate the testing result on one real data set called Wine Recognition data set. It contains the results of a chemical analysis of wines grown in the same region in Italy but derived from three different cultivars. There are 178 instances, each of which has 13 features (dimensions), including alcohol, magnesium, color intensity, etc. The data set has three clusters with the sizes of 59, 71 and 48. We perform the algorithms on the Wine data set. The accuracy rate of PanKNN-scalable is 94.1%, which is higher than the accuracy rate of PanKNN (92.9%).

6 Conclusion

In the paper we present our strategy to improve the similarity search approaches. On each dimension we divide the value range into segments with equal size and merge the segment containing the query point with the neighboring segments to acquire the group of data points closest to data point Q on each dimension. This data processing algorithm can be applied in many fields such as bioinformatics, pattern recognition, data clustering and signal processing.

References

1. White, D.A., Jain, R.: Similarity Indexing with the SS-tree. In: Proceedings of the 12th Intl. Conf. on Data Engineering, New Orleans, Louisiana, pp. 516–523 (February 1996)
2. Achtert, E., Böhm, C., Kröger, P., Kunath, P., Pryakhin, A., Renz, M.: Efficient reverse k-nearest neighbor search in arbitrary metric spaces. In: SIGMOD 2006, pp. 515–526. ACM, New York (2006)
3. Aggarwa, C.C.: Towards meaningful high-dimensional nearest neighbor search by human-computer interaction. In: ICDE (2002)
4. Aggarwal, C.C., Hinneburg, A., Keim, D.A.: On the surprising behavior of distance metrics in high dimensional space. In: Van den Bussche, J., Vianu, V. (eds.) ICDT 2001. LNCS, vol. 1973, p. 420. Springer, Heidelberg (2000), `citeseer.nj.nec.com/aggarwal01surprising.html`
5. Bay, S.D.: The UCI KDD Archive,University of California, Irvine, Department of Information and Computer Science, `http://kdd.ics.uci.edu`

6. Berchtold, D.A., Keim, S., Kriegel, H.-P.: The X-tree: An index structure for high-dimensional data. In: VLDB 1996, Bombay, India, pp. 28–39 (1996)
7. Beyer, K., Goldstein, J., Ramakrishnan, R., Shaft, U.: When is "nearest neighbor" meaningful? In: International Conference on Database Theory 1999, Jerusalem, Israel, pp. 217–235 (1999)
8. Cui, B., Shen, H., Shen, J., Tan, K.: Exploring bit-difference for approximate KNN search in high-dimensional databases. In: Australasian Database Conference (2005)
9. Fagin, R., Kumar, R., Sivakumar, D.: Efficient similarity search and classification via rank aggregation (2003)
10. Gionis, A., Indyk, P., Motwani, R.: Similarity search in high dimensions via hashing. The VLDB Journal, 518–529 (1999)
11. Hinneburg, A., Aggarwal, C.C., Keim, D.A.: What is the nearest neighbor in high dimensional spaces? The VLDB Journal, 506–515 (2000)
12. Seidl, T., Kriegel, H.-P.: Optimal multi-step k-nearest neighbor search. SIGMOD Rec. 27(2), 154–165 (1998)
13. Shi, Y., Zhang, L.: A dimension-wise approach to similarity search problems. In: The 4th International Conference on Data Mining, DMIN 2008 (2008)
14. Weber, R., Schek, H.-J., Blott, S.: A quantitative analysis and performance study for similarity-search methods in high-dimensional spaces. In: Proc. 24th Int. Conf. Very Large Data Bases, VLDB, pp. 194–205, 24–27(1998)

Hospital RFID-Based Patient u-Healthcare Design over Wireless Medical Sensor Network

Randy S. Tolentino[1] and Sungwon Park[2,*]

[1] Graduate School, Hannam University,
[2] Department of Nursing, Hannam University,
133 Ojeong-dong, Daedeok-gu, Daejeon, Korea
daryn2004@yahoo.com, sungwon@hnu.kr

Abstract. Ubiquitous sensor network is drawing a lot of attention as a method for realizing a ubiquitous society. It collects environmental information to realize a variety of functions, through a countless number of compact wireless nodes that are located everywhere to form an ad hoc arrangement, which does not require a communication infrastructure. However, there isn't any flexible and robust communication infrastructure to integrate these devices into an emergency care setting. An efficient wireless communication substrate for medical devices that addresses ad hoc or fixed network for information, naming and discovery, transmission efficiency of the data, data security and authentication, as well as filtration and aggregation of vital sign data need to be studied. We propose a Hospital RFID-based Patient u-Healthcare Management System architecture that possesses the essential elements of each of the future medical applications are integration with existing medical practices and technology, real-time, long-term, remote monitoring, miniature, wearable sensors to assist the elderly and chronic patients, and to amplify the transmission of data by installing routers in the hospital.

Keywords: U-Health Care, RF-ID, Quality Life, Hospital Networks.

1 Introduction

In recent years, in almost every country in the world, substantial financial resources have been allocated to the health care sector. Technological development and modern medicine practices are amongst the outstanding factors triggering this shift. Developed countries like South Korea, Japan, and China are currently facing a middle-and older-aged marketplace from a predominantly youth-driven marketplace. This trend is resulting in a greater demand for health care-related services and greater competition among health care providers.

Achieving a high operational efficiency in the health care sector is an essential goal for organizational performance evaluation. Efficiency uses to be considered as the primary indicator of hospital performance. From a managerial perspective, understanding the hospitals cost structure and their inefficiency in utilizing resources

* Corresponding author.

Y. Zhang et al. (Eds.): DTA/BSBT 2010, CCIS 118, pp. 299–308, 2010.

is crucial for making health care policies and budgeting decisions. The cost of medical services in hospitals is likely control by higher operational efficiency and to provide more affordable care and improved access to the public.

In the last few years' health care providers are making a concerted effort in using information technology for bringing down the spiraling health care cost here in Korea and technologically growing countries. In order to bring down the cost and improve efficiency, intelligent systems can play a significant role in providing intelligently processed and personalized information about patients to doctors, their health care staff (i.e., nurses) and health care administrators. The performance of health care management system is far behind compared to the service and manufacturing industries. Health care organizations nowadays are dealing with greater rank diseases, their cost, quality and delivery has essentially not improved significantly, and even the difference with the other industries seems to have increased further [1].

This paper propose an Architecture of Hospital RFID-based patient u-healthcare system that provides a healthcare monitoring and assistance service without any inconvenience and interference caused by the measuring apparatus, elderly persons and patients can be continuously monitored in the hospital environment by wireless ad-hoc network technology. This technology could also satisfy the requirements of "ubiquitous" wireless health state monitoring, including minimum intervention by medical caregivers. An environment of this type can be constructed using wireless sensor network (WSN) technology, which allows the coverage area of single wireless network to be expanded by ad-hoc network technology, capable of handing over wireless data to neighboring wireless networks.

This paper is structured as follows: Section 2 discusses related works. Section 3 outlines the RFID model used for developing a health care system. Section 4 illustrates the architecture of Hospital RFID-based Patient u-Healthcare Management System. Section 5 illustrates the application of the health care systems architecture using a Hospital RFID-based Patient u-Healthcare Management System. Section 6 concludes the paper.

2 Related Work

Research about u-Healthcare system which enables us to monitor patients' status and receive medical services is being conducted. The most significant limitations of wireless networks are the slow data transfer rate and lack of a single connectivity standard that enables devices to communicate with one another and to exchange data. Other limitations include wireless devices, which are still in their infancy stages. The current medical systems are merely providing medical services when a patient who has already a bad health status visits medical facilities.

However, a u-Healthcare system with sensor network enables patients to receive medical services from caregivers through mobile devices and remote clinic services anytime, anywhere. At the same time, the caregivers can provide medical services to prevent diseases by discovering the symptoms in advance through monitoring the patient continuously before her/his status is worsened. As examples of existing u-Healthcare systems based on sensor networks, there are the Codeblue [2] of Harvard University and the in-home monitoring [3] system of Virginia University.

In order to implement a u-Healthcare system with a sensor network, there are a few issues to be considered. Firstly, correction technique of the vital signs error caused by wireless transmission is needed. Secondly, augmenting signal technique of a sensor node having limited transmission signal strength is needed. Thirdly, since it is difficult to analyze a large number of vital signs collected from sensor network with mobile devices having limited processing ability, a technique to solve that problem is required.

3 RF-ID Model for Developing Healthcare System

Healthcare providers (i.e., hospitals) traditionally use a paper-based 'flow chart' to capture patient information during registration time, which is updated by the on duty nurse and handed over to the incoming staff at the end of each shift. Although, the nurses spent large amount of time on updating the paperwork at the bedside of the patient, it is not always accurate, because this is handwritten.

The nurses play a vital role at the Hospital system in the success of both inpatient and outpatient care. They also play a very important role in bridging to execute clinical orders or to communicate information between the hospital and the patient that motivates to evaluate the potential of RFID (Radio Frequency Identification) technology, and to reinforce the critical job of information handover.

RFID is one of the emerging technologies offering a solution, which can facilitate automating and streamlining safe and accurate patient identification, tracking, and processing important health related information in health care sector such as hospitals. Each RFID tag/wristband is identified by a Unique Identification Number (UIN) that can be programmed either automatically or manually and then password protected to ensure high security.

RFID wristband can be issued to every patient at registration, and then it can be used to identify patients during the entire hospitalization period. It can also be used to store patient important data (such as name, patient ID, drug allergies, drugs that the patient is on today, blood group, and so on) in order to dynamically inform staff before critical. RFID encoded wristband data can be read through bed linens, while patients are sleeping without disturbing them [2].

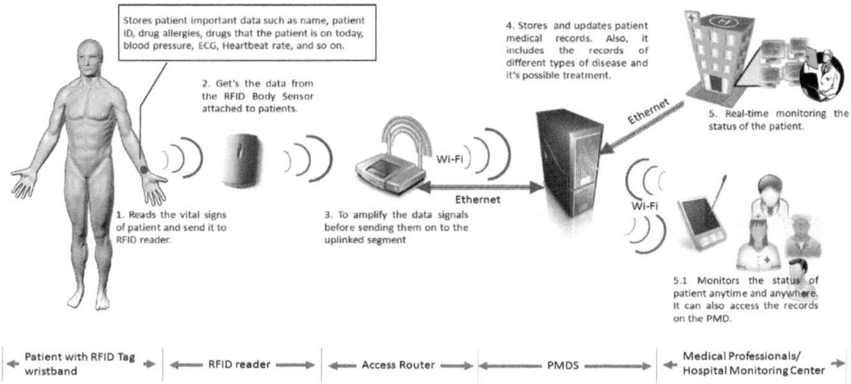

Fig. 1. Major Components of Hospital RFID- based Patient U-Healthcare Management System

Figure 1 shows how RFID technology provides a method to transmit and receive data from a patient to health service provider/medical professionals without human intervention (i.e., wireless communication). It is an automated data-capture technology that can be used to identify, track, and store patient information electronically contained on RFID wristband (i.e., smart tag). Although, medical professionals/consultants can access/update patient's record remotely via WiFi connection using mobile devices such as PDA (Personal Digital Assistant), laptops and other mobile devices. WiFi (wireless fidelity'), wireless local area networks (WLAN) that allows healthcare provider (e.g., hospitals) to deploy a network more quickly, at lower cost, and with greater flexibility than a wired system [3].

4 Proposed Architecture and Methodology of the System

4.1 System Architecture

A key requirement for healthcare system is the ability to operate continuously over a long time periods and to periodically update the data on the database. Generally speaking, the primary objective of a wireless sensor network is to maximize the node/network lifetime, while performance metrics are secondary objectives. On the other hand, the main aim of a wireless healthcare system should be reliable data transfer with minimum delay. The proposed system architecture combines the benefits of wireless sensor networks and RFID technology in the design and implementation of an internal positioning system for tracking the status of patients in a hospital.

The proposed architecture provides location/context specific information and content to a user via a RFID tag/wristband and RFID reader. Figure 2 provides a high level overview of the system architecture.

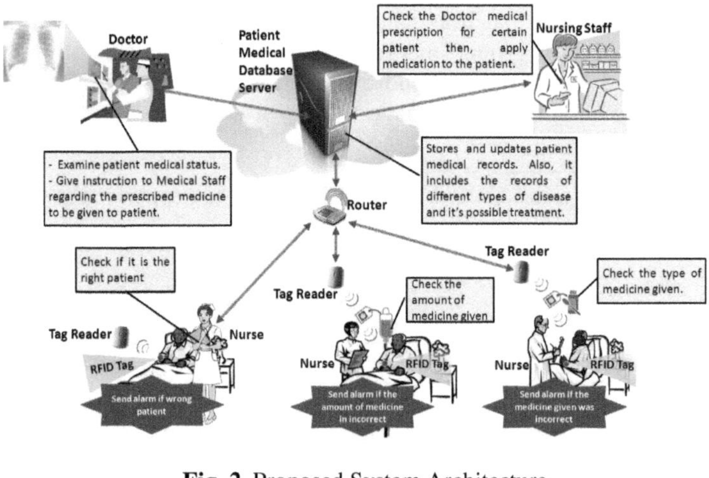

Fig. 2. Proposed System Architecture

The operation of the system is based on strategically located RFID readers placed inside the hospital identifying passive RFID tags/wristband. When a patient with RFID tags/wristbands roam inside the hospital (e.g. at the Lobby, Pharmacy, Physical and Therapy Rooms, Operating Room, Intensive Care Unit, etc.), the attached RFID tag/wristband is read by the RFID reader. The RFID reader communicates the tag ID to the router which then transmits the data through the wireless sensor network to the database server then, the router transmits the data corresponding to the status and location of that particular patient to a database server.

4.2 Benefits of System in Healthcare Sectors

Figure 3 shows the potential benefits of the system such as improve patients safety, eliminates paper-based document (e.g., bed side patient card), cost savings, increases efficiency and productivity, prevent/reduce medical errors, reduce patient waiting time, and so on, of using RFID technology within health care sector (e.g., hospitals) is numerous.

4.3 Multi-layer Architecture of U-Healthcare System

Figure 4 shows the multi-layered (i.e., six layer) RFID-based healthcare systems architecture, namely, physical layer, middleware layer, process layer, data access layer, application layer and user interface layer.

The physical layer consists of the actual hardware components that include RFID tag, antennas, readers, PDA, Smart phone. RFID Middleware is the interface between the RFID reader and healthcare providers (i.e., hospitals) databases and patient management system. It is an important element of RFID systems in healthcare sector.

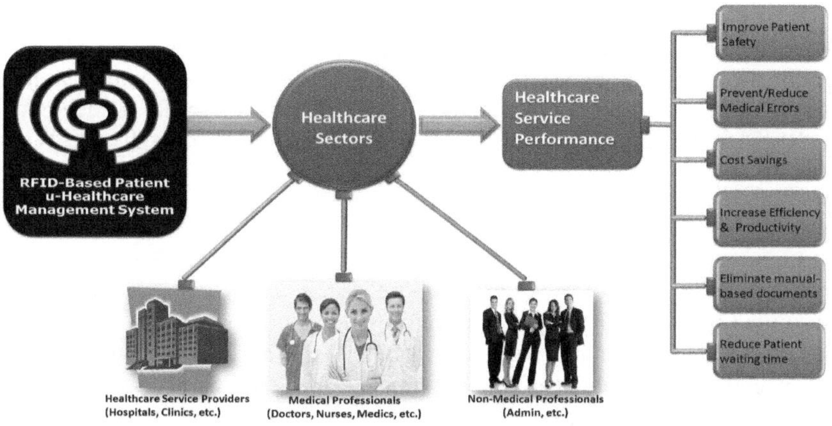

Fig. 3. System benefits on Healthcare Sectors

RFID middleware layer enables healthcare providers (e.g., hospitals) a quick connectivity with readers, lower the volume of information that medical applications need to process by grouping and filtering raw RFID observations from readers, and provide application-level interface for managing readers and process large volumes of RFID data for their medical applications. This layer is also monitoring physical layer components and supports International Organization for Standardization (ISO) standard [4].

The Process Layer drives hospitals to deploy RFID-based healthcare system (business) processes that provide real-time integration into existing systems. This layer enables data mapping, formatting, business rule execution and the service interactions with databases. The data access layer composed of a Relational Database Management System (i.e., normalization, database schema) and applications that allow healthcare providers for the creation of RFID "events". This layer includes a data loading approach that supports high volumes of RFID data, access them using structured query Language (SQL) and enables the data to be easily presented through the use of customized views to healthcare provider (i.e., hospital) to use them [5].

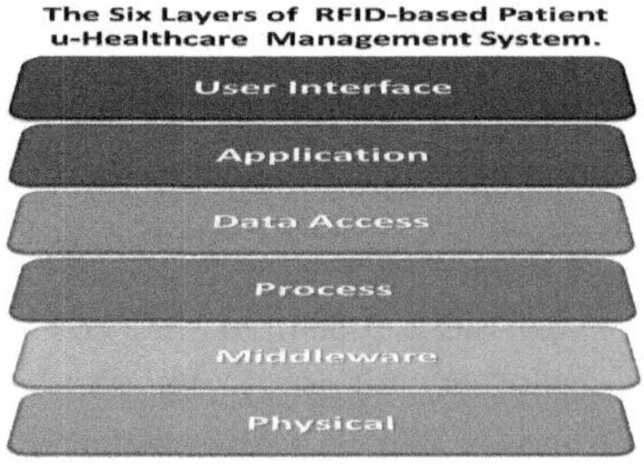

Fig. 4. Six-Layer of RFID-based Patient U-Healthcare Management System

The application layer interacts with multiple patients' wristband in the medical application (i.e., patient management system). Finally, the user interface layer is comprised of an extensible, graphical user interfaces that allow RFID devices (e.g., reader, RFID tag) in a uniform, user-friendly way to work seamlessly in a Windows environment.

4.4 How the System Works

Figure 5 shows the system flow of our propose system architecture. It uses the unique ID transmitted by RFID tags as a key to information stored in database. After reading the data, the RFID reader reads the data from RFID tag/wristband then, the RFID reader authenticate the patient tag ID by transmitting it to the access router for fast transmission of data. The access router sends the data to the database server to authenticate patient tag ID. If the patient tag ID exist, it starts to process the corresponding data of that certain patient. The doctor/caregiver checks the data of the patient and analyze it. If the status of the patient is not stable, the doctor/caregiver sends instruction (e.g. prescribe medicine for patient, check the amount of medicine given) to the nursing staff. After complying the doctor/caregiver instructions. The nurse updates the medical records of the patient on the database server.

In addition, healthcare professionals (e.g., doctors, caregivers) can assign a security password for patient's medical record to increased patient data security.

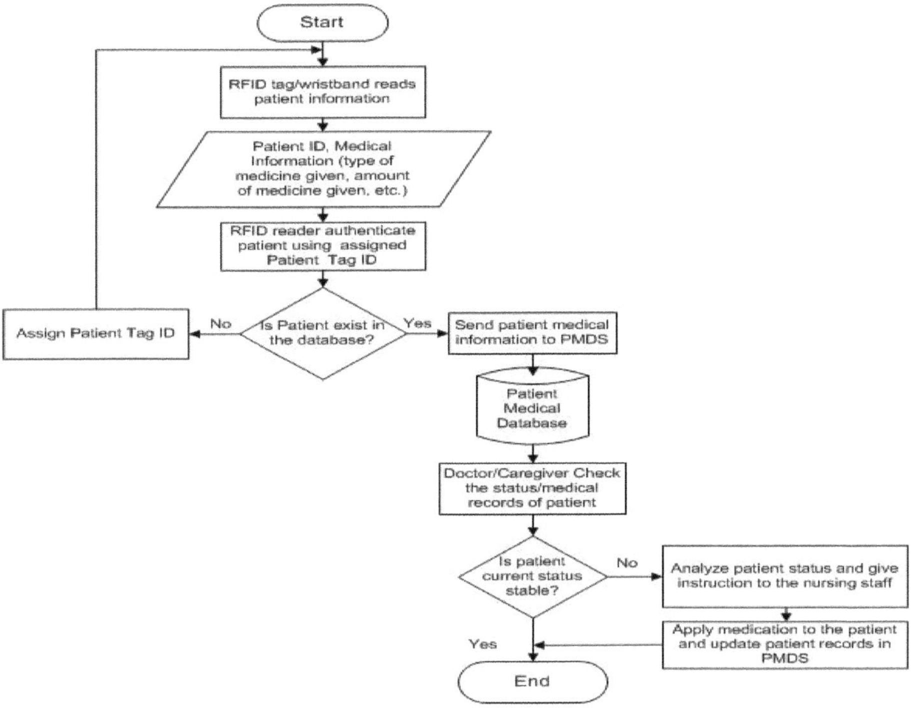

Fig. 5. System Flow Chart

5 Application of the Proposed RFID-Based Patient U-Healthcare System

The application can be implemented in departments (e.g., medicine, surgery, obstetrics and gynecology, pediatrics) in both public and private hospitals for fast and accurate patient identification without human intervention. But, most of the patients at public hospitals get admission into different units (e.g., medicine, intensive care unit, trauma center, etc.) through emergency department. So, our immediate focus is to implement it on hospitals emergency department. The proposed system architecture combines the benefits of wireless sensor networks and RFID technology in the design and implementation of an internal positioning system for tracking the status and location of patients that roaming around the hospital (e.g. here in South Korea most of the patients at the hospital does not stay inside their room. They usually roam around the hospital because they feel bored staying inside their room and they want to exercise to make themselves feel a little bit better.).

As a demonstrative example the architecture will be used to provide location/context specific information and content to a doctor/caregiver via a PDA/Smart phone. Figure 6 provides a high level overview of the system architecture.

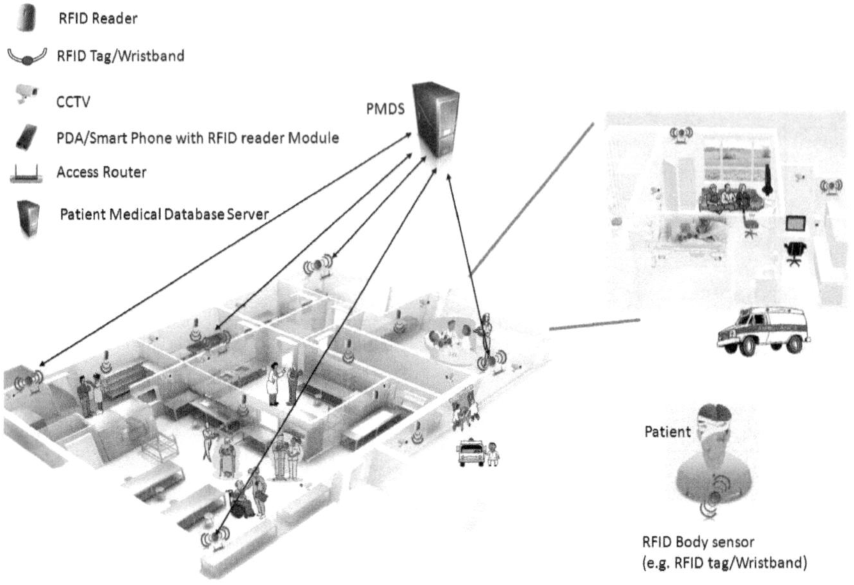

Fig. 6. Hospital-based Wireless Sensor Network Model

The operation of the system, which is based on strategically located RFID readers placed around the hospital identifying passive RFID tags/wristband with unique identifiers. When a patient with RFID tag/wristband passes a tag reader in the wireless sensor network the attached tag is read by the connected RFID reader. The RFID reader communicates the tag ID to the router which then transmits the data through the wireless sensor network to the PMDS. The PDMS stores the data corresponding to the location of that particular patient. Based on the patient medical information that stored on the PMDS, the doctor/caregiver can check and analyze the current status and location of a patient via PDA/Smart phone that connected through wireless network. The doctor/caregiver instructs the nursing staff if the status of the patient is not stable.

6 Conclusion and Future Works

The architecture we designed is to present the new capabilities for both remote healthcare and real time monitoring of patients within hospital healthcare premise. In this paper, we have described the Major components of Hospital RFID-based Patient u-Healthcare Management System, multi-layer health care system architecture, system architecture, System benefits in Healthcare Sectors, Flow of the system, and the application of system. Using the proposed system architecture, health care providers (e.g., hospitals) have a chance to track fast and accurate patient identification, improve patient's safety by capturing basic data (such as patient unique ID, name, blood group, drug allergies, drugs that the patient is on today), prevent/reduce medical errors, increases efficiency and productivity, and cost savings through wireless communication. The system also helps hospitals to build a better, more collaborative environment between different departments, such as the wards, medication, examination, and payment.

This generic architecture of health care system can be applied to other areas such as new born babies and aged-care management in both public and private hospitals. Where, health care providers (i.e., hospitals) will have an opportunity of use smart tag (RFID) for people's babies, which look pretty much alike to ensure that they match the right mother with the right baby. Having RFID tracking means hospitals can use the propose system to track whether infants are even inside the hospital that presumably is useful if they lose track of patients. The similar concept can be applied for the aged-care management.

In a future version of Hospital RFID-based Patient u-Healthcare Management System, we will explore the functionality to access patient's medical record from other health care providers (e.g., hospitals) databases through Internet. We are planning to design an architecture that integrates RFID reader module to PDA/Smart phone that will serve as a tag reader of RFID for reading patients vital signs using body sensor. This RFID reader module has a buffer which stores patient's information temporarily while waiting to be sent to the database for authentication and update.

Acknowledgments. This work has been supported by the 2010 Hannam University Research Fund.

References

1. Correa, F.A., Gil, M.J., Redín, L.B.: Benefits of Connecting RFID and Lean Principles in Health Care., Working Paper 05-4, Business Economics Series 10 (2005)
2. GAO RFID, RFID Solutions for Healthcare Industry, `http://healthcare.gaorfid.com/` (accessed on February10, 2007)
3. Laverty, D.: What is WiFi? An Introduction to Wireless Networks for the Small/Medium Enterprise
4. Glover, B., Bhatt, H.: RFID Essentials, pp. 54–169. O'Reilly Media, Sebastopol (2006)
5. Sybase, Inc., Sybase Radio Frequency Identification Technology Architecture One Sybase Drive Dublin CA, 94568 USA (2005), `http://www.sybase.com/sb_content/1031464/16056_RFID_Arch_L02607_FNL3.pdf` (accessed on February 11, 2007)

Authors

Randy S. Tolentino received the B.S. degree in Information Technology from the Western Visayas College of Science and Technology, Lapaz, Iloilo, Philippines, 2003. Currently, he is on the Integrated Course in Hannam University. His research interests include ubiquitous e-health care, ubiquitous medical sensor network modeling, wireless healthcare network, multimedia system, Multimedia Healthcare, SCADA security and sensor network.

Sungwon Park received the B.S., M.S. and Ph.D. degrees in Nursing from Yonsei University in 1998, 2002 and 2005 respectively. From 2006 to February 28, 2010, she was an assistant professor in the college of nursing at Hyechon University. Since March 1, 2010, he has been a professor in the Department of Nursing at Hannam University, Daejeon, Korea. Her research interests include psychiatric nursing, schizophrenia, social emotion, ubiquitous e-health care, ubiquitous medical sensor network modeling and social cognitive functions.

Author Index

GPSR Compliance

The European Union's (EU) General Product Safety Regulation (GPSR) is a set of rules that requires consumer products to be safe and our obligations to ensure this.

If you have any concerns about our products, you can contact us on ProductSafety@springernature.com

In case Publisher is established outside the EU, the EU authorized representative is:

Springer Nature Customer Service Center GmbH
Europaplatz 3
69115 Heidelberg, Germany

Batch number: 09474011

Printed by Printforce, the Netherlands